Advances in Experimental Medicine and Biology

Volume 922

More information about this series at http://www.springer.com/series/5584

Isabel Moraes
Editor

The Next Generation in Membrane Protein Structure Determination

Springer

Editor
Isabel Moraes
Membrane Protein Laboratory
Diamond Light Source/Imperial College London
Harwell Campus
Didcot, Oxfordshire, UK

ISSN 0065-2598 ISSN 2214-8019 (electronic)
Advances in Experimental Medicine and Biology
ISBN 978-3-319-35070-7 ISBN 978-3-319-35072-1 (eBook)
DOI 10.1007/978-3-319-35072-1

Library of Congress Control Number: 2016950435

Printed on acid-free paper

This Springer imprint is published by Springer Nature
The registered company is Springer International Publishing AG Switzerland

Preface

Over the years membrane proteins have fascinated scientists for playing a fundamental role in many critical biological processes. Located across the native cell membrane or mitochondria wall, integral membrane proteins perform a large diversity of vital functions including energy production, transport of ions and/or molecules across the membrane and signaling. Mutations or improper folding of these proteins are associated with many known diseases such as Alzheimer's, Parkinson's, depression, heart disease, cystic fibrosis, obesity, cancer and many others. It is estimated that more than one quarter of the human genome codes for integral membrane proteins and it is therefore imperative to investigate the role of these proteins in human health and diseases. Today, around 60 % of the drugs on the market target membrane proteins. Although most of the commercially available drugs have been facilitated by conventional drug discovery methods, it is the information provided by the protein atomic structures that discloses details regarding the binding mode of drugs. In addition, atomic structures contribute to a better understanding of the protein function, mechanism, and regulation at the molecular level. Consequently, membrane protein structural information plays a significant role not just in medicine but also in many pharmaceutical drug discovery programs.

More than 30 years have passed since the first atomic structure of an integral membrane protein was solved (Deisenhofer et al. 1985). Nevertheless, the number of membrane protein structures available, when compared with soluble proteins is still very low (http://blanco.biomol.uci.edu/mpstruc/) and the main reason for this has been the many technical challenges associated with protein expression, purification, and the growth of well-ordered crystals for X-ray structure determination. In the last few years, developments in recombinant methods for overexpression of membrane proteins; new detergents/lipids for more efficient extraction and solubilisation; protein engineering through mutations, deletions, fusion partners and monoclonal antibodies to promote diffraction quality crystals; automation/miniaturization and synchrotron/beamline developments have been crucial to recent successes in the field. In addition, developments in computational approaches have been of extremely valuable importance to the link between the protein structure and its physiological function. Molecular dynamics simulations combined with homology modeling has become a powerful tool in the development of novel pharmacological drug targets.

The chapters presented in this book provide a unique coverage of different methods and developments essential to the field of membrane proteins structural biology. The contributor authors are all experts in their respective fields and it is our hope that the material found within the book will provide valuable information to all the researchers whether experts or new.

Oxfordshire, UK
February 2016

Isabel Moraes

Acknowledgments

The editor wish to thank to all the authors who enthusiastically have agreed to be part of this volume. The research of the editor is supported by the Wellcome Trust grant 099165/Z/12/Z and by the EU Marie Curie FP7-PEOPLE-2011-ITN NanoMem.

Contents

Expression Screening of Integral Membrane Proteins by Fusion to Fluorescent Reporters

1

Louise E. Bird, Joanne E. Nettleship, Valtteri Järvinen, Heather Rada, Anil Verma, and Raymond J. Owens

Abstract

The production of recombinant integral membrane proteins for structural and functional studies remains technically challenging due to their relatively low levels of expression. To address this problem, screening strategies have been developed to identify the optimal membrane sequence and expression host for protein production. A common approach is to genetically fuse the membrane protein to a fluorescent reporter, typically Green Fluorescent Protein (GFP) enabling expression levels, localization and detergent solubilisation to be assessed. Initially developed for screening the heterologous expression of bacterial membrane proteins in *Escherichia coli*, the method has been extended to eukaryotic hosts, including insect and mammalian cells. Overall, GFP-based expression screening has made a major impact on the number of membrane protein structures that have been determined in the last few years.

Keywords

Integral membrane protein • Green fluorescent protein • Insect cells • *Escherichia coli* • *Saccharomyces cerevisiae* • *Pichia pastoris* • HEK 293 cells

1.1 Introduction

The production of recombinant integral membrane proteins (IMPs) for structural and functional studies is technical challenging due to low levels of expression often limited by toxicity to the expression host cells. To overcome these limitations screening of sequence variants either engineered or exploiting the natural sequence diversity of orthologues, has been successfully used to improve the production of many mem-

L.E. Bird • J.E. Nettleship • V. Järvinen • H. Rada
A. Verma • R.J. Owens (✉)
OPPF-UK, The Research Complex at Harwell, Rutherford Appleton Laboratory Harwell, Oxford, UK

Division of Structural Biology, Henry Wellcome Building for Genomic Medicine, University of Oxford, Roosevelt Drive, Oxford, UK
e-mail: louise.bird@rc-harwell.ac.uk; joanne.nettleship@rc-harwell.ac.uk; valtteri.jarvinen@rc-harwell.ac.uk; heather.rada@rc-harwell.ac.uk; anil.verma@rc-harwell.ac.uk; ray@strubi.ox.ac.uk

I. Moraes (ed.), *The Next Generation in Membrane Protein Structure Determination*, Advances in Experimental Medicine and Biology 922, DOI 10.1007/978-3-319-35072-1_1

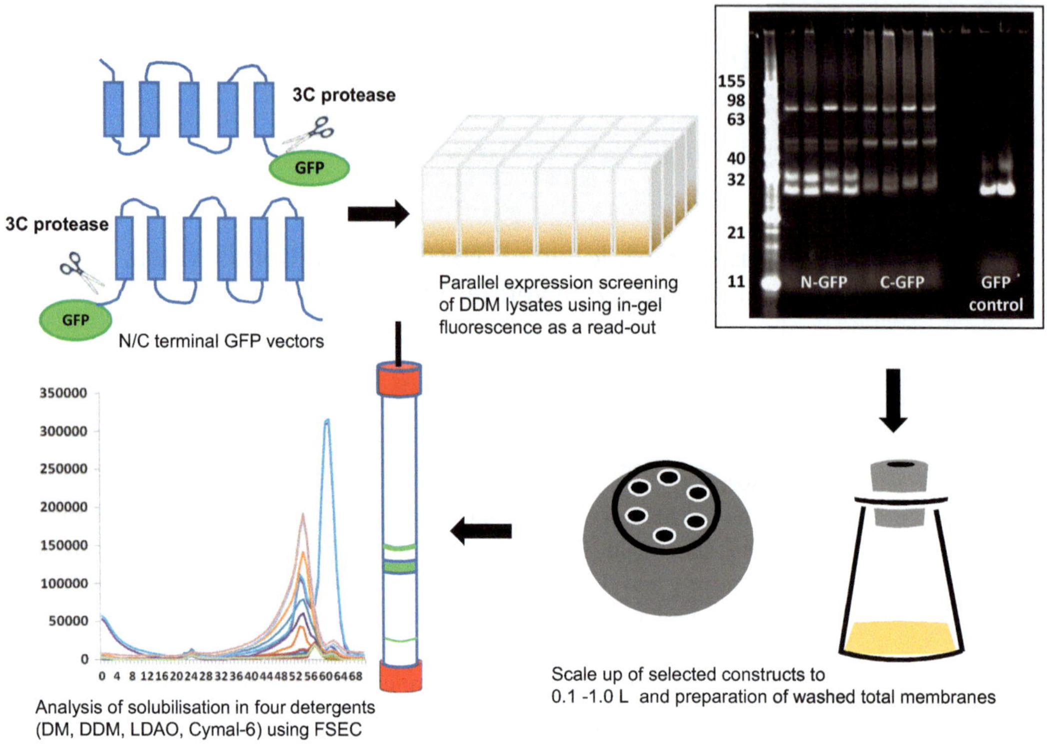

Fig. 1.1 Schematic diagram of workflow for screening for expression of integral membrane proteins

brane proteins. This approach has been greatly facilitated by genetic fusion to a fluorescent reporter protein, typically Green fluorescent protein (GFP). This enables rapid expression screening and hence identification of proteins that are stably inserted into the membrane without the need to purify the membrane protein (Drew et al. 2005). Once a well expressed stable protein is identified the GFP moiety can also be used to monitor purification and for pre-crystallization screening (Drew et al. 2006; Kawate and Gouaux 2006). A generic workflow for this method is shown in Fig. 1.1. In this chapter the use of GFP as a reporter for the expression of membrane proteins in different heterologous hosts will be reviewed.

1.2 Bacteria

Escherichia coli is the most commonly used prokaryotic host for overexpression of IMPs, followed by the Gram positive bacterium, *Lactococcus lactis* (Kunji et al. 2003; Drew et al. 2006; Gordon et al. 2008; Frelet-Barrand et al. 2010; Chen 2012; King et al. 2015). Bacterial hosts have obvious advantages for the over-expression of recombinant proteins with rapid growth rates, inexpensive growth media and the ease of genetic manipulation. Moreover, the biology of transcription, translation and insertion into membranes are also well characterised, allowing manipulation of the host cell to facilitate heterologous expression of proteins. Nevertheless, the expression of membrane proteins in bacteria can be problematical for a number of reasons. The expressed protein may prove to be toxic to the host cell (Kunji et al. 2003) or saturate the membrane insertion machinery (Loll 2003; Wagner et al. 2006). Rare codons in the protein or insufficient amino acid availability (Angov et al. 2008; Marreddy et al. 2010; Bill et al. 2011) or insufficient membrane capacity (Arechaga et al. 2000) may all limit the expression of membrane proteins in bacteria. Therefore, screening for correctly folded protein is critical, with fusion to GFP at either the N or C-terminus now being widely used as a reporter of insertion into the bacterial membrane (Drew et al. 2001; Sonoda et al. 2011; Lee et al. 2014a, b, c). The combination of (1) high-throughput cloning strategies to construct fusion GFP fusion vectors with (2) screening in *E. coli* using in gel-fluorescence of detergent lysates of whole cells, enables the expression of large numbers of IMPs to be evaluated at small scale (Sonoda et al. 2011; Schlegel et al. 2012; Lee et al. 2014a; Bird et al. 2015). For example, in one study, 47 orthologues of bacterial SEDS (shape, elongation, division, and sporulation) proteins were cloned and candidate proteins rapidly identified for further analysis (Bird et al. 2015). Typically an affinity purification tag, for example octa-histidine, is included with the GFP reporter so that fluorescence can be used to monitor the mono-dispersity and integrity of the membrane proteins during purification by size exclusion chromatography (Fluorescence detected Size Exclusion Chromatography, FSEC) (Drew et al. 2006; Bird et al. 2015). Thus, fusion to GFP has facilitated purification to homogeneity and subsequent crystallization of many IMPs expressed in *E. coli*, for example, *Pseudomonas aeruginosa* lysP, *E. coli* sodium-proton NhaA and the *Streptococcus thermophilus* peptide transporter $PepT_{St}$ (Lee et al. 2014b; Nji et al. 2014).

Fusion of IMPs to GFP is useful for comparing expression in different strains of bacteria (see Fig. 1.2 for an example). The *E. coli* strain BL21(DE3) and related strains are most commonly used for heterologous protein production. In these strains, the bacteriophage T7 RNA polymerase is expressed from the mutant *lac*UV5 promoter resulting in high-level expression of a polymerase that is more processive than the native *E. coli* RNA polymerase (Iost et al. 1992). Driving transcription generally leads to higher levels of heterologous protein production. How-

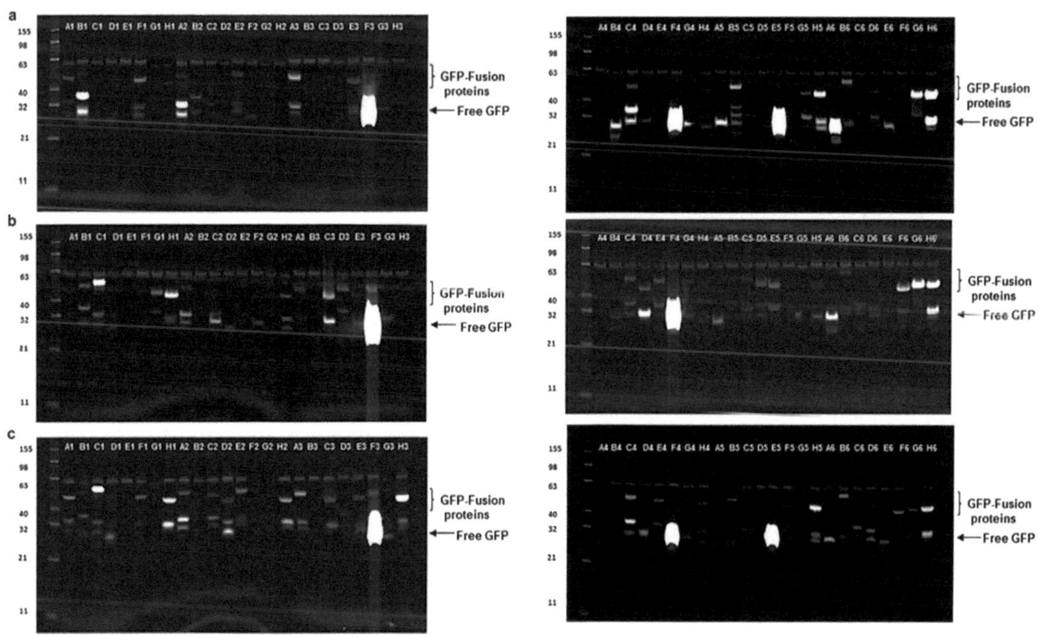

Fig. 1.2 Screening expression in *E.coli* of 47 SED (Sporulation Elongation Division) proteins from a wide range of bacteria, by in-gel fluorescence. Strains were grown in Powerbroth (Molecular Dimensions) and expression induced at 20 °C overnight. (**a**) C41(DE3) plysS, induced with 1 mM IPTG. (**b**) Lemo21 (DE3), grown in the presence of 0.625 mM rhamnose and induced with 1 mM IPTG. (**c**) KRX, induced with 2.5 mM rhamnose and 1 mM IPTG. Detergent lysates of *E. coli* cells were analysed by SDS-PAGE and gels imaged using Blue Epi illumination and a 530/28 filter. A GFP control is shown in lane F3 and the numbers to the *left* refer to the sizes in kDa of molecular weight markers run in parallel

ever, for membrane proteins this can result in saturation of the *Sec* translocon and subsequent misfolding of much of the expressed membrane protein (Wagner et al. 2006, 2007; Klepsch et al. 2011). To avoid this problem, Miroux and Walker isolated strains of BL21(DE3) that survived the over-expression of membrane proteins by an unknown mechanism (Miroux and Walker 1996). These strains, C41(DE3) and C43(DE3), known as the Walker strains, are used pragmatically to express a membrane proteins, though high levels of expression are not seen for all membrane proteins (Miroux and Walker 1996; Wagner et al. 2008). Analyses of the Walker strains, using the bacterial membrane protein YidC fused to GFP (Wagner et al. 2007), showed that mutations in the *lac*UV5 promoter are responsible for the often improved membrane protein expression (Drews et al. 1973; Wagner et al. 2008). The mutations that were found, result in lower levels of mRNA production and hence a slower rate of protein synthesis. This presumably ensures that membrane protein translocation machinery is not saturated.

These data suggested that to optimize expression levels of folded and functional inserted IMPs, it is important to match the rate of transcription /translation with the capacity of the *Sec* translocon. The Lemo21(DE3) strain has been specifically engineered according to this principal and incorporates the gene for T7 lysozyme on a plasmid under the control of the highly titratable rhamnose promoter (Giacalone et al. 2006; Wagner et al. 2008). T7 lysozyme is an inhibitor of T7RNA polymerase, and Schlegel et al. showed that the expression level of a number of membrane proteins could be optimised by varying the level of rhamnose in the cell media (Schlegel et al. 2012). However, not all IMPs express well in Lemo21(DE3) and screening *E. coli* strains with different expression kinetics is important for achieving expression (Schlegel et al. 2012; Bird et al. 2015).

Fusion of IMPs with GFP at the C-terminus of the protein in tandem with the erythromycin resistance protein (23S ribosomal RNA adenine *N*-6 methyl transferase, ErmC) has been used to evolve both *E coli* and *L. lactis* strains for improved production of membrane proteins (Linares et al. 2010; Gul et al. 2014). In both cases the protein is under the regulation of a titratable promoter, the arabinose inducible pBAD promoter in *E. coli* and the NICE (nisin-inducible controlled gene expression) promoter in *L. lactis*. In this approach, the optimum inducer concentration, induction time and temperature of induction are established using readout from the GFP reporter. The cells are then exposed, under these conditions, to increasing levels of erythromycin, since the GFP and ErmC are at the C-terminus, cells that have evolved to express higher levels of the functional protein will be resistant to a higher concentration of erythromycin. The strains are then plated on erythromycin at the highest concentration used and the most fluorescent colonies are analysed. The strains can be cured of the selection plasmid and it was shown that expression is increased for proteins other than the test plasmid (Linares et al. 2010; Gul et al. 2014). The evolved *E. coli* when compared with the parental strain showed up to a tenfold increase in fluorescence levels and when compared to the Walker strains had increased levels of expression per unit of biomass (Gul et al. 2014). Interestingly, deep sequencing of four evolved *E. coli* strains revealed that all had mutations were in the gene encoding DNA-binding protein, H-NS, which is involved in chromosome organization and transcriptional silencing, although the exact mechanism causing the elevated expression is unclear (Gul et al. 2014). In *L. lactis* the strain selection led to a two to eightfold increases in the expression levels of a variety of proteins. In contrast to *E. coli*, deep sequencing of the genome of the evolved strains identified point mutations in a single gene, *nisK*, which is the histidine kinase sensor protein of the two component regulatory system that directs nisin-A mediated expression. It seems likely that the mutations enhance phosphoryl transfer to NisR and increase transcription from the nisin-A promoter (Linares et al. 2010).

Most IMPs have been produced in *E. coli*, which reflects its popularity as a host for heterologous expression of soluble proteins. However other bacterial species may be more suitable for IMP production. For example, Gram positive bacteria, such as *L. lactis*, express two copies of the IMP chaperone YiDC and thus may be better than *E. coli* at translocating heterologous proteins and hence may be less susceptible to saturation of the integration machinery (Zweers et al. 2008; Funes et al. 2009; Funes et al. 2011; Schlegel et al. 2014) A number of other features of *L. lactis*, like the slower growth rate and reduced proteolytic activity when compared to *E. coli*, may also facilitate IMP production in this bacterium (Schlegel et al. 2014).

1.3 Yeast

Like *E. coli*, yeast require relatively low cost of media, have fast growth rates and can be easily genetically modified, making them attractive expression host for IMP production. Moreover, the post translational modifications and lipid environment of yeast cells may be more appropriate for the expression of eukaryotic IMPs. The two yeast strains that have been widely used for IMP production are *Saccharomyces cerevisiae* and *Pichia pastoris* and less commonly, *Schizosaccharomyces pombe* (Yang and Murphy 2009; Yang et al. 2009; He et al. 2014). It is important to note that protein glycosylation in yeast is not typical of higher eukaryotic cells with N-linked glycosylation sites in *S. cerevisiae* hyper-glycosylated with high mannose glycoforms. In *P. pastoris*, the N-linked glycans are shorter than in *S. cerevisiae* and strains have been engineered that add glycoforms more typical of human glycoproteins (Hamilton et al. 2006; Darby et al. 2012).

The GFP screening pipeline used with *E. coli* has been adapted to both *S. cerevisiae* and *P. pastoris* (Drew et al. 2008; Drew and Kim 2012b; Brooks et al. 2013; Scharff-Poulsen and Pedersen 2013). There are, however, some differences, for example, as part of the screening process it can be useful to include a confocal microscope image to confirm the localization of the IMP-GFP fusion protein (Newstead et al. 2007; Drew et al. 2008) (Fig. 1.3). Additionally, *S. cerevisiae* cloning can be carried out by *in vivo* homologous recombination of PCR products into 2 μ based episomal vectors (Drew and Kim 2012a; Scharff-Poulsen and Pedersen 2013). The inducible *GAL1* promoter is often used to drive expression as the yields are generally higher compared to constitutive promoters (Newstead et al. 2007). The induction of the IMP-GFP fusion can be optimized by varying parameters, such as, the timing of induction, using non-selective media, the addition of chemical chaperones such as DMSO, glycerol

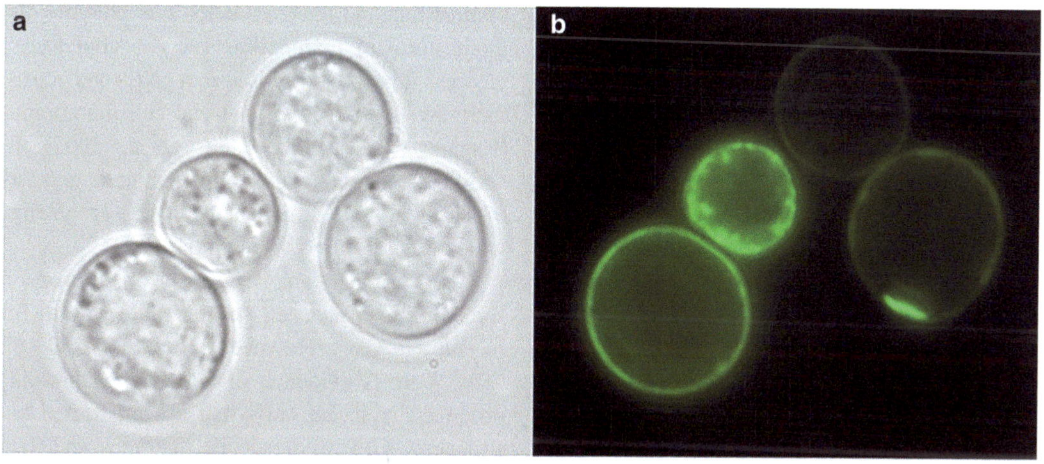

Fig. 1.3 *S. cerevisiae* expressing a recombinant *Candida albicans* TOK1 GFP fusion protein observed under (**a**) white light (**b**) fluorescence optics (Image courtesy of Prof. Per Pedersen, University of Copenhagen)

and histidine and also by lowering the temperature (Drew and Kim 2012c). Furthermore, the levels of expression of IMP-GFP fusions can be improved by the choice of strain and by plasmid engineering (Pedersen et al. 1996; Drew and Kim 2012a; Scharff-Poulsen and Pedersen 2013; Molbaek et al. 2015). For example, Molbaek *et al.* produced functional full-length human ERG K^+-GFP fusions by utilizing the strain PAP1500, which overexpresses the *GAL4* transcriptional activator. This was combined with a vector that has a strong hybrid CYC-GAL promoter and the compromised *leu2-d* gene, which elevates the episomal copy number to between 200 and 400 plasmids per cell in response to leucine starvation (Romanos et al. 1992; Molbaek et al. 2015).

For *P. pastoris*, strain development is more complicated. Since genes to be expressed have to be integrated into the yeast genome using a resistance marker such as zeocin and typically use the methanol inducible AOX1 promoter (Logez et al. 2012). This means that a shuttle vector has to be constructed and different *P. pastoris* transformants have to be characterised to identify the best recombinant strain for IMP expression. Again, fusion to GFP enables the expression screening of integrated clones using a plate based assay. For example, using this methodology Brooks et al. isolated a clone of mouse PEMT (ER associated phosphatidyl ethanolamine N-methyl transferase) that gave a final yield of 5 mg/L of purified protein (Brooks et al. 2013). In an interesting development, Parcej et al. reported the use of fusions to different fluorophores to monitor the expression of the human heterodimeric ATP binding cassette (ABC) transporter associated with antigen processing (TAP) in *P. pastoris*. The subunits were tagged with either monomeric venus and a HIS_{10} tag or monomeric cerulean with a strepII tag, dual wavelength monitoring was then used to monitor expression of individual subunits and purification of the complex (Parcej et al. 2013). This approach could clearly be applied to the expression of multi-subunit IMPs in other cell hosts.

Yeast is clearly a very useful host for expression of IMPs, however in a study of 43 eukaryotic membrane proteins Newstead et al. showed that while 25 out of 29 yeast membrane proteins were produced to greater than 1 mg/L in *S. cerevisiae*, only 4 of the 14 membrane proteins from higher eukaryotic organisms were produced at this level, suggesting that a higher eukaryotic heterologous expression systems is often necessary for higher eukaryotic proteins (Newstead et al. 2007).

1.4 Insect and Mammalian Cells

Insect cells are widely used for the production of eukaryotic recombinant proteins, including IMPs. The cells are easy to handle and in general give higher yields of recombinant proteins than transfected mammalian cells. The main cell lines in use are from *Spodoptera frugiperda* (Sf9 and Sf21) and *Trichoplusia ni* (High Five) with the gene of interest typically introduced using the baculovirus expression vector system (BEVS) (Zhang et al. 2008; Mus-Veteau 2010; Milic and Veprintsev 2015). Transient transfection with plasmid vectors has also been reported for rapid screening of IMP expression using GFP fusion proteins (Chen et al. 2013). In addition, *Drosophilia melanogaster* S2 cells in combination with inducible plasmid vectors have been used for the expression of recombinant IMPs (Brillet et al. 2010). However, it is important to note that the lipid composition of insect cell membranes differs from those of mammalian and bacterial cells. For example, the main sterol in mammalian cells is cholesterol, whereas it is ergosterol in insect cells (and yeast): there are no sterols in bacterial cell membranes (Lagane et al. 2000; Eifler et al. 2007). In addition, N-glycosylation in insect cells consists of short so-called pauci-mannose glycoforms, which are not found on mammalian IMPs.

GFP-tagging can be used for expression screening in insect cells in the same way as for *E. coli* and yeast. However in contrast to *E. coli* cells, there is evidence of GFP-tagged proteins produced in insect cells that are misfolded but still show GFP fluorescence (Thomas and Tate 2014). Fusion to GFP remains a convenient way for screening many constructs in parallel at

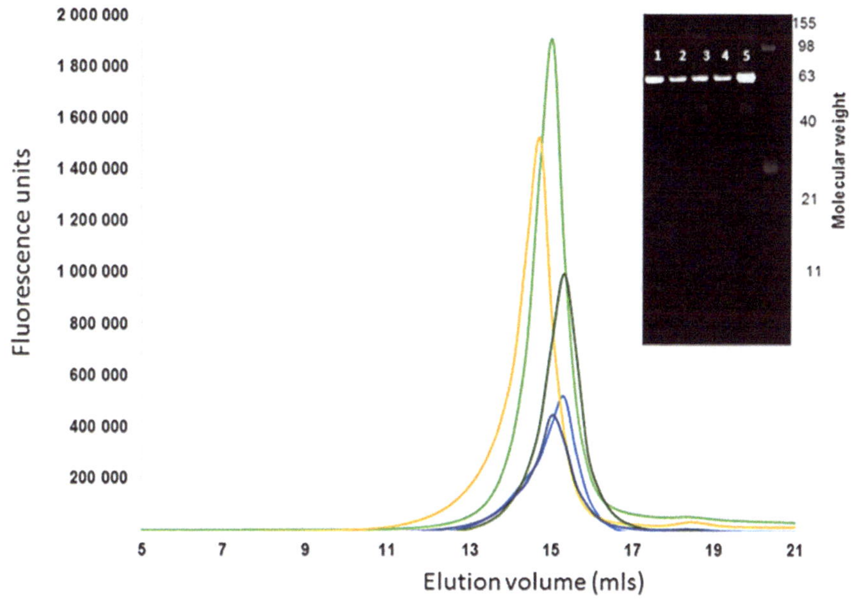

Fig. 1.4 Fluorescence detected size exclusion profiles (FSEC) and in-gel fluorescence (inset) of detergent extracts of the total membrane fraction from SF9 insect cells expressing *Caenorhabditis elegans* GTG1 fused to GFP. Membranes were extracted in the following detergents (1 % final concentration plus 0.2 % cholesterol): n-Decyl-β-D-Maltoside (DM: lane 1, *dark blue trace*); n-Dodecyl-β-D-Maltoside (DDM: lane 2, *dark green trace*); Lauryldimethylamine-N-Oxide (LDAO: lane 3, *yellow trace*); 6-Cyclohexyl-1-Hexyl-β-D-Maltoside (cymal-6: lane 4, *blue trace*); n-Dodecylphosphocholine (FC12; lane 5, *green trace*)

small scale, particularly different orthologues, in order to identify the best expressed candidate for purification and crystallization (Lee and Stroud 2010; He et al. 2014; Hu et al. 2015). Analysis of the subsequent products by FSEC (see Fig. 1.4 for an example) enables the optimal detergent for solubilisation to be identified and any misfolded fusion proteins to be detected.

Transient expression in Human Embryonic Kidney cells (HEK293) provides a rapid way of screening protein expression, including IMPs and has become the system of choice for the production of secreted/cell surface glycoproteins for structural biology (Aricescu and Owens 2013). In particular HEK-293 cells deficient in N-acetylglucosamine tranferase I (HEK Gnt1 −/−) are used to produce proteins containing only a high mannose glycoform, which can be removed by endoglycosidase treatment following purification. Simplifying the N-glycosylation of proteins appears to favour crystallization since sample heterogeneity is reduced (Chang et al. 2007). This approach is equally relevant for modifying the N-glycans of IMPs which may in turn aid crystallization.

The use of GFP fusions in combination with transient expression in HEK cells was introduced by Gouaux and co-workers (Kawate and Gouaux 2006) for optimizing the expression of the ATP-gated ion channel P2X4. Protein production for crystallization was subsequently transferred to insect cells (Kawate et al. 2009). For IMP production in mammalian cells, inducible stable cell lines are usually required to generate sufficient biomass without the problem of toxicity from constitutive expression (Chaudhary et al. 2011, 2012). Although this requires more time and effort than using insect cells, there are now a number of structures of membrane proteins produced in this way. In all cases, multiple constructs were initially screened by transient expression using fusion to GFP as a reporter of protein expression and stability by FSEC analysis. Although recombinant protein yields from mammalian cells are generally lower than either microbial or insect cell over-expression systems, there may be a significant advantage in using mammalian cells for the production of human/mammalian IMPs. The proteins will be produced in a cellular context with native post-translational modifications and lipid environment, it is becoming increasingly apparent that this leads to improved protein quality due to lower levels of misfolded aggregates (Yamashita et al. 2005; Chaudhary et al. 2011).

An alternative to the production of stable cell lines for IMP production is the use of baculovirus mediated gene transduction for large-scale production of IMPs in mammalian cells, typically HEK Gnt1 −/− (Goehring et al. 2014). The so-called BacMam system (Dukkipati et al. 2008) involves the inclusion of a mammalian cell transcription unit(s) within a baculovirus transfer vector so that on generation of a recombinant virus, the inserted gene can be expressed in mammalian cells. The same plasmid vector can be used for small-scale transient transfection of HEK cells to identify the optimal construct and then to generate a BacMam baculovirus for scaling up of protein production by bulk transduction of HEK cells for further characterization (Goehring et al. 2014). Using this protocol, sample preparation can be accomplished in 4–6 weeks, which is at least half the time required to generate and scale-up stable cell lines. The approach has been used by the Gouaux group to produce a number of IMPs for structural determination (Althoff et al. 2014; Baconguis et al. 2014; Dürr et al. 2014; Lee et al. 2014c; Wang et al. 2015).

1.5 Summary and Conclusions

Initially developed for screening the expression of bacterial membrane proteins in *Escherichia coli*, the use of GFP fusions has been successfully extended to eukaryotic hosts, including insect and mammalian cells. Although *E. coli* and yeast are useful tools for the over-expression of recombinant membrane proteins, there is a marked difference in the lipid compositions of membranes from prokaryotes and eukaryotes. This in turn may affect the quality and quantity of heterologous proteins inserted into the host membrane. Given that the host cell determines the nature of post-translational modifications, such as glycosylation and phosphorylation, in choosing an expression host for screening, it may the appropriate to match the host cell to the recombinant product for example, human IMPs in mammalian cells.

Acknowledgments The OPPF-UK is funded by the Medical Research Council, UK (grant MR/K018779/1).

References

Althoff T, Hibbs RE, Banerjee S, Gouaux E (2014) X-ray structures of GluCl in apo states reveal a gating mechanism of Cys-loop receptors. Nature 512(7514):333–337

Angov E, Hillier CJ, Kincaid RL, Lyon JA (2008) Heterologous protein expression is enhanced by harmonizing the codon usage frequencies of the target gene with those of the expression host. PLoS ONE 3(5):e2189

Arechaga I, Miroux B, Karrasch S, Huijbregts R, de Kruijff B, Runswick MJ, Walker JE (2000) Characterisation of new intracellular membranes in Escherichia coli accompanying large scale over-production of the b subunit of F(1)F(o) ATP synthase. FEBS Lett 482(3):215–219

Aricescu AR, Owens RJ (2013) Expression of recombinant glycoproteins in mammalian cells: towards an integrative approach to structural biology. Curr Opin Struct Biol 23(3):345–356

Baconguis I, Bohlen CJ, Goehring A, Julius D, Gouaux E (2014) X-ray structure of acid-sensing ion channel 1-snake toxin complex reveals open state of a Na(+)-selective channel. Cell 156(4):717–729

Bill RM, Henderson PJF, Iwata S, Kunji ERS, Michel H et al (2011) Overcoming barriers to membrane protein structure determination. Nat Biotechnol 29(4):335–340

Bird LE, Rada H, Verma A, Gasper R, Birch J, Jennions M et al (2015) Green fluorescent protein-based expression screening of membrane proteins in Escherichia coli. J Vis Exp 95:e52357

Brillet K, Pereira CA, Wagner R (2010) Expression of membrane proteins in Drosophila Melanogaster S2 cells: production and analysis of a EGFP-fused G protein-coupled receptor as a model. Methods Mol Biol 601:119–133

Brooks CL, Morrison M, Joanne Lemieux M (2013) Rapid expression screening of eukaryotic membrane proteins in Pichia pastoris. Protein Sci 22(4):425–433

Chang VT, Crispin M, Aricescu AR, Harvey DJ, Nettleship JE et al (2007) Glycoprotein structural genomics: solving the glycosylation problem. Structure 15(3):267–273

Chaudhary S, Pak JE, Pedersen BP, Bang LJ, Zhang LB et al (2011) Efficient expression screening of human membrane proteins in transiently transfected human embryonic kidney 293S cells. Methods 55(4): 273–280

Chaudhary S, Pak JE, Gruswitz F, Sharma V, Stroud RM (2012) Overexpressing human membrane proteins in stably transfected and clonal human embryonic kidney 293S cells. Nat Protoc 7(3):453–466

Chen R (2012) Bacterial expression systems for recombinant protein production: E. coli and beyond. Biotechnol Adv 30(5):1102–1107

Chen H, Shaffer PL, Huang X, Rose PE (2013) Rapid screening of membrane protein expression in transiently transfected insect cells. Protein Expr Purif 88(1):134–142

Darby RA, Cartwright SP, Dilworth MV, Bill RM (2012) Which yeast species shall I choose? Saccharomyces cerevisiae versus Pichia pastoris (review). Methods Mol Biol 866:11–23

Drew D, Kim H (2012a) Preparation of Saccharomyces cerevisiae expression plasmids. Methods Mol Biol 866:41–46

Drew D, Kim H (2012b) Screening for high-yielding Saccharomyces cerevisiae clones: using a green fluorescent protein fusion strategy in the production of membrane proteins. Methods Mol Biol 866:75–86

Drew D, Kim H (2012c) Optimizing Saccharomyces cerevisiae induction regimes. Methods Mol Biol 866:191–195

Drew D, von Heijne G, Nordlund P, de Gier JWL (2001) Green fluorescent protein as an indicator to monitor membrane protein overexpression in Escherichia coli. FEBS Lett 507(2):220–224

Drew D, Slotboom DJ, Friso G, Reda T, Genevaux P et al (2005) A scalable, GFP-based pipeline for membrane protein overexpression screening and purification. Protein Sci 14(8):2011–2017

Drew D, Lerch M, Kunji E, Slotboom DJ, de Gier JW (2006) Optimization of membrane protein overexpression and purification using GFP fusions. Nat Methods 3(4):303–313

Drew D, Newstead S, Sonoda Y, Kim H, von Heijne G, Iwata S (2008) GFP-based optimization scheme for the overexpression and purification of eukaryotic membrane proteins in Saccharomyces cerevisiae. Nat Protoc 3(5):784–798

Drews J, Grasmuk H, Unger FM (1973) Peptide chain initiation with chemically formylated Met-tRNAs from E. coli and yeast. Biochem Biophys Res Commun 51(3):804–812

Dukkipati A, Park HH, Waghray D, Fischer S, Garcia KC (2008) BacMam system for high-level expression of recombinant soluble and membrane glycoproteins for structural studies. Protein Expr Purif 62(2): 160–170

Dürr KL, Chen L, Stein RA, De Zorzi R, Folea IM, Walz T, Gouaux E (2014) Structure and dynamics of AMPA receptor GluA2 in resting, pre-open, and desensitized states. Cell 158(4):778–792

Eifler N, Duckely M, Sumanovski LT, Egan TM, Oksche A et al (2007) Functional expression of mammalian receptors and membrane channels in different cells. J Struct Biol 159(2):179–193

Frelet-Barrand A, Boutigny S, Kunji ER, Rolland N (2010) Membrane protein expression in Lactococcus lactis. Methods Mol Biol 601:67–85

Funes S, Hasona A, Bauerschmitt H, Grubbauer C, Kauff F et al (2009) Independent gene duplications of the YidC/Oxa/Alb3 family enabled a specialized cotranslational function. Proc Natl Acad Sci U S A 106(16):6656–6661

Funes S, Kauff F, van der Sluis EO, Ott M, Herrmann JM (2011) Evolution of YidC/Oxa1/Alb3 insertases: three independent gene duplications followed by functional specialization in bacteria, mitochondria and chloroplasts. Biol Chem 392(1–2):13–19

Giacalone MJ, Gentile AM, Lovitt BT, Berkley NL, Gunderson CW, Surber MW (2006) Toxic protein expression in Escherichia coli using a rhamnose-based tightly regulated and tunable promoter system. Biotechniques 40(3):355–364

Goehring A, Lee CH, Wang KH, Michel JC, Claxton DP et al (2014) Screening and large-scale expression of membrane proteins in mammalian cells for structural studies. Nat Protoc 9(11):2574–2585

Gordon E, Horsefield R, Swarts HG, de Pont JJH, Neutze R, Snijder A (2008) Effective high-throughput overproduction of membrane proteins in Escherichia coli. Protein Expr Purif 62(1):1–8

Gul N, Linares DM, Ho FY, Poolman B (2014) Evolved Escherichia coli strains for amplified, functional expression of membrane proteins. J Mol Biol 426(1):136–149

Hamilton SR, Davidson RC, Sethuraman N, Nett JH, Jiang Y et al (2006) Humanization of yeast to produce complex terminally sialylated glycoproteins. Science 313(5792):1441–1443

He Y, Wang K, Yan N (2014) The recombinant expression systems for structure determination of eukaryotic membrane proteins. Protein Cell 5(9):658–672

Hu NJ, Rada H, Rahman N, Nettleship JE, Bird L et al (2015) GFP-based expression screening of membrane proteins in insect cells using the baculovirus system. Methods Mol Biol 1261:197–209

Iost I, Guillerez J, Dreyfus M (1992) Bacteriophage T7 RNA polymerase travels far ahead of ribosomes in vivo. J Bacteriol 174(2):619–622

Kawate T, Gouaux E (2006) Fluorescence-detection size-exclusion chromatography for precrystallization screening of integral membrane proteins. Structure 14(4):673–681

Kawate T, Michel JC, Birdsong WT, Gouaux E (2009) Crystal structure of the ATP-gated P2X(4) ion channel in the closed state. Nature 460(7255):592–598

King MS, Boes C, Kunji ER (2015) Membrane protein expression in Lactococcus lactis. Methods Enzymol 556:77–97

Klepsch MM, Persson JO, De Gier JWL (2011) Consequences of the overexpression of a eukaryotic membrane protein, the human KDEL receptor, in Escherichia coli. J Mol Biol 407(4):532–542

Kunji ER, Slotboom DJ, Poolman B (2003) Lactococcus lactis as host for overproduction of functional membrane proteins. Biochim Biophys Acta 1610(1):97–108

Lagane B, Gaibelet G, Meilhoc E, Masson JM, Cézanne L, Lopez A (2000) Role of sterols in modulating the human mu-opioid receptor function in Saccharomyces cerevisiae. J Biol Chem 275(43):33197–33200

Lee JK, Stroud RM (2010) Unlocking the eukaryotic membrane protein structural proteome. Curr Opin Struct Biol 20(4):464–470

Lee C, Kang HJ, Hjelm A, Qureshi AA, Nji E, Choudhury H et al (2014a) MemStar: a one-shot Escherichia coli-based approach for high-level bacterial membrane protein production. FEBS Lett 588(20):3761–3769

Lee C, Yashiro S, Dotson DL, Uzdavinys P, Iwata S et al (2014b) Crystal structure of the sodium-proton antiporter NhaA dimer and new mechanistic insights. J Gen Physiol 144(6):529–544

Lee CH, Lü W, Michel JC, Goehring A, Du J, Song X, Gouaux E (2014c) NMDA receptor structures reveal subunit arrangement and pore architecture. Nature 511(7508):191–197

Linares DM, Geertsma ER, Poolman B (2010) Evolved Lactococcus lactis strains for enhanced expression of recombinant membrane proteins. J Mol Biol 401(1):45–55

Logez C, Alkhalfioui F, Byrne B, Wagner R (2012) Preparation of Pichia pastoris expression plasmids. Methods Mol Biol 866:25–40

Loll PJ (2003) Membrane protein structural biology: the high throughput challenge. J Struct Biol 142(1):144–153

Marreddy RK, Geertsma ER, Permentier HP, Pinto JP, Kok J, Poolman B (2010) Amino acid accumulation limits the overexpression of proteins in Lactococcus lactis. PLoS ONE 5(4):e10317

Milic D, Veprintsev DB (2015) Large-scale production and protein engineering of G protein-coupled receptors for structural studies. Front Pharmacol 6:66

Miroux B, Walker JE (1996) Over-production of proteins in Escherichia coli: mutant hosts that allow synthesis of some membrane proteins and globular proteins at high levels. J Mol Biol 260(3):289–298

Molbaek K, Scharff-Poulsen P, Helix-Nielsen C, Klaerke DA, Pedersen PA (2015) High yield purification of full-length functional hERG K+ channels produced in Saccharomyces cerevisiae. Microb Cell Fact 14:15

Mus-Veteau I (2010) Heterologous expression of membrane proteins for structural analysis. Methods Mol Biol 601:1–16

Newstead S, Kim H, von Heijne G, Iwata S, Drew D (2007) High-throughput fluorescent-based optimization of eukaryotic membrane protein overexpression and purification in Saccharomyces cerevisiae. Proc Natl Acad Sci U S A 104(35):13936–13941

Nji E, Li D, Doyle DA, Caffrey M (2014) Cloning, expression, purification, crystallization and preliminary X-ray diffraction of a lysine-specific permease from Pseudomonas aeruginosa. Acta Crystallogr F Struct Biol Commun 70(10):1362–1367

Parcej D, Guntrum R, Schmidt S, Hinz A, Tampé R (2013) Multicolour fluorescence-detection size-exclusion chromatography for structural genomics of membrane multiprotein complexes. PLoS ONE 8(6):e67112

Pedersen PA, Rasmussen JH, Jørgensen PL (1996) Expression in high yield of pig alpha 1 beta 1 Na, K-ATPase and inactive mutants D369N and D807N in Saccharomyces cerevisiae. J Biol Chem 271(5):2514–2522

Romanos MA, Scorer CA, Clare JJ (1992) Foreign gene expression in yeast: a review. Yeast 8(6):423–488

Scharff-Poulsen P, Pedersen PA (2013) Saccharomyces cerevisiae-based platform for rapid production and evaluation of eukaryotic nutrient transporters and transceptors for biochemical studies and crystallography. PLoS ONE 8(10):e76851

Schlegel S, Löfblom J, Lee C, Hjelm A, Klepsch M et al (2012) Optimizing membrane protein overexpression in the Escherichia coli strain Lemo21(DE3). J Mol Biol 423(4):648–659

Schlegel S, Hjelm A, Baumgarten T, Vikström D, de Gier JW (2014) Bacterial-based membrane protein production. Biochim Biophys Acta 1843(8):1739–1749

Sonoda Y, Newstead S, Hu NJ, Alguel Y, Nji E et al (2011) Benchmarking membrane protein detergent stability for improving throughput of high-resolution X-ray structures. Structure 19(1):17–25

Thomas J, Tate CG (2014) Quality control in eukaryotic membrane protein overproduction. J Mol Biol 426(24):4139–4154

Wagner S, Bader ML, Drew D, de Gier JW (2006) Rationalizing membrane protein overexpression. Trends Biotechnol 24(8):364–371

Wagner S, Baars L, Ytterberg AJ, Klussmeier A, Wagner CS et al (2007) Consequences of membrane protein overexpression in Escherichia coli. Mol Cell Proteomics 6(9):1527–1550

Wagner S, Bader ML, Drew D, de Gier JW (2008) Tuning Escherichia coli for membrane protein overexpression. Proc Natl Acad Sci U S A 105(38):14371–14376

Wang KH, Penmatsa A, Gouaux E (2015) Neurotransmitter and psychostimulant recognition by the dopamine transporter. Nature 521(7552):322–327

Yamashita A, Singh SK, Kawate T, Jin Y, Gouaux E (2005) Crystal structure of a bacterial homologue of Na+/Cl–dependent neurotransmitter transporters. Nature 437(7056):215–223

Yang H, Murphy AS (2009) Functional expression and characterization of Arabidopsis ABCB, AUX 1 and PIN auxin transporters in Schizosaccharomyces pombe. Plant J 59(1):179–191

Yang Y, Hu Z, Liu Z, Wang Y, Chen X, Chen G (2009) High human GLUT1, GLUT2, and GLUT3 expression in Schizosaccharomyces pombe. Biochemistry (Mosc) 74(1):75–80

Zhang F, Manzan MA, Peplinski HM, Thiem SM et al (2008) A new Trichoplusia ni cell line for membrane protein expression using a baculovirus expression vector system. Vitro Cell Dev Biol Anim 44(7): 214–223

Zweers JC, Barák I, Becher D, Driessen AJ, Hecker M et al (2008) Towards the development of Bacillus subtilis as a cell factory for membrane proteins and protein complexes. Microb Cell Fact 7:10

Detergents in Membrane Protein Purification and Crystallisation

2

Anandhi Anandan and Alice Vrielink

Abstract

Detergents play a significant role in structural and functional characterisation of integral membrane proteins (IMPs). IMPs reside in the biological membranes and exhibit a great variation in their structural and physical properties. For *in vitro* biophysical studies, structural and functional analyses, IMPs need to be extracted from the membrane lipid bilayer environment in which they are found and purified to homogeneity while maintaining a folded and functionally active state. Detergents are capable of successfully solubilising and extracting the IMPs from the membrane bilayers. A number of detergents with varying structure and physicochemical properties are commercially available and can be applied for this purpose. Nevertheless, it is important to choose a detergent that is not only able to extract the membrane protein but also provide an optimal environment while retaining the correct structural and physical properties of the protein molecule. Choosing the best detergent for this task can be made possible by understanding the physical and chemical properties of the different detergents and their interaction with the IMPs. In addition, understanding the mechanism of membrane solubilisation and protein extraction along with crystallisation requirements, if crystallographic studies are going to be undertaken, can help in choosing the best detergent for the purpose. This chapter aims to present the fundamental properties of detergents and highlight information relevant to IMP crystallisation. The first section of the chapter reviews the physicochemical properties of detergents and parameters essential for predicting their behaviour in solution. The second section covers the interaction of detergents with the biologic membranes and proteins followed by their role in membrane

A. Anandan • A. Vrielink (✉)
School of Chemistry and Biochemistry, University of Western Australia, 35 Stirling Highway, Crawley, WA 6009, Australia
e-mail: anandhi.anandan@uwa.edu.au; alice.vrielink@uwa.edu.au

I. Moraes (ed.), *The Next Generation in Membrane Protein Structure Determination*,
Advances in Experimental Medicine and Biology 922, DOI 10.1007/978-3-319-35072-1_2

protein crystallisation. The last section will briefly cover the types of detergent and their properties focusing on custom designed detergents for membrane protein studies.

Keywords
Detergents • Lipids • Micelles • Membrane proteins • Protein purification • Crystallisation

2.1 Physicochemical Properties of Detergents

Detergents are surfactants (surface acting reagents) that decrease the interfacial tension between two immiscible liquids. The overall molecular structure of detergents consists of a hydrophilic polar head group and a hydrophobic non-polar tail group (Fig. 2.1a) that renders them amphiphilic. The polar head group of a detergent can be ionic, non-ionic or zwitterionic and usually has a strong attraction for aqueous solvent molecules whereas the detergent non-polar tail is generally repelled from the aqueous solvent. Consequently, in an aqueous medium, the hydrophobic tail of detergent molecules usually orients itself to minimize contact with water while the hydrophilic head interacts with the water molecules. As a result, the detergent monomers align themselves as a single layer at the hydrophilic-hydrophobic interface, reducing the surface tension of the solvent (Fig. 2.1b). This alignment not only reduces the interaction of the hydrophobic tail with water molecules, it also allows the interaction between the detergent head group and the solvent, facilitating the detergent molecules to stay soluble in aqueous media (Rosen and Kunjappu 2012).

Detergent molecules persist as monomers in solution up to a particular concentration. As the detergent concentration increases, detergent molecules assemble into complex structures called micelles. The hydrophobic tails of the detergent molecules pack together, forming the core of the micelle and reducing their interaction with the water molecules. In contrast, the polar head groups orient themselves outwards from the micelle core, enabling interaction with the aqueous solvent (Fig. 2.1c). The minimal detergent concentration required for the formation of micelles is called the 'critical micelle concentration ' (CMC) and the number of detergent monomers required to form a micelle is called the 'aggregation number' (Helenius

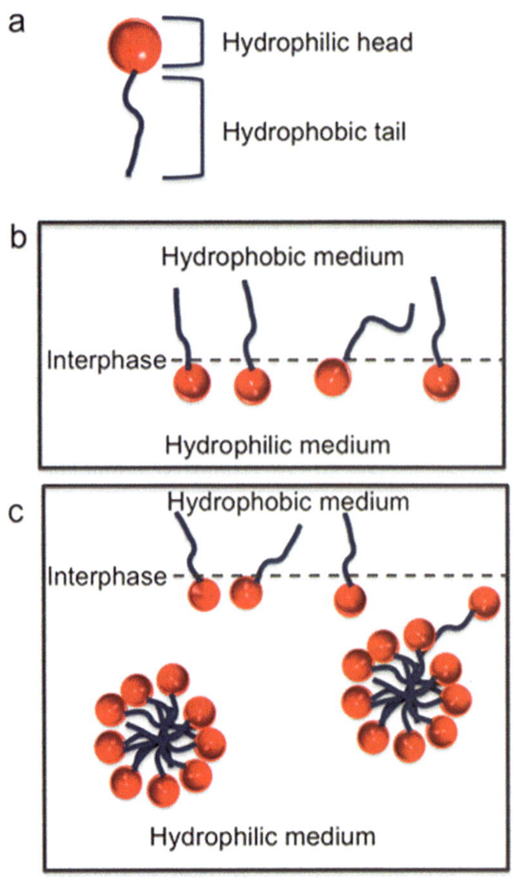

Fig. 2.1 (**a**) Schematic representation of the overall molecular structure of detergents. (**b**) Alignment of detergent molecules at hydrophobic and hydrophilic interface and (**c**) detergent micelles at CMC

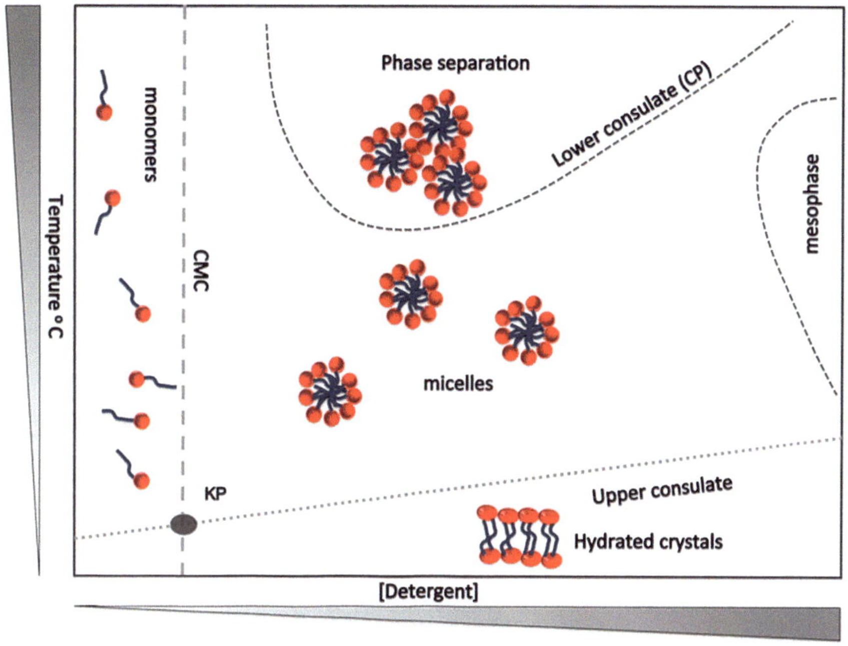

Fig. 2.2 A general phase diagram showing the various phases and their boundaries at varying detergent concentration and temperature. *KP* represents the Krafft point and *CP* represents the cloud point

et al. 1979; Neugebauer 1990). The CMC is of great importance when extracting and solubilising membrane proteins for structural and functional studies. The detergent CMC is dependent on the detergent alkyl chain size and its saturation. For example, the CMC value decreases with the length of the alkyl chain and increases with the addition of double bonds. It is thus understandable that it is the CMC value that determines the micelle size of a detergent. While detergents with lower CMC values form large micelles and exchanging them with other detergents is difficult, detergents with high CMC values require a higher concentrations for extraction and purification (Keyes et al. 2003). Detergents with a CMC between 0.5 and 50 mM have been reported to be suitable for IMP solubilisation and purification. Finally, experimental conditions such as buffer composition and temperature also have a profound influence on the CMC and aggregation number of the detergent.

Above the CMC the detergent molecules co-exist as both monomers and micelles in solution. Detergent solutions are also dynamic systems undergoing a constant exchange of detergent molecules between monomeric and micellar state (le Maire et al. 2000). A further increase in the detergent concentration might result in aggregation of the detergent micelles leading to phase separation. The two phase-system comprises a detergent rich phase and detergent poor phase (Arnold and Linke 2007). In addition to the influence of the detergent concentration on micelle formation and phase separation, the temperature, pH, ionic strength and type of detergent also play an important role. The temperature at which detergent monomers form micelles is called the Krafft point or upper consolute temperature (Gu and Sjöblom 1992). The temperature at which phase separation occurs is called the cloud point or lower consolute temperature (Arnold and Linke 2007). Figure 2.2 shows a simplified general phase diagram of a detergent displaying the solubility of the detergent as a function of concentration and temperature.

Detergent micelles are asymmetric in structure with rough surfaces and disorganised clumps of alkyl tails within the hydrophobic core region (Garavito and Ferguson-Miller 2001). Micelle

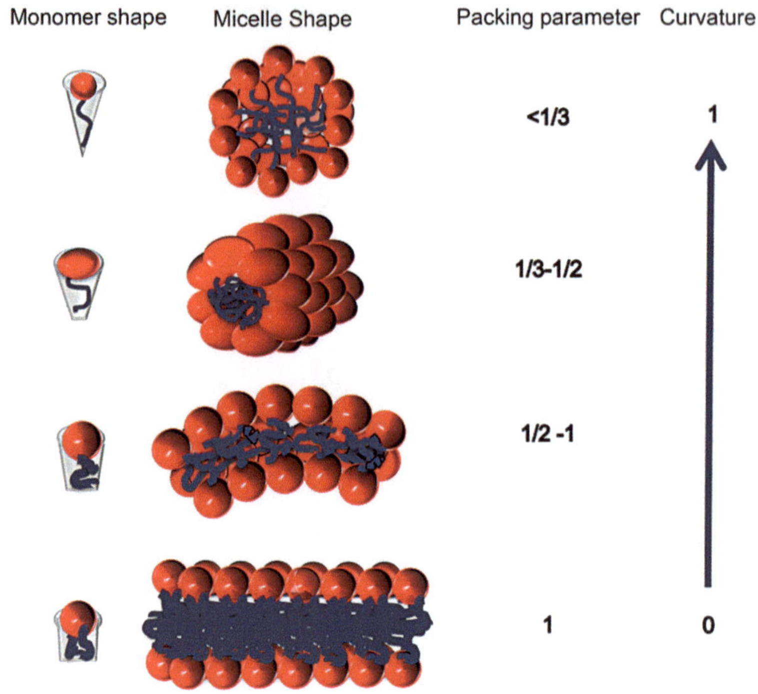

Fig. 2.3 Representation how detergent monomers shape influences the micelle shape, packing parameter and the curvature of detergent micelles

shapes vary from spheres to bilayers, as shown in Fig. 2.3, depending on the overall size and shape of the detergent monomers. Furthermore, the size of the detergent molecule determines the packing ability while the shape of the detergent molecule determines the curvature of the micelle formed (Lichtenberg et al. 2013). The packing ability or packing parameter (P) of detergent monomers is defined as:

$$P = \frac{v}{l\,a}$$

where 'v' is the volume of the hydrocarbon chain, 'l'is the length of the hydrocarbon chain that is extended in fluid and 'a' is the optimal area per molecule. Spherical micelles are formed when the P value is between 0 and 1/3, cylindrical when the P value is between 1/3 and 1/2 and bilayers occur when the P value is between1/2 and 1 (Nagarajan 2002). The packing parameter predicts the assembly and shape of the micelles. For example, detergents with a large head group or which undergo strong head group repulsion will have a smaller packing parameter, which results in spherical micelles. Likewise, detergents with double tails will have a packing parameter twice as large as the corresponding single tail molecule when the tail length and head group are constant. As a result the double tail molecules form a bilayer vesicle while single tail molecules with the same head group form spherical or globular micelles (Nagarajan 2011).

Increasing the length of the hydrophobic chain not only affects the packing parameter, but also the diameter of the micelles formed. In general, the size of micelles increases by 1.2–1.5 Å for every hydrocarbon added to the alkyl chain (Oliver et al. 2013). In addition, the overall shape of the detergent monomer determines the spontaneous curvature of these amphiphilic molecules in solution. Detergent monomers that are cylindrical in shape are curvophobic and assemble into a bilayer conformation with a spontaneous curvature of 0. On the other hand, molecules that are conical in shape are curvophilic and have a negative or positive spontaneous curvature, resulting in a spherical or tubular micelle, respectively, as shown in the Fig. 2.3 (Lichtenberg et al. 2013). The spontaneous curvature of the detergent micelle is inversely proportional to the radius of the surface along which a given amphiphile assembles (Lichtenberg et al. 2000).

Finally, the packing parameter is not only influenced by the size and shape of the detergent, but also by the solvent composition. The presence of electrolytes decreases the electrostatic repulsion between head groups in the case of ionic detergents. In contrast, for non-ionic detergents, the electrolytes indirectly affect the hydrophobic tail packing by affecting hydrogen bonding of water molecules. Kosmotropic electrolytes favour the hydrogen bonding of water molecules and stabilize the hydrophobic tail packing, decreasing the CMC of the detergent (Ray and Nemethy 1971; Damodaran and Song 1990). Additionally, an increase in temperature causes a decrease in the steric repulsion effects between the head groups in non-ionic detergents, allowing more detergent monomers to pack into the micelles (Nagarajan 2011). Also, the presence of polar solvents decreases the interfacial tension of the solvent, leading to tighter packing of detergent molecules. For example, a transition from cylindrical micelles to spherical micelles can be achieved by the addition of low molecular weight alcohols such ethanol or isopropanol (Nagarajan 2011).

The hydrophilic-lipophilic balance (HLB) is a calculated parameter used to describe the surface activity of detergents in solution. It is represented as a scale ranging from 1 to 40 and is dependent on the size and ratio of the hydrophilic to lipophilic part of the detergent. The HLB value is proportional to the solubility of the detergent in water (Neugebauer 1990), where the HLB number increases as the solubility of the detergent in water increases. For example, detergents with HLB values lower than 10 are usually of low solubility in water, while detergents with HLB values higher than 10 are easily soluble in water or buffer. The most common detergents used in membrane protein research have a HLB number between 12 and 15. In addition, the HLB scale is useful for comparing detergents within the same family (Linke 2009).

The structures of detergent micelles mimic the arrangement of lipids in native membrane bilayers, in which membrane proteins are normally embedded or anchored. The ability of detergents to form micelles and be fairly soluble in aqueous media makes them suitable to extract and solubilise membrane proteins. In addition, the micellar environment must be able to reconstitute the extracted IMPs without affecting their structural and functional properties. Though the detergent micelle structure is similar to the membrane bilayer, their head group and the hydrophobic core are packed loosely with constant exchange of detergent monomers between the aqueous medium and micelles resulting in a dynamic environment. Detergents tend to have bulkier head groups when compared to bilayer forming lipids, a result of this difference is that the inner core of a micelle is less shielded by the bulky head groups, allowing water molecules to access the hydrophobic core (Goyal and Aswal 2001; Bordag and Keller 2010).

In the native state, the conformation and function of IMPs are influenced and maintained by the lateral pressure provided by the membrane and also by the presence of lipid molecules bound to the hydrophobic surface of the protein (Anglin and Conboy 2008). Detergent micelles do not exert a similar lateral pressure to the extracted IMP and in some instances, bound lipids necessary for protein stability and function are removed from the protein periphery (Bae et al. 2015). In spite of all these differences, specific detergents have been successfully used for the extraction, purification, and structural characterization of IMP.

2.2 Role of Detergents in Membrane Protein Solubilisation

The first step in membrane protein solubilisation is the disruption of the membranes. An understanding of detergent-lipid interactions is of practical importance to extract the protein from the membrane and to reconstitute it in a suitable environment for functional studies (Lasch 1995). Biological membranes are composed of a variety of loosely packed lipids, differing in their hydrophobicity, along with proteins and peptides embedded in between them. Saturated lipids with higher hydrophobicity pack more tightly than unsaturated lipids with lower hydrophobicity. This

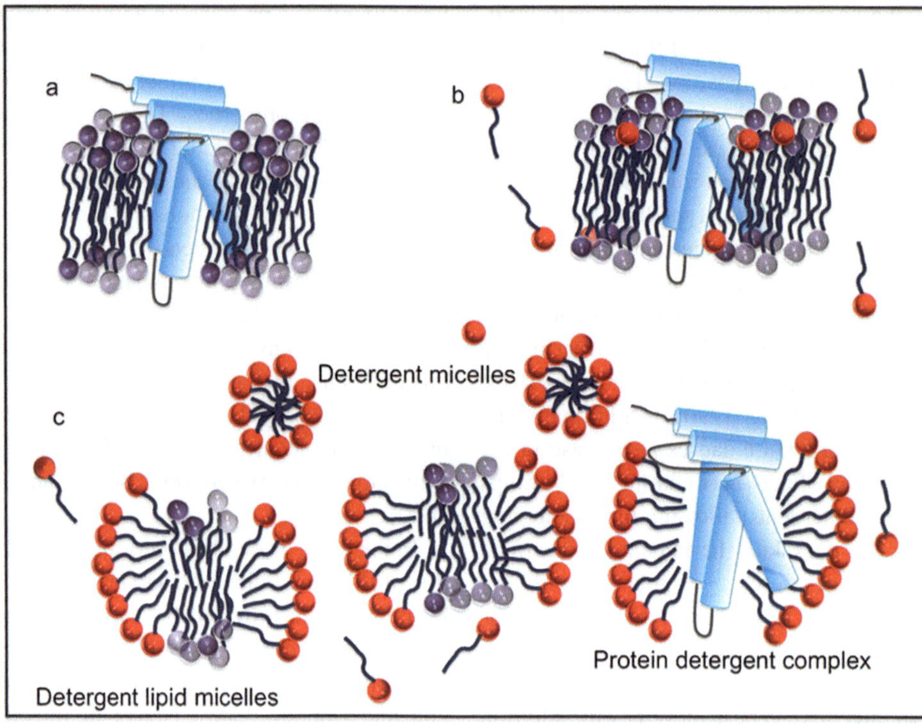

Fig. 2.4 Representation of membrane protein solubilisation. (**a**) Membrane protein embedded in the lipid bilayer. (**b**) Insertion of detergent monomers disrupting the lipid bilayer. (**c**) Solubilisation of the lipid bilayer and extraction of membrane protein

arrangement leads to lipid driven compartmentalisation in membranes. Thus the ability of the detergent to solubilise such membranes depends both on the detergent and the lipid properties (Schuck et al. 2003).

The detergent structure, it's concentration and the micelle structure play important roles in the penetration and disruption of the membrane. Detergent molecules, even at lower concentrations, are capable of interacting with and permeabilising the membrane (Kragh-Hansen et al. 1993). The interaction between the detergent and the membrane starts as non-cooperative binding to the outer layer of the biological membrane. As the concentration of the detergent in the solution increases, the interaction with the membrane becomes cooperative. The concentration at which cooperative binding begins is called the C^{sat}, and the cooperative and non-cooperative binding of detergent to the biological membrane is called the transbilayer mechanism. The bound detergent molecules start flipping from the outer bilayer to the inner bilayer. This is a dynamic action and is described as the flip-flop mechanism (Kragh-Hansen et al. 1998, 1993). The ability of the detergent to enter into the flip-flop mechanism depends on the HLB and on the size of the detergent, which also determines the speed of solubilisation (Lichtenberg et al. 2013). The more detergent monomers are incorporated cooperatively into the membrane, the more disrupted the membrane becomes. Further incorporation of detergent leads to the formation of lipid-detergent mixed micelles. The membrane falls apart during this process (Fig. 2.4). Detergents that are capable of flip-flopping and solubilise membranes faster are commonly called "strong detergents". Others, particularly those with large polar head groups, which lack the capability of flip-flopping across the non-polar membrane bilayer, are called "mild detergents". The micellar mechanism is an alternate proposed process, where detergent micelles, not individual detergent molecules, are involved in solubilising membranes. Detergent micelles interact with the outer layer of the membrane and extract the membrane lipids. This leads to a rearrangement of the remaining phospholipids from the inner to the outer leaflet of the membrane, resulting in fragmentation, followed by solubilisation of the membrane (Kragh-Hansen et al. 1998).

Membrane-embedded proteins can perturb the orderly arrangement of lipids in the membrane, promoting easy access for detergent micelles to insert and solubilise the bilayer (Kragh-Hansen et al. 1998). Some tightly packed membrane bilayers can be resistant to the detergent used, resulting in extracted proteins that are still associated with lipid moieties (Ilgu et al. 2014). Therefore, an optimal detergent/protein ratio and detergent/lipid ratio is required for efficient solubilisation of such membranes and for complete protein extraction (Lórenz-Fonfría et al. 2011). The concentration of detergent dictates the amount of micelles in solution that is essential for accommodating the solubilised protein molecules. The balance between the hydrophobic and hydrophilic detergent moieties also influences the ability of a detergent to solubilise the membrane. Detergents with moderate hydrophobicity are capable of perforating the membrane, resulting in efficient solubilisation. Detergents with a high HLB value are more hydrophilic and their micelles tend to accumulate in solution leading to inefficient solubilisation. In contrast, detergents with a low HLB value are more hydrophobic, they incorporate easily into the bilayer, inducing vesicle growth, however not interfering with the protein solubilisation (Lin et al. 2011; Lichtenberg et al. 2013). Amphiphile compounds derived from pentaerythritol with different ratios of hydrophilic and hydrophobic composition, have been used to study the HLB characteristics of detergents and the solubility of membranes. Amphiphiles with single polar head groups and single alkyl chains or detergents with two polar heads and di-alkyl chains were shown to be more efficient in solubilisation and protein stabilisation than detergents with larger polar head groups or tri-alkyl chains (Zhang et al. 2011).

It is also important to keep in mind that the type of detergent used for the solubilisation of biological membranes depends on the type of membrane targeted for solubilisation. For example, detergents like Sarkosyl and Triton X-100 are very suitable for the solubilisation of the bacterial inner membrane, but not for the outer membrane (Linke 2009). Wiseman et al. in their efforts to solubilise and purify the ABC transporter BmrA, showed that the solubilisation profile from different detergents resulted in different levels of protein contaminants from the same cell line. For example, OmpF, an outer membrane protein, was solubilised in excess when extracting BmrA with n-Dodecylphosphocholine (Fos-choline-12) but the same did not happened with Lauryldimethylamine N-oxide (LDAO) or maltose derivatives (Wiseman et al. 2014). Fos-choline based detergents are zwitterionic with a head group similar to phospholipids in native membranes and with a single hydrophobic tail. This similarity allows them to pack and disrupt the *E. coli* membranes efficiently. Sometimes harsher detergents with ionic head groups may be used for effective solubilisation of membranes and the removal of the desired IMP. Furthermore, the concentration of detergent used for solubilisation should always be higher than the concentration required to maintain the protein in a soluble and stable state. However, high concentrations of detergent may also be detrimental to the stability of the protein immediately after the extraction/solubilisation. Hence, detergent reduction and/or detergent exchange are often essential. This is discussed in the following section.

2.3 Detergent Exchange or Removal

Different detergents may be required for the different steps involved in the structural and functional studies of membrane proteins. Additionally, as mentioned above, the high concentration of detergent required for the protein extraction/solubilisation is typically not suitable for downstream applications. Hence, detergent exchange or removal might be necessary for purification and crystallisation. This can be achieved either by dialysis, size exclusion chromatography or adsorption on hydrophobic beads depending on the properties of the detergent used. Dialysis can be used for detergents that form small micelles and can pass through the pores of the dialysis tubing. During dialysis, free detergent monomers traverse the dialysis membrane reducing the detergent concentration in the protein solution. Over a period of time the concentration of the detergent decreases in the solution (Lorch and Batchelor 2011). Size exclusion chromatography is efficient for detergent removal or exchange if the micellar size is significantly different from the size of the membrane protein (Furth et al. 1984). Hydrophobic beads can also be used to adsorb the excess detergent for downstream applications. Furthermore, affinity tagged membrane proteins can be bound to the corresponding affinity column, washed and eluted with buffers containing lower concentrations of detergent or a different detergent (Arnold and Linke 2008; Seddon et al. 2004).

2.4 Role of Detergents in Membrane Protein Stability

IMPs usually have one or more segments that are mainly made up of hydrophobic amino acid residues that span the membrane bilayer. They may also have hydrophilic segments that extend into either the cytoplasm or the periplasm. Interspersed between the membrane spanning hydrophobic segments are loops comprised of amino acid residues with either polar or charged side chains that do not penetrate the hydrophobic core of the membrane (Bowie 2005). The hydrophobicity of the membrane spanning segments and the polar residues within the loops influence both the insertion of the protein into membrane and their stability within the membrane. Additional factors such as the membrane lipid composition, the low dielectric constant of membrane interior, the lateral pressure exerted by the membrane and the presence of other proteins and peptides contributes to the structural stability and conformational flexibility of IMPs (Gohon and Popot 2003; Bordag and Keller 2010). Stability and conformational flexibility are essential for biological function of IMPs.

The detergent used for purification and storage of the protein must be able to maintain this integrity for downstream studies. Not all detergents are suitable for this purpose and certain detergents can cause denaturation of the native conformation or unfolding of the protein. Detergents used for protein extraction from the membrane may be harsh and a different detergent may be required for purification. Understanding how detergents assemble around the protein and interact with the macromolecule may assist the user in choosing a suitable detergent to maintain protein stability.

Generally, detergent micelles are highly curved when compared to the membrane bilayer, also exhibiting a less orderly arrangement of detergent monomers as a result of their bulky head groups. This loose arrangement, along with the increased solubility of the detergents in the aqueous medium, enables a constant exchange of detergent monomers between the micelles and the newly formed protein-detergent complex (PDC). As a result, the shielding effect of the interfacial head group becomes weak and water molecules penetrate more easily into the hydrophobic core of the micelles compared to the membrane lipid bilayer (Bordag and Keller 2010). In spite of these differences, many specific detergents have been reported to maintain the native conformation of proteins, successfully making them suitable for IMP research studies (Franzin et al. 2007; Tulumello and Deber 2012).

The detergent concentration, size and shape of the micelles, are examples of parameters that affect the formation of the PDC. The solubilised protein has a considerable amount of detergent bound to it (from 40 to 200 % of its weight). The amount of detergent bound to the protein is dictated by the concentration of detergent used (Garavito et al. 1996) and the exposed hydrophobic surface area of the protein (Tulumello and Deber 2012; Ilgu et al. 2014).

Large hydrophobic sections of IMPs are usually covered with detergent monomers forming an oblate protein detergent micelle. The hydrophobic part of the detergent monomer aligns around hydrophobic portion of the IMP such that they are both shielded from the aqueous medium (le Maire et al. 2000). On the other hand the projecting loops of IMPs with their hydrophilic residues can impose a steric hindrance and reduce the packing of the detergent molecules around the hydrophobic area (Melnyk et al. 2003). The cross sectional area of the PDC depends on the packing parameter of the detergent around the exposed hydrophobic surface of the protein and this in turn depends on the size of the non-polar tail and polar head group portions of the detergent monomers. Detergents with longer alkyl chains have been reported to be milder to the protein compared to those with shorter alkyl chains (Wiener 2004). The diameter of the micelles formed by detergents with 7–12 carbon atoms in their non-polar tail matches the thickness of most native biological membranes. The micelles formed by longer alkyl chains exhibit lower water permeability into the hydrophobic core compared to micelles formed by shorter chain detergents. This absence of water molecules is essential for stabilising the intrahelical hydrogen bonds of the transmembrane helices. Also, larger micelles formed by detergents with longer alkyl chains are capable of exerting a higher lateral pressure on the protein transmembrane helices. While these properties of detergent micelles help to maintain the stability of the IMP structure *in vitro* (Santonicola et al. 2008) the hydrophilic residues in the loops interact with the head group of the detergent playing an addition important role in the stabilisation of the protein within the detergent micelle. Insufficient packing of detergent molecules around the hydrophobic region of the protein can lead to protein aggregation in an effort to sequester their hydrophobic surfaces from the water molecules in the aqueous medium (Prive 2007). This is a particularly serious problem for detergents with short (C7 – C10) alkyl chains such as octylglucoside (OG).

Contrary to non-ionic detergents, ionic detergents have been reported to bind differently to the soluble domains versus the transmembrane domains of IMPs. Monomers of ionic detergents bind to residues with charged side chains, initially stabilising the protein. As the concentration of the detergent increases, more cooperative binding occurs, especially in the soluble domain, resulting in destabilisation of the protein structure (Otzen 2011). Repulsion between the charged detergent head groups is the main reason for the unfolding of the protein (Otzen 2002; Tulumello and Deber 2012). A few ionic detergents capable of binding to the helices in the transmembrane domain also cause a destabilisation of the helix – helix interactions, resulting in the unravelling of the protein (Renthal 2006). Lactose transporter protein from *Streptococcus thermophiles* adopted different conformations in two different non-ionic detergents. It had a random conformation lacking activity in n-Dodecyl β-D-maltoside (DDM) while in Triton X-100 the protein adopted a native conformation and regained activity (Knol et al. 1998).

A number of additional factors require careful attention while purifying IMP in detergent micelles. One of them is the total molecular weight of the PDC. The molecular weight and shape of the PDC is different to the molecular weight of the protein alone and can affect the separation of the protein from impurities by size exclusion chromatography methods. Smaller IMPs bind more detergent per gram of protein than larger proteins thus masking the size of the protein moiety. Hydrophobic interaction chromatography is unsuitable for all types of detergents while ion exchange chromatography should not be used for detergents with charged head groups (Schagger et al. 2003).

2.5 Detergents for Membrane Protein Purification and Crystallisation

2.5.1 Conventional Detergents

Detergents commonly used for IMP purification and crystallisation are broadly classified as ionic, non-ionic and zwitterionic depending

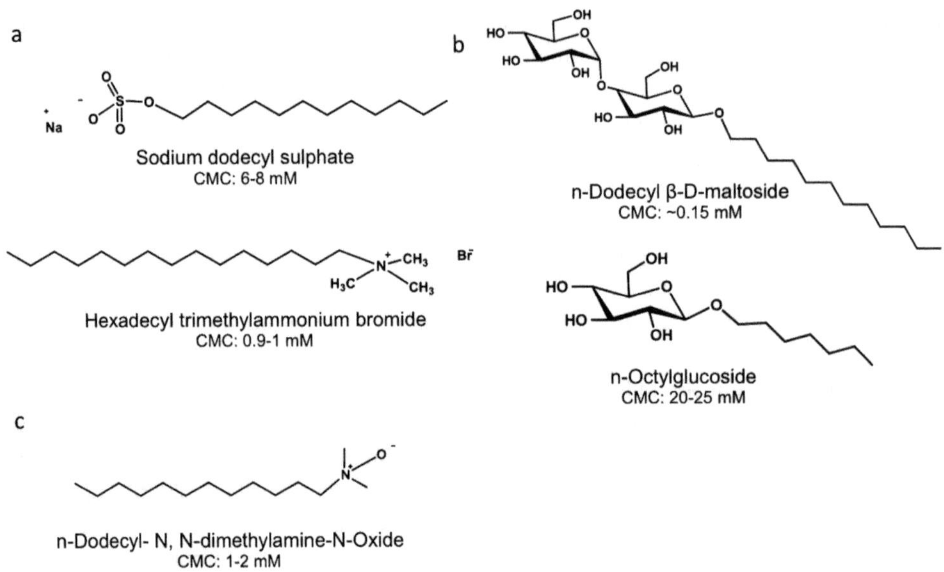

Fig. 2.5 Chemical structures of conventional detergents. (**a**) Ionic detergents; e.g., sodium dodecyl sulphate (anionic detergent) and hexadecyl trimethylammonium bromide (cationic detergent). (**b**) Non-ionic detergents; e.g., n-dodecyl-β-D-maltoside (DDM) and n-octylglucoside (OG) and (**c**) Zwitterionic detergent; e.g., n-dodecyl-N,N-dimethylamine-N-oxide (LDAO)

on the charge of their hydrophilic head group. Ionic detergents have either an anionic or a cationic charged head group. The hydrophobic tail can be composed of either a straight chain hydrocarbon, as in sodium dodecyl sulphate (SDS) and cetyltrimethylammonium bromide (CTAB) (Fig. 2.5a) or exhibit a steroidal structure as in sodium deoxycholate. On the other hand, zwiterionic detergents have head groups that are both anionic and cationic as in the case for lauryldimethylamine-N-oxide (LDAO) (Fig. 2.5b). In general, zwiterionic detergents are milder than the ionic detergents but still efficient in the solubilisation of IMP. Non-ionic detergents have an uncharged head group, such as a glycosidic group or polyoxyethylene, and are classified as mild detergents. Examples of non-ionic detergents are n-dodecyl-β-D-maltoside (DDM), n-decyl-β-D-maltoside (DM) and n-octyl-glucoside (OG) (Fig. 2.5c). So far DDM, DM, OG and LDAO have been the most common detergents that have contributed to various MP structures that has been successfully solved and still remain the first choice when working with membrane proteins (Parker and Newstead 2012).

2.5.2 New Detergents

A number of new generation detergents have emerged in the market for membrane protein characterisation. They have been purposefully designed to combat the problems associated with conventional detergents in terms of solubilisation, stabilisation and crystallisation of membrane proteins. One problem associated with the instability of solubilised membrane proteins is the highly flexible alkyl chain of the detergent (Zhang et al. 2011). The newer detergents are designed to have enhanced rigidity. Their alkyl chain structures are altered by:

- Substituting hydrogen atoms with fluorine atoms (Durand et al. 2014)
- Incorporating cyclic molecules (e.g. cyclic maltosides)
- Branching the alkyl chain (e.g. tripods)

Examples and chemical structures of these alkyl modified detergents are shown in Fig. 2.6.

Some newer detergents have also been designed to mimic the lipids in the membrane

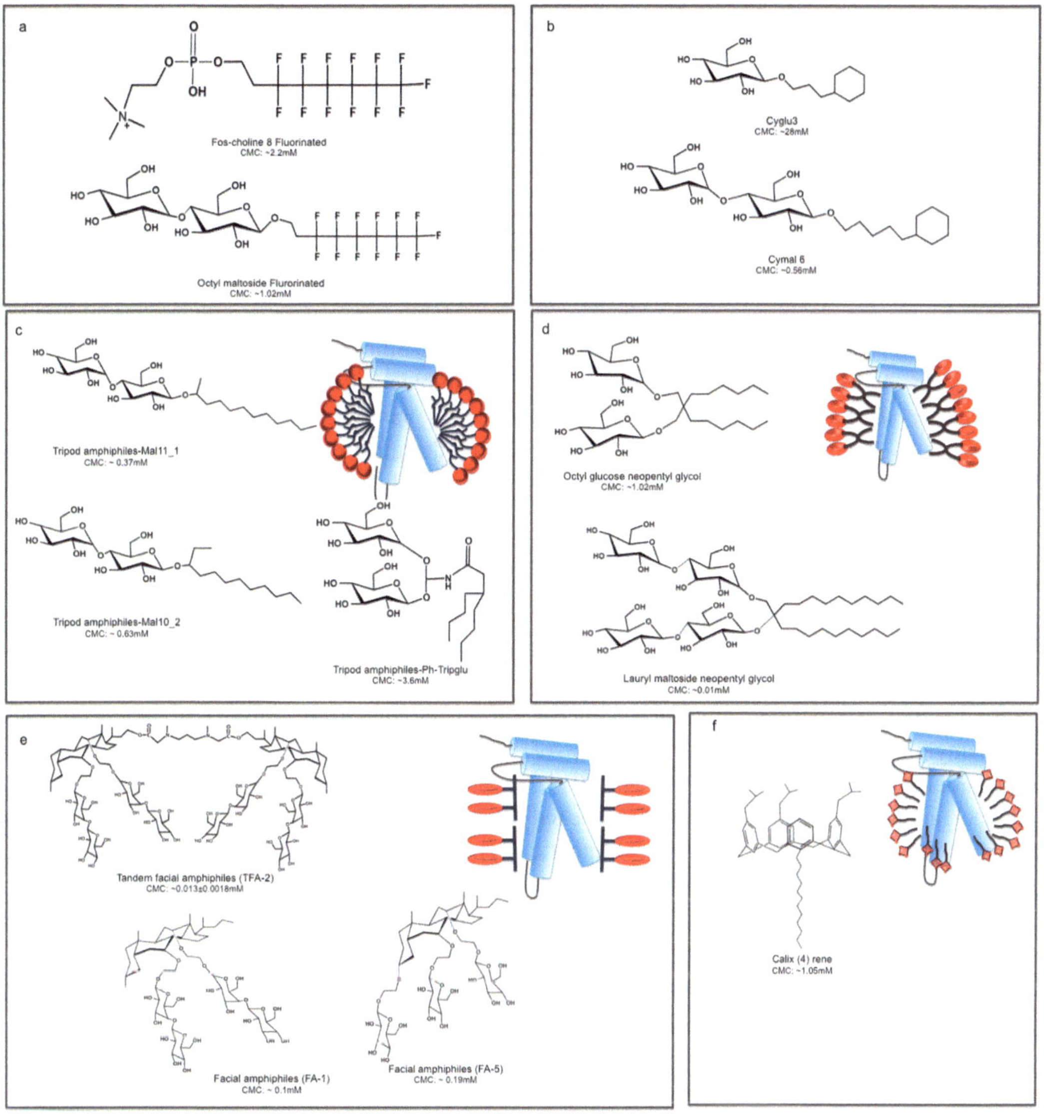

Fig. 2.6 Chemical structure of "new" detergents. Examples of hydrophobic tail modified detergents (**a**) fluorine substituted alkyl chain, (**b**) cyclic molecules incorporated in the tail (**c**) branched alkyl chain e.g. tripods. Example of lipid like detergents (**d**) and steroid based facial amphiphiles (**e**). Structure of calix[4]arene (**f**). DetergentCMC values in water are listed under the respective chemical structures. Cartoon representation of hypothetical micelle shape for different classes of detergents is also shown

bilayer, for example the maltoside-neopentyl glycols (MNGs), see Fig. 2.6d. The MNGs include maltose derivatives of neopentyl glycol amphiphiles with two units of maltose as the hydrophilic head group and two alkyl chains as the hydrophobic tail group. MNGs have reduced flexibility and increased hydrophobicity in the interior of the micelle. These properties contribute to the stability of the extracted IMP and assist in the crystallisation. Furthermore, their hydrophilicity and hydrophobicity characteristics are well balanced, enhancing the efficiency of protein solubilisation (Chae et al. 2010b). MNG amphiphiles have been used to extract and purify the human agonist β2 adrenoreceptor complex, which yielded diffraction quality crystals in

DDM (Rosenbaum et al. 2011). The TatC core of the twin arginine protein transport system is another example of a membrane protein that was solubilised, purified and crystallised in the presence of MNG amphiphiles (Rollauer et al. 2012).

Crystal contacts between protein molecules are absolutely required for crystallisation. A smaller detergent "belt" around the hydrophobic domain of the protein improves the chances of crystal contacts and enhances crystal quality (Garavito et al. 1996). Conventional detergents with shorter alkyl chains and smaller micelles have given high resolution diffracting crystals, unfortunately IMPs tend to become more denatured in these detergents due to a lack of sufficient hydrophobicity. This issue was overwhelmed in the new detergent design by including two short alkyl chains that provide sufficient hydrophobicity to maintain the stability of IMPs while forming a smaller hydrophobic "belt" around the protein molecule. Glucose neo-pentyl glucoside is a good example of this series of detergents (Cho et al. 2014). Another approach was the design of branched chain maltoside detergents. A short aliphatic branch chain was introduced between the hydrophilic and hydrophobic interface. This new detergent design mimics the second aliphatic chain of conventional lipid molecules, reducing the penetration of water molecules into the hydrophobic core. The performance of these new detergents was reported to be comparable to DDM, yielding crystals for Human connex 26 in different crystal forms (Hong et al. 2010)

Facial amphiphiles (FAs) are a different class of detergents that are steroid based and form small protein detergent complexes favouring crystallisation. They are structurally distinct from other detergents with hydrophilic maltoside residues attached to a steroid backbone instead of an alkyl chain. This provides a large hydrophobic surface, facilitating better packing around the hydrophobic region of the IMP, resulting in a small protein-detergent complex. The FAs have promoted successful crystal formation for a number of proteins in test conditions (Lee et al. 2013). An extension of the FAs is the tandem facial amphiphile (TFA), where two deoxycholate bis maltoside units are linked with a diaminopropane moiety (Fig. 2.6e). TFAs have been reported to have a similar width to native biological membranes and form smaller micelles compared to conventional DDM (Chae et al. 2010a). Deoxycholate based glucosides (DCG) have branched diglucoside head groups linked to deoxycholate by amide bonds. These detergents are reported to solubilize membranes efficiently and provide better stability to the IMP (Bae et al. 2015).

Anionic calix[4]arene based detergents are another family of ionic detergents that have been designed to improve the stability of IMPs. This family of detergents exploits the fact that IMPs display a high level of basic residues at the cytosol-membrane interface. Calix[4]arenes are made up of building blocks composed of 4 aromatic rings. Three of these rings are substituted with methylene carboxyl groups while an alkyl tail is attached to the fourth aromatic ring (Fig. 2.6f). The alkyl tail bonds to the hydrophobic region of the protein and the charged methylene carboxyl group forms salt bridges with the basic residues of the protein. These detergents are reported to be milder than Fos-choline-12 or SDS in spite of having 3 negative charges and have been shown to extract IMPs efficiently (Matar-Merheb et al. 2011).

2.5.3 The Role of Detergents in Membrane Protein Crystallisation

Crystals of membrane proteins are obtained by subjecting the PDC to different crystallisation conditions with the aim of the protein molecules pack in a well-ordered 3D array once the solution becomes supersaturated. The structural integrity of solubilised and purified membrane protein in detergent micelles are absolutely essential for protein molecules to assemble into building blocks, which is required for successful crystallisation (Rummel and Rosenbusch 2003). Another important criteria required for crystallisation is the homogeneity of the PDC (Kang et al. 2013). Depending on the type of crystal packing, crystals are classified into type I or type II 3D crystals. Usually type I 3D crystals are obtained by *in meso* methods such as lipidic cubic phase and bicelles (Caffrey 2003). Here the detergent micelle surrounding the protein becomes absorbed into a lipid making up the cubic phase, thus allowing the hydrophobic portion of the protein to form crystal contacts (Bill et al. 2011). Type II crystals are obtained by *in surfo* methods such as vapor diffusion. The crystal contacts are predominantly mediated by interactions between the hydrophilic segments of the PDC. Hence, detergent choice plays an important role in determining the ability of the membrane protein to crystallise especially for *in surfo* methods (Moraes et al. 2014).

Detergents with alkyl chains as their hydrophobic tails have so far been the most successful detergents used for membrane protein crystallisation. An analysis of successful crystallisation conditions for α-helical IMP structures by Parker *et al* revealed that the non-ionic alkyl maltopyranosides detergents, mainly DDM and DM, have been the most successful ones followed by n-octyl-β-D-glucopyranoside (OG) and LDAO. Crystals obtained in the presence of OG have been reported to yield high resolution diffraction (Parker and Newstead 2012; Sonoda et al. 2011). Though DDM, DM, OG and LDAO have been recommended as good first choice detergents for crystallisation, the ability to crystallise a protein varies depending on the properties of the PDC. Small variations in the chemical structure of the detergent, such differences in their chain length, can affect the protein crystallisation ability and the diffraction quality of the crystals. Ostermeier *et al* have reported that crystals of two subunits of cytochrome c oxidase (from soil bacterium *Paracoccus denitrificans*) in n-undecyl-β-D-maltoside (UDM) diffracted to 2.6 Å while crystals in DDM diffracted to 8 Å. In contrast, DM failed to yield crystals (Ostermeier and Michel 1997). Fos-choline series are zwitterionic detergents that have been widely used for the extraction of IMPs, however only very few IMPs have crystallised in their presence. An exception to this is the porin OmpF, which was solubilised, purified and crystallised in the presence of Fos-choline-12 (Kefala et al. 2010). Therefore, although a particular detergent may not previously have resulted in successful crystallisation trials it should not be excluded from consideration when undertaking crystallisation of membrane proteins.

Polar interactions between neighbouring detergent micelles within the crystal packing have proved to be significant contributors to the stability of the lattice (Ostermeier and Michel 1997). Neutron diffraction studies of crystals from the outer membrane phospholipase A have revealed rings of OG micelles covering the β-barrel surface of the protein molecule. These rings are fused with one another within the crystal lattice, stabilising crystal contacts (Snijder et al. 2003). In addition, attraction between detergent micelles can aid in forming a detergent rich phase conducive to crystal formation (Arnold and Linke 2007).

Furthermore, it is important to bear in mind that precipitants used for crystallisation also influence the detergent system. Precipitants affect the detergent phase, inducing nucleation of crystals (Hitscherich et al. 2001). Finally, purity and homogeneity of the detergent is another important factor affecting crystallisation. Trace amount of impurities in the detergent are capable of hampering the crystallisation process and crystal quality (Prive 2007).

2.5.3.1 Mixed Micelles

Using a combination of detergents or detergent-lipid mixtures to yield mixed micelles is another approach to combat the problem of providing a suitable environment for the isolated IMP. An important criteria for successful membrane protein crystallisation is obtaining a small PDC that would facilitate formation of the crystal contacts, promoting an orderly packing of the molecules, eventually leading to crystals with a reduced solvent content and better ordered lattices, and thus to a better diffraction quality (Tate 2010; Sonoda et al. 2011). Usually small PDCs are obtained by using detergents with short alkyl chains, however short chain detergents are usually more denaturing to the protein molecules when compared to the long chain detergents (Bill et al. 2011).

A systematic comparison of binary mixtures of commonly used detergents in IMP studies revealed that they formed micelles of different shape and size rather than mixture of pure micelles of the individual detergent. Mixing detergents with different head groups (e.g. mixing an ionic detergent with non-ionic detergent) modulate the surface potential of the micelles which in turn can play a key role in stabilising membrane proteins in solution and facilitating crystallisation. Thus, micelles with desired properties can be designed by systematically mixing detergents with different properties in varying ratios (Oliver et al. 2014). Another approach is to work with lipid detergent mixed micelles. The use of lipids can aid in re-establishing the needed native lipids that were lost during solubilisation, thereby enhancing the stability of the protein. The addition of lipids has proved to be very effective for a number of membrane proteins, in particular bacterial transporters and G protein coupled receptors (GPCRs). As an example, the stability and activity of G protein rhodopsin complexes were improved by using asolectin along with DDM (Jastrzebska et al. 2009). It is crucial to pay attention to buffer conditions such as pH, ionic strength and temperature for either detergents or detergent-lipid mixtures used in mixed micelles. A practical observation has shown that using a mismatch combination can lead to precipitation of the detergents (Keyes et al. 2003).

2.6 Conclusions

Given the great variation in structure and physical properties of membrane proteins, no single detergent or class of detergents will be effective for all the different cases. Each detergent displays a unique variety of physical and chemical properties, such as ionic charge, molecular size, and degree of hydrophobicity. Understanding these properties and the factors that influence them is a key step to the successful utilisation of detergents for membrane protein studies. As it is not possible to "guess" the correct detergent to be used for a particular membrane protein, knowledge about the detergent properties will aid in strategic planning of the experiments.

Acknowledgments The authors wish to thank Dr Isabel Moraes for helpful discussions. Additionally the authors acknowledge funding by grants from the National Health and Medical Research Council of Australia (APP1003697 and APP1078642).

References

Anglin TC, Conboy JC (2008) Lateral pressure dependence of the phospholipid transmembrane diffusion rate in planar-supported lipid bilayers. Biophys J 95:186–193

Arnold T, Linke D (2007) Phase separation in the isolation and purification of membrane proteins. Biotechniques 43:427–434

Arnold T, Linke D (2008) The use of detergents to purify membrane proteins. Curr Protoc Protein Sci. Chapter 4:Unit 4.8.1–4.8.30. doi:10.1002/0471140864.ps0408s53

Bae HE, Gotfryd K, Thomas J, Hussain H, Ehsan M et al (2015) Deoxycholate-based glycosides (DCGs) for membrane protein stabilisation. ChemBioChem 16:1454–1459

Bill RM, Henderson PJ, Iwata S, Kunji ER, Michel H et al (2011) Overcoming barriers to membrane protein structure determination. Nat Biotechnol 29:335–340

Bordag N, Keller S (2010) Alpha-helical transmembrane peptides: a "divide and conquer" approach to membrane proteins. Chem Phys Lipids 163:1–26

Bowie JU (2005) Solving the membrane protein folding problem. Nature 438:581–589

Caffrey M (2003) Membrane protein crystallization. J Struct Biol 142:108–132

Chae PS, Gotfryd K, Pacyna J, Miercke LJ, Rasmussen SG et al (2010a) Tandem facial amphiphiles for membrane protein stabilization. J Am Chem Soc 132:16750–16752

Chae PS, Rasmussen SG, Rana RR, Gotfryd K, Chandra R et al (2010b) Maltose-neopentyl glycol (MNG) amphiphiles for solubilization, stabilization and crystallization of membrane proteins. Nat Methods 7:1003–1008

Cho KH, Bae HE, Das M, Gellman SH, Chae PS (2014) Improved glucose-neopentyl glycol (GNG) amphiphiles for membrane protein solubilization and stabilization. Chem Asian J 9:632–638

Damodaran S, Song KB (1990) Effect of water structure makers and breakers on the adsorption of β-casein at the air—water interface. Colloids Surf 50:75–86

Durand G, Abla M, Ebel C, Breyton C (2014) New amphiphiles to handle membrane proteins: "Ménage à Trois" between chemistry, physical chemistry, and

biochemistry. In: Membrane proteins production for structural analysis. Springer, New York, pp 205–251
Franzin CM, Teriete P, Marassi FM (2007) Structural similarity of a membrane protein in micelles and membranes. J Am Chem Soc 129:8078–8079
Furth AJ, Bolton H, Potter J, Priddle JD (1984) Separating detergent from proteins. Methods Enzymol 104:318–328
Garavito RM, Ferguson-Miller S (2001) Detergents as tools in membrane biochemistry. J Biol Chem 276:32403–32406
Garavito RM, Picot D, Loll PJ (1996) Strategies for crystallizing membrane proteins. J Bioenerg Biomembr 28:13–27
Gohon Y, Popot J-L (2003) Membrane protein–surfactant complexes. Curr Opin Colloid 8:15–22
Goyal PS, Aswal VK (2001) Micellar structure and inter-micelle interactions in micellar solutions: results of small angle neutron scattering studies. Curr Sci 80:972–979
Gu T, Sjöblom J (1992) Surfactant structure and its relation to the Krafft point, cloud point and micellization: some empirical relationships. Colloids Surf 64:39–46
Helenius A, McCaslin DR, Fries E, Tanford C (1979) Properties of detergents. Method Enzymol 56:734–749
Hitscherich C, Aseyev V, Wiencek J, Loll PJ (2001) Effects of PEG on detergent micelles: implications for the crystallization of integral membrane proteins. Acta Crystallogr D Biol Crystallogr 57(7):1020–1029
Hong WX, Baker KA, Ma X, Stevens RC, Yeager M, Zhang Q (2010) Design, synthesis, and properties of branch-chained maltoside detergents for stabilization and crystallization of integral membrane proteins: human connexin 26. Langmuir 26:8690–8696
Ilgu H, Jeckelmann JM, Gachet MS, Boggavarapu R, Ucurum Z, Gertsch J, Fotiadis D (2014) Variation of the detergent-binding capacity and phospholipid content of membrane proteins when purified in different detergents. Biophys J 106:1660–1670
Jastrzebska B, Goc A, Golczak M, Palczewski K (2009) Phospholipids are needed for the proper formation, stability, and function of the photoactivated rhodopsin-transducin complex. Biochemistry 48:5159–5170
Kang HJ, Lee C, Drew D (2013) Breaking the barriers in membrane protein crystallography. Int J Biochem Cell Biol 45:636–644
Kefala G, Ahn C, Krupa M, Esquivies L, Maslennikov I, Kwiatkowski W, Choe S (2010) Structures of the OmpF porin crystallized in the presence of foscholine-12. Protein Sci 19:1117–1125
Keyes M, Gray D, Kreh K, Sanders C (2003) Solubilizing detergents for membrane proteins. Methods and results in crystallization of membrane proteins. International University Line, La Jolla, pp 15–33
Knol J, Sjollema K, Poolman B (1998) Detergent-mediated reconstitution of membrane proteins. Biochemistry 37:16410–16415
Kragh-Hansen U, le Maire M, Noel JP, Gulik-Krzywicki Tand Moller JV (1993) Transitional steps in the solubilization of protein-containing membranes and liposomes by nonionic detergent. Biochemistry 32:1648–1656
Kragh-Hansen U, le Maire M, Moller JV (1998) The mechanism of detergent solubilization of liposomes and protein-containing membranes. Biophys J 75:2932–2946
Lasch J (1995) Interaction of detergents with lipid vesicles. Biochim Biophys Acta 1241:269–292
le Maire M, Champeil P, Moller JV (2000) Interaction of membrane proteins and lipids with solubilizing detergents. Biochim Biophys Acta 1508: 86–111
Lee SC, Bennett BC, Hong WX, Fu Y, Baker KA et al (2013) Steroid-based facial amphiphiles for stabilization and crystallization of membrane proteins. Proc Natl Acad Sci U S A 110:E1203–E1211
Lichtenberg D, Opatowski E, Kozlov MM (2000) Phase boundaries in mixtures of membrane-forming amphiphiles and micelle-forming amphiphiles. Biochim Biophys Acta 1508:1–19
Lichtenberg D, Ahyayauch H, Goni FM (2013) The mechanism of detergent solubilization of lipid bilayers. Biophys J 105:289–299
Lin CM, Chang GP, Tsao HK, Sheng YJ (2011) Solubilization mechanism of vesicles by surfactants: effect of hydrophobicity. J Chem Phys 135(4):045102
Linke D (2009) Detergents: an overview. Methods Enzymol 463:603–617, Elsevier Science
Lorch M, Batchelor R (2011) Stabilizing membrane proteins in detergent and lipid systems. In: Production of membrane proteins: strategies for expression and isolation. Wiley-VCH Verlg GmbH & Co. KGaA, Weinheim, pp 361–390
Lórenz-Fonfría V, Perálvarez-Marín A, Padrós E, Lazarova T (2011) Solubilization, purification, and characterization of integral membrane proteins. In: Production of membrane proteins: strategies for expression and isolation. Wiley-VCH Verlag GmbH & Co. KGaA, Weinheim, pp 317–360
Matar-Merheb R, Rhimi M, Leydier A, Huche F, Galian C et al (2011) Structuring detergents for extracting and stabilizing functional membrane proteins. PLoS ONE 6(3):e18036
Melnyk RA, Partridge AW, Yip J, Wu Y, Goto NK, Deber CM (2003) Polar residue tagging of transmembrane peptides. Biopolymers 71:675–685
Moraes I, Evans G, Sanchez-Weatherby J, Newstead S, Stewart PD (2014) Membrane protein structure determination – the next generation. Biochim Biophys Acta 1838:78–87
Nagarajan R (2002) Molecular packing parameter and surfactant self-assembly: the neglected role of the surfactant tail. Langmuir 18:31–38
Nagarajan R (2011) Amphiphilic surfactants and amphiphilic polymers: principles of molecular assembly. In: Amphiphiles: molecular assembly and applications. American Chemical Society, Oxford University Press, Washington, DC, pp 1–22
Neugebauer JM (1990) Detergents: an overview. Methods Enzymol 182:239–253

Oliver RC, Lipfert J, Fox DA, Lo RH, Doniach S, Columbus L (2013) Dependence of micelle size and shape on detergent alkyl chain length and head group. PLoS ONE 8:e62488
Oliver RC, Lipfert J, Fox DA, Lo RH, Kim JJ et al (2014) Tuning micelle dimensions and properties with binary surfactant mixtures. Langmuir 30:13353–13361
Ostermeier C, Michel H (1997) Crystallization of membrane proteins. Curr Opin Struct Biol 7:697–701
Otzen DE (2002) Protein unfolding in detergents: effect of micelle structure, ionic strength, pH, and temperature. Biophys J 83:2219–2230
Otzen D (2011) Protein-surfactant interactions: a tale of many states. Biochim Biophys Acta 1814:562–591
Parker JL, Newstead S (2012) Current trends in alpha-helical membrane protein crystallization: an update. Protein Sci 21:1358–1365
Prive GG (2007) Detergents for the stabilization and crystallization of membrane proteins. Methods 41:388–397
Ray A, Nemethy G (1971) Effects of ionic protein denaturants on micelle formation by nonionic detergents. J Am Chem Soc 93:6787–6793
Renthal R (2006) An unfolding story of helical transmembrane proteins. Biochemistry 45:14559–14566
Rollauer SE, Tarry MJ, Graham JE, Jaaskelainen M, Jager F et al (2012) Structure of the TatC core of the twin-arginine protein transport system. Nature 492:210–214
Rosen MJ, Kunjappu JT (2012) Characteristic features of surfactants. In: Surfactants and interfacial phenomena. Wiley, Hoboken, pp 1–38
Rosenbaum DM, Zhang C, Lyons JA, Holl R, Aragao D et al (2011) Structure and function of an irreversible agonist-beta(2) adrenoceptor complex. Nature 469:236–240
Rummel G, Rosenbusch J (2003) Crystallization of bacterial outer membrane proteins from detergent solutions: porin as a model. Methods and results in crystallization of membrane proteins. International University Line, La Jolla, pp 101–129
Santonicola MG, Lenhoff AM, Kaler EW (2008) Binding of alkyl polyglucoside surfactants to bacteriorhodopsin and its relation to protein stability. Biophys J 94:3647–3658
Schagger H, Link TA, Jagow G (2003) Purification strategies for membrane proteins. In: Membrane protein purification and crystallization: a practical guide. Academic press, Amsterdam/Boston, pp 3–21
Schuck S, Honsho M, Ekroos K, Shevchenko A, Simons K (2003) Resistance of cell membranes to different detergents. Proc Natl Acad Sci U S A 100:5795–5800
Seddon AM, Curnow P, Booth PJ (2004) Membrane proteins, lipids and detergents: not just a soap opera. Biochim Biophys Acta 1666:05–117
Snijder HJ, Timmins PA, Kalk KH, Dijkstra BW (2003) Detergent organisation in crystals of monomeric outer membrane phospholipase A. J Struct Biol 141:122–131
Sonoda Y, Newstead S, Hu NJ, Alguel Y, Nji E et al (2011) Benchmarking membrane protein detergent stability for improving throughput of high-resolution X-ray structures. Structure 19:17–25
Tate CG (2010) Practical considerations of membrane protein instability during purification and crystallisation. In: Heterologous expression of membrane proteins. Springer, New York, pp 187–203
Tulumello DV, Deber CM (2012) Efficiency of detergents at maintaining membrane protein structures in their biologically relevant forms. Biochim Biophys Acta 1818:1351–1358
Wiener MC (2004) A pedestrian guide to membrane protein crystallization. Methods 34:364–372
Wiseman B, Kilburg A, Chaptal V, Reyes-Mejia GC, Sarwan J, Falson P, Jault JM (2014) Stubborn contaminants: influence of detergents on the purity of the multidrug ABC transporter BmrA. PLoS ONE 9(12):e114864
Zhang Q, Tao H, Hong WX (2011) New amphiphiles for membrane protein structural biology. Methods 55:318–323

NMR of Membrane Proteins: Beyond Crystals

3

Sundaresan Rajesh, Michael Overduin, and Boyan B. Bonev

Abstract

Membrane proteins are essential for the flow of signals, nutrients and energy between cells and between compartments of the cell. Their mechanisms can only be fully understood once the precise structures, dynamics and interactions involved are defined at atomic resolution. Through advances in solution and solid state NMR spectroscopy, this information is now available, as demonstrated by recent studies of stable peripheral and transmembrane proteins. Here we highlight recent cases of G-protein coupled receptors, outer membrane proteins, such as VDAC, phosphoinositide sensors, such as the FAPP-1 pleckstrin homology domain, and enzymes including the metalloproteinase MMP-12. The studies highlighted have resulted in the determination of the 3D structures, dynamical properties and interaction surfaces for membrane-associated proteins using advanced isotope labelling strategies, solubilisation systems and NMR experiments designed for very high field magnets. Solid state NMR offers further insights into the structure and multimeric assembly of membrane proteins in lipid bilayers, as well as into interactions with ligands and targets. Remaining challenges for wider application of NMR to membrane structural biology include the need for overexpression and purification systems for the production of isotope-labelled proteins with fragile folds, and the availability of only a few expensive perdeuterated detergents.

S. Rajesh
Henry Wellcome Building for Biomolecular NMR Spectroscopy, School of Cancer Sciences, University of Birmingham, Edgbaston, Birmingham B15 2TT, UK
e-mail: m.oveduin@bham.ac.uk

M. Overduin (✉)
Faculty of Medicine & Dentistry, Department of Biochemistry, University of Alberta, Edmonton, AB T6G 2H7, Canada
e-mail: overduin@ualberta.ca

B.B. Bonev (✉)
School of Life Sciences, University of Nottingham, Queen's Medical Centre, Nottingham NG7 2UH, UK
e-mail: boyan.bonev@nottingham.ac.uk

I. Moraes (ed.), *The Next Generation in Membrane Protein Structure Determination*,
Advances in Experimental Medicine and Biology 922, DOI 10.1007/978-3-319-35072-1_3

Step changes that may transform the field include polymers, such as styrene maleic acid, which obviate the need for detergent altogether, and allow direct high yield purification from cells or membranes. Broader demand for NMR may be facilitated by MODA software, which instantly predicts membrane interactive residues that can subsequently be validated by NMR. In addition, recent developments in dynamic nuclear polarization NMR instrumentation offer a remarkable sensitivity enhancement from low molarity samples and cell surfaces. These advances illustrate the current capabilities and future potential of NMR for membrane protein structural biology and ligand discovery.

Keywords
High resolution NMR • Solid state NMR • Protein structure • Protein interactions • Membrane targets

3.1 Introduction

Membranes of cells, organelles and some viral envelopes are asymmetric topologically planar bilayer assemblies composed at approximately equal weight fractions of proteins and lipids with diverse structure and function. They present permeability barriers, which support electrical potentials and chemical gradients, and are vital to cells. Membranes contain a significant fraction of specialised transmembrane proteins, which actively maintain an asymmetric electrical and solute environment. Other proteins may be loosely associated with the membrane or persistently anchored to the bilayer, which permits localised functionality with substrates at or near the membrane surface. A complex and dynamic lateral organisation exists within the proteolipid membrane, where sphingolipids and cholesterol-rich domains play an important role in membrane transport and signalling (Simons and Ikonen 1997). A structural variation exists between proteins within the plasma membranes of cells, which are largely made of helical transmembrane segments, and proteins from bacterial outer membranes, which predominantly follow β-barrel folds in contact with outer leaflet lipopolysaccharides.

The challenges in high resolution structural analysis of proteins are greater for membrane proteins, which require a hydrophobic membrane to support their folded state. Model membranes are often used and mimetic systems have been developed, which provide hydrophobic support to maintain the protein fold in the absence of a bilayer, while allowing for the 3D arrangement of isotropically mobile protein suspensions. Mimetics include detergent micelles, detergent/lipid bicelles, lipidic cubic phases and protein/lipid nanodisks (Warschawski et al. 2011) which can accommodate biologically important lipids alongside the proteins of interest. While this approach has provided the opportunity for structural characterisation of membrane peptides and proteins, studying the interactions of membrane proteins with each other or with lipids or looking at the regulation by ligands requires, in most cases, the presence of an intact membrane or a liposome membrane model. This is also required for functional characterisation in cases where an asymmetric solute environment is required, such as in transport or voltage-dependent ionic conductance.

Membrane protein studies by crystallography have yielded a set of key structures, which are bringing our understanding of their fundamental architecture towards a basis of high resolution templates for building a more comprehensive picture of membrane protein folding and function. Yet, studying membrane protein structure under unrestricted, hydrated and functional conditions remains possible by nuclear magnetic resonance

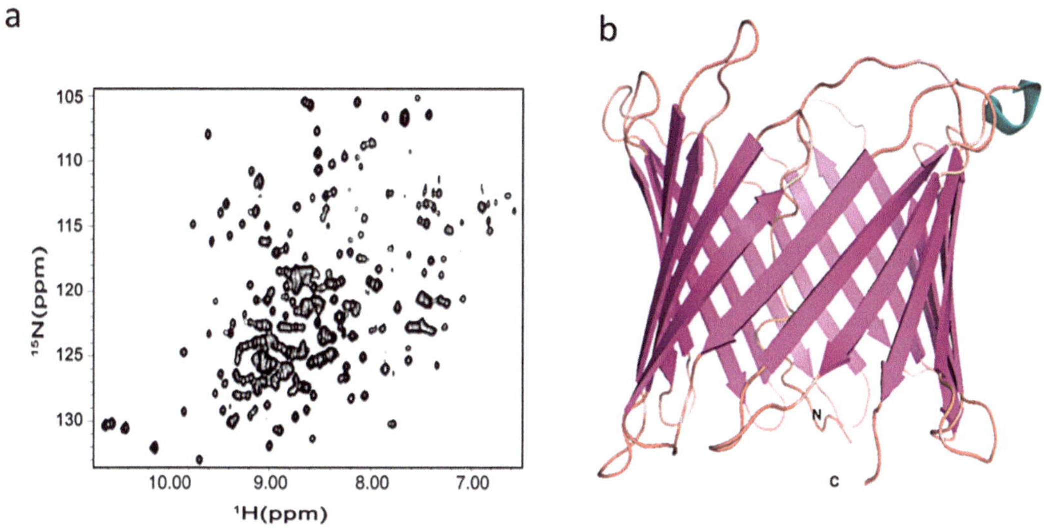

Fig. 3.1 (**a**) Two-dimensional ^{15}N-^{1}H- TROSYNMR spectrum of 0.5 mM [u-^{2}H,^{15}N] VDAC-1 in 300 mM detergentLDAO micelles at 30 °C. (**b**) Cartoon representation of the three-dimensional solution-NMR structure of VDAC-1 in LDAO micelles with the intracellular N- and C-terminus, as labelled (Adapted from (Hiller et al. 2008))

(NMR) alone. Here, we discuss representative applications of NMR to structural studies of membrane proteins and peptides, as well as their interactions with lipids, targets and ligands.

3.2 Barrel Folds

The mitochondrial outer membrane contains a protein known as the voltage-dependent anion channel (VDAC), which exchanges ions and molecules between mitochondria and cytosol. This is the largest β-barrel structure solved by NMR to date, and hence represents the state-of-the-art and also illustrates the technical challenges involved (Fig. 3.1). The VDAC fold consists of 19 β-strands that cross the membrane at a tilted angle and form a central pore that becomes occluded by the flexible N-terminus to mediate channel gating. The assignment involved a suite of 3D TROSY experiments performed at 900 MHz. The structure of the 283 residue VDAC-1 protein in micelles composed of perdeuterated LDAO was solved using 4D NOESY spectra collected with non-uniform sampling (Hiller et al. 2008). The binding sites are evident by ligand-induced perturbations in amide chemical shifts, which map the surfaces that interact with cholesterol, the metabolite β-NADH, and its anti-apoptotic partner Bcl-X_L, respectively. The experimental strategy benefitted from very high expression, purification and refolding yields of isotope labelled protein from *E. coli*, which were essential for cost-effective perdeuteration, selective ^{13}C-methyl labelling and amide proton exchange. Unfortunately, these requirements represent insurmountable barriers for the determination of the majority of eukaryotic membrane proteins, and point to the need for alternative means of producing pure, stable and labelled proteins that are functionally intact. Nonetheless, there is now a growing number of solution structures of bacterial outer membrane proteins, the stable β-barrel folds of which have been determined in detergent micelles composed of phosphocholine (PC)-based detergents and octyl glucoside. This demonstrates that modern solution NMR methods can be used to elucidate novel solution structures of 75–90 kDa complexes of stable β-barrel proteins in micelles if a suitable expression systems can be found.

The challenges of studying membrane proteins within lipid bilayers can be addressed through the applications of solid state NMR, which is often used in conjunction with solution NMR. Spectral resolution in membrane samples can be enhanced by magic angle sample spinning (MAS) (Haeberlen and Waugh 1968), which permits the determination of distance and torsion angle constraints, similar to those obtained in solution. Using a CP-HSQC approach, resonance assignment under fast MAS has been shown for the monomeric outer membrane poring OmpG from *E. coli* (Barbet-Massin et al. 2014).

3.3 Helical Targets

The diverse array of α-helical membrane proteins is challenging to study using solution NMR and this is reflected in the low number of structures resulved using this technique. Complete structure determination of larger seven-helical transmembrane GPCR proteins are yet to be attempted using solution NMR methods. However solution structures of proteins with multiple transmembrane regions have recently been solved, indicating that helical bundles in micelles can now be characterized if they are sufficiently stable. The successes include the bacterial enzymes DsbB and DAGK (Zhou et al. 2008; Van Horn et al. 2009), proteorhodopsin (Reckel et al. 2011), sensory rhodopsin II (Gautier et al. 2010), mitochondrial uncoupling protein 2 (Berardi et al. 2011), six human proteins with 2 or 3 transmembrane helices (Klammt et al. 2012), YgaP (Eichmann et al. 2014), LAMP2A (Rout et al. 2014), the Cl^- channel of human Glycine receptor α1 (Mowrey et al. 2013), and the voltage sensor-domain of voltage-dependent K^+ channel KvAP (Butterwick and MacKinnon 2010). The typical difficulties in solution NMR studies of helical membrane proteins include insufficient sample production, choice of membrane mimics, spectral overlap, limited spectral dispersion, shortage of long-range distance information and increased NMR data acquisition time.

Most membrane protein structures that have been determined by solution NMR methods were obtained from recombinant proteins expressed in *E.coli*. However not all proteins (e.g. GPCRs) readily express in prokaryotes, and some lack activity due to altered folding pathways and the absence of post-translational modifications. In such cases, eukaryotic expression hosts such as *Pichia pastoris*, *Drosophila* S2 and Sf9 cells serve as alternatives to producing uniformly ^{13}C,^{15}N-labeled membrane proteins for NMR studies (Fan et al. 2011). Recently, cell free expression systems using *E.coli* (Klammt et al. 2012), wheat-germ (Madono et al. 2011) or insect-cell extracts (Sobhanifar et al. 2010) have been successfully developed to produce membrane proteins for NMR. These cell free systems offer the advantage of amino acid selective (^{13}C methyl groups of isoleucine, leucine and valine) or targeted labelling, as well as the combinatorial dual-labelling scheme that help to address problems in NMR assignment, such as spectral overlap and fast relaxation in helical transmembrane proteins. Suitable membrane mimetics are screened for their ability to solubilise proteins from native membranes, mimic the native lipid environment, confer stability to proteins over a period of time at temperatures up to 50 °C and are sufficiently small for NMR studies. Further optimization by varying protein-detergent ratio to improve spectral quality can be done by monitoring the 2D ^{1}H-^{15}N TROSY spectra or by measuring rotational diffusion (Gautier and Nietlispach 2012; Horst et al. 2012). The use of smaller nanodisks can improve the relaxation properties of reconstituted membrane proteins and aided in the high resolution NMR structure calculation of bacteriorhodopsin (Hagn et al. 2013). The spectral overlap, due to limited dispersion of helical membrane proteins, is addressed by measurements at high temperatures, use of selectively methyl-labeled amino acids (ILV) and combinatorial-dual labelling. Unlike β-barrel proteins, helical transmembrane domain (TMD) proteins lack sufficient long range NOEs and even when available, the problem is compounded by the difficulty in obtaining complete side chain assignment. Paramagnetic relaxation enhancement (PRE) allows distance restraints to be measured using spin label attached to selected

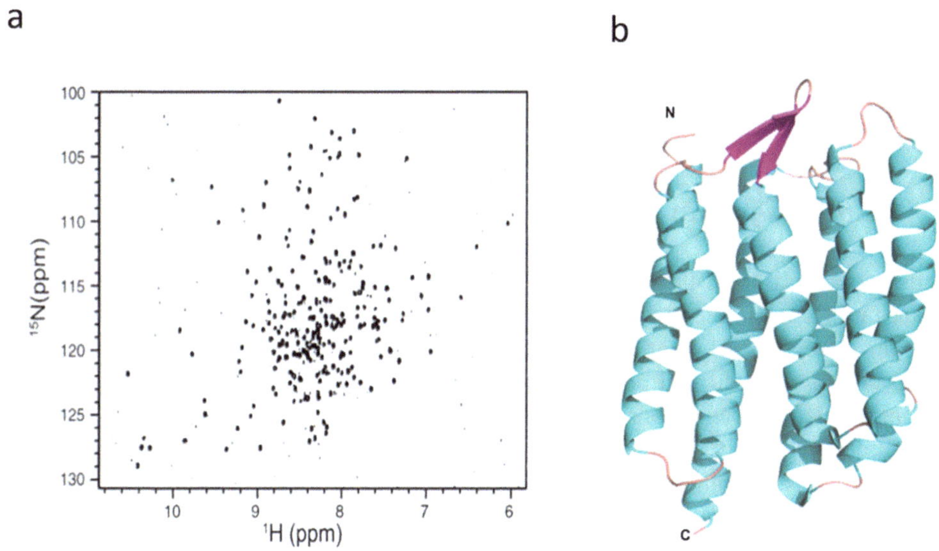

Fig. 3.2 (**a**) Two-dimensional ^{15}N-^{1}H- TROSYNMR spectrum of refolded 0.5 mM pSRII in 60 mM detergent c7-DHPC micelles at 50 °C. (**b**) Cartoon representation of the three-dimensional solution-NMR structure of pSRII in DHPC micelles with the extracellular N- and intracellular C-terminus, as labelled (Adapted from (Gautier et al. 2010))

protein sites. The PRE restraints provide distance information between 12 and 25 Å, and have contributed significantly to a number of large membrane protein structures obtained by NMR (Van Horn et al. 2008). The advantages of a non-uniform sampling scheme in reducing NMR acquisition times while increasing sensitivity of higher dimensionality experiments has been beautifully exemplified in the high resolution structure calculation of pSRII and holds promise for solution NMR studies of membrane proteins in general (Fig. 3.2).

Selective isotope labelling of GPCRs using ^{19}F probes (on cysteines), ^{13}C-methylated methionine (methyl-^{13}C-Met) or dimethylated lysines [Methyl-^{13}C-Lys] was successfully used to define population states and to probe conformational changes in transmembrane helices during receptor activation of β2 adrenergic and μ-opiod class of GPCRs (Kofuku et al. 2012; Liu et al. 2012; Sounier et al. 2015). These studies highlight the versatility of NMR methods in determining conformational dynamics and population states in the protein energy landscape, complementing the static X-ray structures. Understanding multiple conformations during receptor activation is important for designing small molecule inhibitors or monoclonal antibodies targeting these GPCRs.

In some cases, preparations for crystallographic studies can yield low-diffracting crystals, which can be investigated by solid state Magic Angle Spinning (MAS) NMR to obtain protein structures of significantly enhanced resolution. In one example, a combination of solid state MAS NMR and X-ray crystallography has provided structures of the bacterial disulfide-processing proteins DsbB/A (Tang et al. 2011; Shahid et al. 2012) by combining intermediate resolution crystallographic data with solid state NMR from the protein in the microcrystalline state. This was achieved using carefully calculated sets of dihedral angles in combination with selective ^{13}C-labelleing. Fast MAS solid state NMR has been used to obtain spectral assignment of the M2 ion channel from influenza (Barbet-Massin et al. 2014). An alternative to MAS permits angular constraints that can be obtained from proteins in macroscopically aligned membranes, which relies on correlations between heteronuclear dipolar couplings with ^{15}N chemical shifts (Wu et al. 1994). A study of pentameric phosphorylated phospholamban in lipid membranes revealed an L-shaped conformation including a TMD and a surface-associated helix within each monomer (Vostrikov et al. 2013).

3.4 Lipid Recognition Domains

A significant proportion of the proteome binds membranes using small, often terminal domains that recognize specific sets of lipids. With each biological membrane in the cell offering unique sets of lipids, this provides a basis for selective recruitment of thousands of proteins to particular organelle surfaces (Overduin and Cheever 2001). The largest superfamily of membrane interactive proteins is defined by the presence of pleckstrin homology (PH) domains, which is the 11th most common domain in the proteome (Lenoir et al. 2015a). Of these, those of the COF subfamily are distinguished by their ability to traffic ceramide, sterols and glycosphingolipids at the trans-Golgi network, and by recognizing phosphatidylinositol 4-phosphate (PI4P), which is the most abundant monophosphorylated inositol lipid. The human FAPP1 protein is the paradigm of this subfamily, and its mechanism has been most intensely studied by solution NMR methods. The complexes of its 108 residue PH domain with micelles and bicelles in the presence and absence of PI4P have been characterized (Lenoir et al. 2015b) (Fig. 3.3a). This revealed one of the largest intracellular membrane binding sites (~1220 Å), as well as convergent PI4P-specific and nonspecific bilayer insertion angles. The contacts with the phosphoinositide as well as several neighboring phospholipid molecules are evident, as are the residues that insert into the bilayer, from NOE and PRE patterns. The resulting distance restraints, along with the associated chemical shift changes that occur during interactions, allow the structures of assemblies to be calculated using high ambiguity-driven biomolecular docking (HADDOCK) (Dominguez et al. 2003) A novel aspect was the recognition of disordered bilayer regions mediated by an extensive and predominantly ordered region of the protein, and the exclusion of ordered bilayer binding. NMR afforded detailed insights not only due to the high expression and stability of the protein, even in detergents, but also to the development of optimized bicelle and mixed micelle formulations. These advances allowed collection of NMR restraints from large

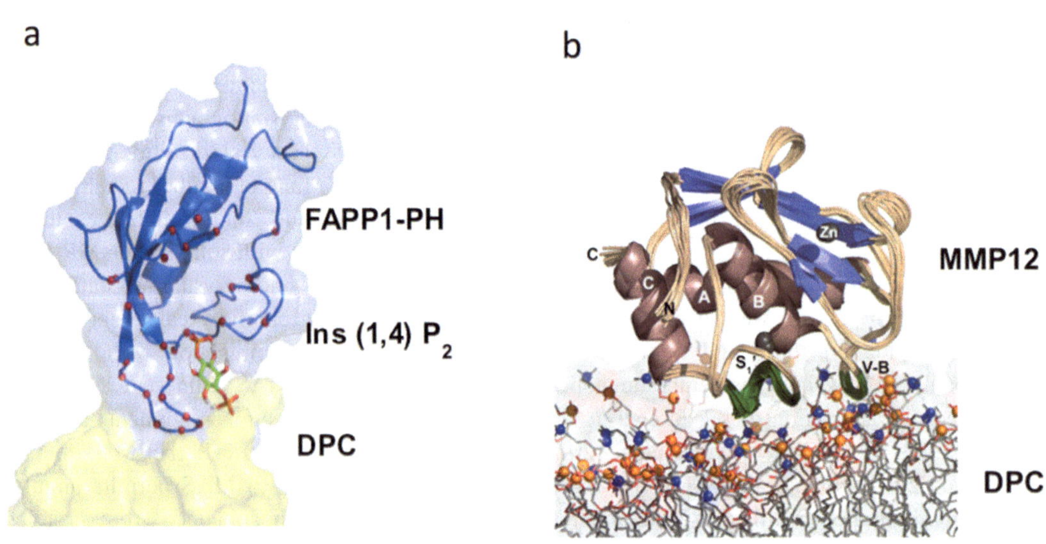

Fig. 3.3 (**a**) Solution-NMR structure of the FAPP1-PH domain (in *blue*) in complex with DPC (*yellow*) micelles in the presence of Inositol(1,4) phosphate (sticks). (**b**) Ensemble of solution-NMR structures of MMP12 (*cartoon*) bound to DPC micelles (*sticks*) with the membrane insertion loop (V-B) and helix (S_1') in *green* (adapted from (Koppisetti et al. 2014))

complexes, as did the use of soluble and lipidic spin labels, ^{13}C-resolved PREs and ultrahigh field magnets.

The membrane interactions of soluble enzymes and receptor domains can be similarly characterized by solution NMR methods, as can those of a variety of lipid binding domains such as C1b, C2, ENTH, FYVE and PX domains (Lemmon 2008). The metalloproteinase MMP-12 is one of the largest such peripheral membrane protein to be studied by solution NMR methods. The addition of bicelles loaded with PC spin labels revealed two opposite surfaces which interact reversibly with the membrane surface (Koppisetti et al. 2014) (Fig. 3.3b). This pair of interfaces is responsible for localising and concentrating the enzyme near the exterior membrane surface of the cell, they also control access to the proximal active site and hinder binding of protein inhibitors such as TIMP2, thus controlling activity and specificity. The alternating membrane-bound orientations can be simulated by molecular dynamics, showing that they differ by 137° and bury 2070 and 2530 Å^2 of surface area, respectively. Despite the large size of the bicelle complex (144 kDa), the PRE values induced by PC spin labels could be measured from relaxation rates to model the complexed structure. Together, this indicates that the reversible membrane interactions of a diversity of folded protein structures that recognize lipids, catalyse reactions, and tubulate and vesiculate membranes can now be characterized by NMR, illuminating fundamental events not readily visible by any other method.

3.5 Protein Structure in Membranes

A further step in developing structural understanding of membrane proteins towards more realistic, *in vivo* conditions is the use of solid state NMR to study proteins reconstituted into lipid membranes. Protein structure in membranes is often sought through a combination of NMR and Small Angle X-ray Scattering (SAXS) in solution or in membrane mimetics, in conjunction with crystallography, or with solid state NMR in membranes (Sanghera et al. 2011). Structural analysis of a microbial retinal-binding photoreceptor ASR from *Anabaena* has been carried out by solid state MAS NMR using PDSD and CHHC from sparsely ^{13}C-labelled protein along with relaxation enhancement (Opella 2013) and in conjunction with existing crystallographic and biochemical data. A hybrid solution/solid state NMR approach has been used in combination with molecular dynamics (MD) simulations to obtain the structure and topology of phospholamban in membranes (Zech et al. 2004). Mechanically oriented samples were used with Polarization Inversion Spin Exchange at the Magic Angle (PISEMA) (Wu et al. 1994) in this study and the approach has since been extended to the analysis of the biologically relevant, pentameric state of phospholamban in conjunction with HSQC in detergent micelles and Dipolar Assisted Rotational Resonance (DARR) MAS NMR from ^{13}C-labelled systems in DOPC (Anderson et al. 2014). The structure and molecular mechanism of a helix-loop-helix 2TMD bacterial mercury transporter, MerF, have been investigated using orientationally restricted sample geometry and in conjunction with multidimensional heteronuclear correlation NMR (Lu et al. 2013) (Fig. 3.4).

Following the successful crystallisation of modified and conformationally stabilised human chemokine receptor CXCR1 from lipidic cubic phases, separated local field solid state NMR has been used to obtain the structure of this GPCR in lipid membranes and to highlight the IL8 binding site near the membrane surface (Park et al. 2012). More challenging proteins amenable to solid state NMR investigation include an 82 kDa bacterial beta-barrel assembly protein called BamA (Renault et al. 2011). Although complete structural characterisation remains outside the capabilities of NMR technologies due to the large size and varied segmental mobility, homology modelling template on crystal structure along with molecular dynamics offer comprehensive insights into its domain dynamics and function. In some cases, sample preparation for membrane protein crystallography can yield

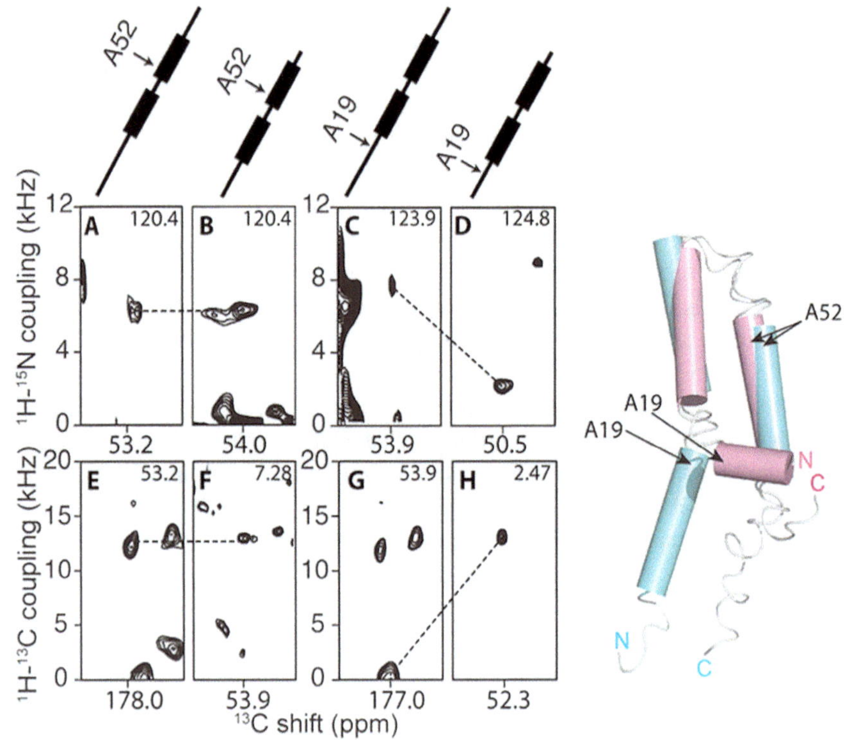

Fig. 3.4 (*Left*) Observed changes in orientationally dependent frequencies demonstrate drastic changes in the structure of MerF (PDB ref. 2 M67) caused by truncation of residues at the N-terminus. All slices are extracted from three-dimensional spectra at the noted values of the third dimension. (*Right*) The structure of the truncated 60-residue protein (magenta) is superimposed on the structure of the full-length 81-residue protein (aqua) (Lu et al. 2013)

crystals, in which segmental dynamics and insufficiently restrictive packing do not permit high resolution crystallographic characterisation. Such weakly diffracting crystals are amenable to characterisation by solid state NMR and offer superior resolution to membrane preparations due to conformational monodispersity, imposed by the crystalline order. In one example, solid state MAS NMR analysis has revealed the structure of the TMD form a Yersinial adhesin, YadA (Tang et al. 2011). Conformational changes in membrane protein ligands can drive changes in the protein functional state and solid state NMR offers high resolution insights. The effects of point mutations in proteorhodopsin on retinal binding and protein conformation have been examined by solid state MAS NMR and the associated retinal shape change, responsible for the green to blue colour shift, has been assessed by Dynamic Nuclear Polarisation (DNP) MAS NMR methods (Mao et al. 2014). The conformation of acetylcholine within the nicotinic acetylcholine receptor has been characterised using ^{13}C-labelled ligand and ligand-receptor contacts have been highlighted (Williamson et al. 2007). Using symmetry-based recoupling (Carravetta et al. 2000) the mature bacterial peptidoglycan intermediate lipid II has been investigated in its membrane complex with antibiotic nisin (Bonev 2013).

Lateral compartmentalisation in membranes allows co-localisation of proteins/targets. In studies of such laterally separated membranes, solid state MAS NMR offers insights into molecular interactions within membrane domains and lipid rafts, such as the specific recognition of raft-embedded ganglioside GM1 by the prion protein (Sanghera et al. 2011) (Fig. 3.5).

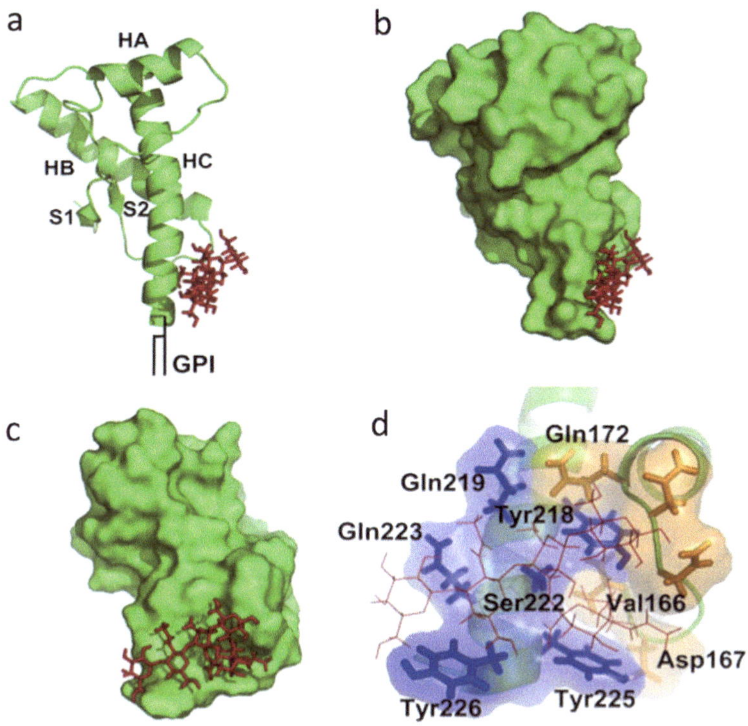

Fig. 3.5 Model of the PrP/GM1os complex: (**a**) Ribbon diagram of the structure of the PrP globular domain (*green*) showing α helices HA, HB, and HC and β sheet strands S1 and S2. The location of the glycosylphosphatidylinositol lipid anchor is highlighted in black; C-terminal end of helix C and the loop between strand S2 and helix B. (**b**) Same as (**a**) but PrP is drawn in surface representation; (**c**) Same as (**b**) but the orientation of this panel is a 90° rotation about the vertical axis of (**b**); (**d**) The putative binding site of GM1os on PrP. The region coloured in orange represents the loop region between strand S2 and helix B, and the region in blue represents the C-terminal end of helix C. The *thin red line* represents GM1os (Sanghera et al. 2011)

3.6 Technologies

Although *E. coli* expression systems have been used to produce proteins, most membrane protein samples for NMR studies have used cell-free expression systems, which can also deliver sufficient yields of protein for structural analysis. These systems rely on yeast or bacterial cell extracts that allow the target mRNA to be transcribed and translated within a reaction chamber. The cell-free approach allows specific amino acid labelling by, for example, the cell-free stereo-array isotope labeling (SAIL) method, and simplifies NMR resonance assignments and enhances resolution (Kainosho et al. 2006). The resulting stereo-specific assignments also improve the quality of the structures determined, including those of eukaryotic membrane proteins (Vinarov et al. 2006). The application of cell-free expression and selective isotope labelling to six helical transmembrane proteins yielded backbone structures using PRE restraints and chemical shift information (Klammt et al. 2012). Elucidating topologies of transmembrane helical bundles can benefit from optimal placement of a minimum number of spin labels (Chen et al. 2011).

In order to avoid the inherent problems associated with detergents, a novel method has been developed to solubilise membrane proteins directly from whole cells or native membranes using an amphipathic polymer. The addition of styrene maleic acid (SMA) results in spontaneous formation of stable bilayer discs that contain the intact protein and complexed lipids. The approach was first demonstrated with bacteriorhodopsin and the PagP β-barrel, which retain activity and stability within the resulting 11 nm diameter discs (Knowles et al. 2009). The method has been successfully applied to a variety of mammalian ABC transporters, yielding thermally stabilized pure protein with native ligand binding properties and structures which can be rapidly characterized by cryo-EM analysis (Gulati et al. 2014). More recently, GPCRs, including the human adenosine A2A receptor, have been purified directly from Pichia pastoris and mammalian cells into SMA lipid particles (SMALPs), resulting in thermally stable proteins with physiologically relevant binding and structural properties (Jamshad et al. 2015). While the polydispersity of the synthetic polymer limits the resolution of the contained protein signals, the observation of native-like binding and stability indicates the value of these nanoparticles for studying the physiological states and interactions of membrane-associated proteins.

The utility of NMR for validating novel sites of membrane interaction discovered within protein structures was demonstrated recently (Kufareva et al. 2014). A novel software tool, MODA (for "Membrane Optimal Docking Area") accurately detects such sites, having been trained on a set of established membrane interactive proteins. This approach was validated with five proteins: the Alix Bro1 domain, Arf1 GTPase, ATF2 acetyltransferase, the von Willebrand factor A3 domain, and the Neisseria gonorrhoeae MsrB protein, which all exhibited potential sites of lipid interaction. The predictions were borne out by generating point mutations and monitoring amide chemical shift changes when micelles were added, showing that the experimentally defined and computationally predicted sites converged. This technique could significantly expand the universe of protein/lipid interaction sites, with NMR being used as the most reliable and data-rich method for confirming such lipid interactions on tractable proteins, as well as for profiling their specificities and induced structural changes. Already the MODA method has been successfully applied to MMP-12, to map its multiple membrane binding surfaces, as confirmed by NMR analysis (Koppisetti et al. 2014). Together, this indicates the general utility of NMR for identifying novel sites, structures and ligands involved in membrane interaction for even challenging systems, underscoring its importance for membrane structural biology.

The Collaborative Computing Project for NMR (CCPN) has emerged to link high throughput data and existing NMR software via common standards and has provided a forum within the community for the discussion of new methods (Fogh et al. 2002). It aims to offer a pipeline for integration of data processing from solution and solid state NMR of proteins with computational structure analysis.

Structural analysis by solid state NMR has been applied commonly to membrane-incorporated peptides and small proteins. Challenges arise from polydispersity in the sample, which leads to inhomogeneously broadening of the spectral lines, strong dipolar coupling networks leading to homogeneous broadening with both effects hindering the assignment of larger and more complex systems (Bonev 2013). In addition, larger proteins within proteoliposomal preparations are present at low molarity, which leads to comparatively low sensitivity even in isotope-enriched samples. Orientational constraints can be obtained from non-rotating aligned samples (Park et al. 2012). The use of aligned samples and microcrystalline preparations improves monodispersity, while ultrafast MAS can be used to reduce dipolar couplings. Significant sensitivity enhancement in the order of 10–100 s can be achieved by combining dynamic nuclear polarization with MAS NMR (Bajaj et al. 2007).

3.7 DNP Sensitivity Enhancement

Solid state NMR delivers high information content at atomic detail on chemically distinct molecular components but at comparatively moderate sensitivity. The application of DNP in conjunc-

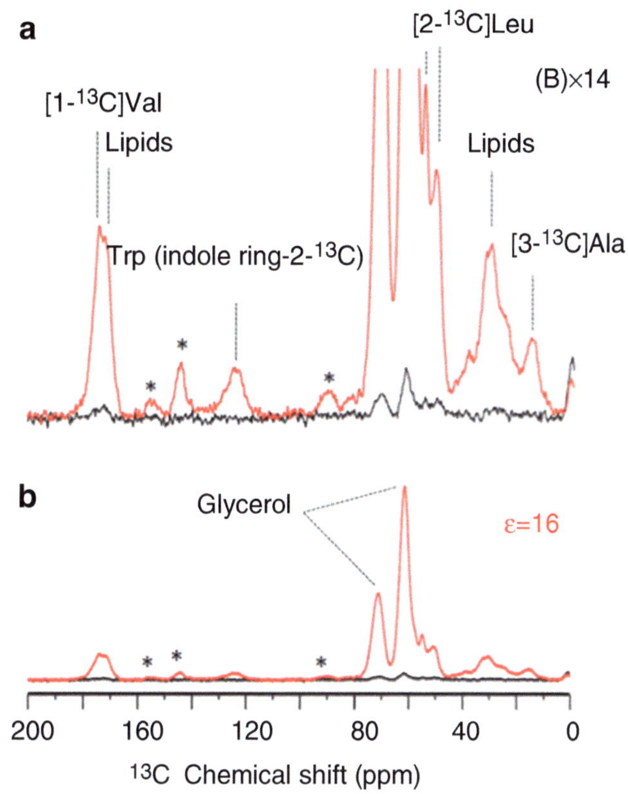

Fig. 3.6 DNP-enhanced solid-state NMR spectroscopy of E. coli cells. One-dimensional in-cell ^{13}C CPMAS NMR spectra of cytochrome b5 in E. coli cells using CPMAS with microwave irradiation on (*red*), and off (*black*); magnified (**a**) and full height (**b**) (Yamamoto et al. 2015)

tion with MAS NMR can boost sensitivity of current solid-state NMR tools by a factor ε 30–90 (Bajaj et al. 2007) and in some cases above 150, which opens the possibility of studying membrane protein structures within native membrane environments (Barbet-Massin et al. 2014) (Fig. 3.6). Photoinduced CDNP studies of intermediate states in bacterial photoreaction centre in *Heliobacter* (Thamarath et al. 2012) provide information on transition states and electron transport. The electronic structure of the electron donor of purple bacteria reaction centres has yielded atomic resolution detail from photo-CIDNP MAS NMR studies (Daviso et al. 2009). Reaction photocentre has also been studied in sulphur bacteria by DNP MAS NMR (Roy et al. 2007) and information on the electron transport and intermediate states has been obtain from photosystem II by DNP MAS NMR (Diller et al. 2007). Recent use of DNP has offered new insights into electron transport in plants (Janssen et al. 2012). A major qualitative advancement in spectroscopy, delivered through the DNP signal enhancement, is the possibility of deriving spectroscopic information from isolated membranes, cell envelopes and from intact cells, such as in studies of cyanobacteria (Janssen et al. 2010) and *E. coli* (Renault et al. 2012; Yamamoto et al. 2015). Substrate binding to multidrug resistance pump EmrE has been observed by DNP MAS NMR, allowing further analysis of the drug binding site (Ong et al. 2013). Using DNP MAS NMR, it has been possible to study protein export by the SecYEG system and to obtain two dimensional spectra from as low as 40 nmoles of bound cargo peptide (Reggie et al. 2011). And, in drug discovery, DNP signal enhancement has made it possible to observe the binding of an inhibitor to the proton transporter from influenza A (Andreas et al. 2013).

Acknowledgments The authors would like to thank the Biotechnology an Biological Sciences Research Council, Campus Alberta Innovates Program, the Engineering and Physical Sciences Research Council and the Wellcome Trust for funding.

References

Anderson TM, Clay MC, Cioffi AG, Diaz KA, Hisao GS et al (2014) Amphotericin forms an extramembranous and fungicidal sterol sponge. Nat Chem Biol 10(5):400–406

Andreas LB, Barnes AB, Corzilius B, Chou JJ, Miller EA et al (2013) Dynamic nuclear polarization study of inhibitor binding to the M2(18–60) proton transporter from influenza A. Biochemistry 52(16): 2774–2782

Bajaj VS, Hornstein MK, Kreischer KE, Sirigiri JR, Woskov PP et al (2007) 250 GHz CW gyrotron oscillator for dynamic nuclear polarization in biological solid state NMR. J Magn Reson 189(2): 251–279

Barbet-Massin E, Pell AJ, Retel JS, Andreas LB, Jaudzems K, Franks WT, Nieuwkoop AJ, Hiller M, Higman V et al (2014) Rapid proton-detected NMR assignment for proteins with fast magic angle spinning. J Am Chem Soc 136(35):12489–12497

Berardi MJ, Shih WM, Harrison SC, Chou JJ (2011) Mitochondrial uncoupling protein 2 structure determined by NMR molecular fragment searching. Nature 476(7358):109–113

Bonev BB (2013) High-resolution solid-state NMR of lipid membranes. Adv Planar Lipid Bilayers Liposomes 17:299–329

Butterwick JA, MacKinnon R (2010) Solution structure and phospholipid interactions of the isolated voltage-sensor domain from KvAP. J Mol Biol 403(4):591–606

Carravetta M, Eden M, Zhao X, Brinkmann A, Levitt MH (2000) Symmetry principles for the design of radiofrequency pulse sequences in the nuclear magnetic resonance of rotating solids. Chem Phys Lett 321(3–4):205–215

Chen H, Ji F, Olman V, Charles KM, Liu Y et al (2011) Optimal mutation sites for PRE data collection and membrane protein structure prediction. Structure 19(4):484–495

Daviso E, Prakash S, Alia A, Gast P, Neugebauer J, Jeschke G, Matysik J (2009) The electronic structure of the primary electron donor of reaction centers of purple bacteria at atomic resolution as observed by photo-CIDNP C-13 NMR. Proc Natl Acad Sci U S A 106(52):22281–22286

Diller A, Roy E, Gast P, van Gorkom HJ, de Groot HJM et al (2007) N-15 photochemically induced dynamic nuclear polarization magic-angle spinning NMR analysis of the electron donor of photosystem II. Proc Natl Acad Sci U S A 104(31):12767–12771

Dominguez C, Boelens R, Bonvin A (2003) HADDOCK: a protein-protein docking approach based on biochemical or biophysical information. J Am Chem Soc 125:1731–1737

Eichmann C, Tzitzilonis C, Bordignon E, Maslennikov I, Choe S, Riek R (2014) Solution NMR structure and functional analysis of the integral membrane protein YgaP from Escherichia coli. J Biol Chem 289(34):23482–23503

Fan Y, Shi L, Ladizhansky V, Brown L (2011) Uniform isotope labeling of a eukaryotic seven-transmembrane helical protein in yeast enables high-resolution solid-state NMR studies in the lipid environment. J Biomol NMR 49(2):151–161

Fogh R, Ionides J, Ulrich E, Boucher W, Vranken W, Linge JP et al (2002) The CCPN project: an interim report on a data model for the NMR community. Nat Struct Mol Biol 9(6):416–418

Gautier A, Nietlispach D (2012) Solution NMR studies of integral polytopic α-helical membrane proteins: the structure determination of the seven-helix transmembrane receptor sensory rhodopsin II, pSRII. In: Membrane protein structure and dynamics. Humana Press, New York, pp 25–45

Gautier A, Mott HR, Bostock MJ, Kirkpatrick JP, Nietlispach D (2010) Structure determination of the seven-helical transmembrane receptor sensory rhodopsin II by solution NMR spectroscopy. Nat Struct Mol Biol 17(6):768–774

Gulati S, Jamshad M, Knowles TJ, Morrison KA, Downing R et al (2014) Detergent-free purification of ABC (ATP-binding-cassette) transporters. Biochem J 461(2):269–278

Haeberlen U, Waugh JS (1968) Coherent averaging effects in magnetic resonance. Phys Rev 175(2): 453–467

Hagn F, Etzkorn M, Raschle T, Wagner G (2013) Optimized phospholipid bilayer nanodiscs facilitate high-resolution structure determination of membrane proteins. J Am Chem Soc 135(5): 1919–1925

Hiller S, Garces RG, Malia TJ, Orekhov VY, Colombini M, Wagner G (2008) Solution structure of the integral human membrane protein VDAC-1 in detergent micelles. Science 321(5893):1206–1210

Horst R, Stanczak P, Serrano P, Wüthrich K (2012) Translational diffusion measurements by micro-coil NMR in aqueous solutions of the Fos-10 detergent-solubilized membrane protein OmpX. J Phys Chem B 116(23):6775–6780

Jamshad M, Charlton J, Lin Y-P, Routledge SJ, Bawa Z et al (2015) G-protein coupled receptor solubilization and purification for biophysical analysis and functional studies, in the total absence of detergent. Biosci Rep 35(2):e00188

Janssen GJ, Daviso E, van Son M, de Groot HJM, Alia A, Matysik J (2010) Observation of the solid-state photo-CIDNP effect in entire cells of cyanobacteria Synechocystis. Photosynth Res 104(2–3):275–282

Janssen GJ, Roy E, Matysik J, Alia A (2012) N-15 photo-CIDNP MAS NMR to reveal functional heterogeneity in electron donor of different plant organisms. Appl Magn Reson 42(1):57–67

Kainosho M, Torizawa T, Iwashita Y, Terauchi T, Ono AM, Güntert P (2006) Optimal isotope labelling for NMR protein structure determinations. Nature 440(7080):52–57

Klammt C, Maslennikov I, Bayrhuber M, Eichmann C, Vajpai N et al (2012) Facile backbone structure determination of human membrane proteins by NMR spectroscopy. Nat Methods 9(8):834–839

Knowles TJ, Finka R, Smith C, Lin Y-P, Dafforn T, Overduin M (2009) Membrane proteins solubilized intact in lipid containing nanoparticles bounded by styrene maleic acid copolymer. J Am Chem Soc 131(22):7484–7485

Kofuku Y, Ueda T, Okude J, Shiraishi Y, Kondo K, Maeda M, Tsujishita H, Shimada I (2012) Efficacy of the β(2)-adrenergic receptor is determined by conformational equilibrium in the transmembrane region. Nat Commun 3:1045

Koppisetti RK, Fulcher YG, Jurkevich A, Prior SH, Xu J et al (2014) Ambidextrous binding of cell and membrane bilayers by soluble matrix metalloproteinase-12. Nat Commun 5:5552

Kufareva I, Lenoir M, Dancea F, Sridhar P, Raush E et al (2014) Discovery of novel membrane binding structures and functions. Biochem Cell Biol 92(6):555–563

Lemmon MA (2008) Membrane recognition by phospholipid-binding domains. Nat Rev Mol Cell Biol 9:99–111

Lenoir M, Grzybek M, Majkowski M, Rajesh S, Kaur J et al (2015a) Structural basis of dynamic membrane recognition by trans-Golgi network specific FAPP proteins. J Mol Biol 427(4):966–981

Lenoir M, Kufareva I, Abagyan R, Overduin M (2015b) Membrane and protein interactions of the pleckstrin homology domain superfamily. Membranes (Basel) 5(4):646–663

Liu W, Chun E, Thompson AA, Chubukov P, Xu F, Katritch V et al (2012) Structural basis for allosteric regulation of GPCRs by sodium ions. Science 337(6091):232–236

Lu GJ, Tian Y, Vora N, Marassi FM, Opella SJ (2013) The structure of the mercury transporter MerF in phospholipid bilayers: a large conformational rearrangement results from N-terminal truncation. J Am Chem Soc 135(25):9299–9302

Madono M, Sawasaki T, Morishita R, Endo Y (2011) Wheat germ cell-free protein production system for post-genomic research. Nat Biotechnol 28(3):211–217

Mao J, Do NN, Scholz F, Reggie L, Mehler M et al (2014) Structural basis of the green-blue color switching in proteorhodopsin as determined by NMR spectroscopy. J Am Chem Soc 136(50):17578–17590

Mowrey D, Cui T, Jia Y, Ma D, Makhov AM et al (2013) Open-channel structures of the human glycine receptor α1 full-length transmembrane domain. Structure 21(10):1897–1904

Ong YS, Lakatos A, Becker-Baldus J, Pos KM, Glaubitz C (2013) Detecting substrates bound to the secondary multidrug efflux pump EmrE by DNP-enhanced solid-state NMR. J Am Chem Soc 135(42):15754–15762

Opella SJ (2013) Structure determination of membrane proteins in their native phospholipid bilayer environment by rotationally aligned solid-state NMR spectroscopy. Acc Chem Res 46(9):2145–2153

Overduin M, Cheever ML (2001) Signaling with phosphoinositides: better than binary. Mol Interv 3:10

Park SH, Das BB, Casagrande F, Tian Y, Nothnagel HJ, Chu M et al (2012) Structure of the chemokine receptor CXCR1 in phospholipid bilayers. Nature 491(7426):779–783

Reckel S, Gottstein D, Stehle J, Löhr F, Verhoefen MK et al (2011) Solution NMR structure of proteorhodopsin. Angew Chem Int Ed 50(50):11942–11946

Reggie L, Lopez JJ, Collinson I, Glaubitz C, Lorch M (2011) Dynamic nuclear polarization-enhanced solid-state NMR of a C-13-labeled signal peptide bound to lipid-reconstituted sec translocon. J Am Chem Soc 133(47):19084–19086

Renault M, Bos MP, Tommassen J, Baldus M (2011) Solid-state NMR on a large multidomain integral membrane protein: the outer membrane protein assembly factor BamA. J Am Chem Soc 133(12):4175–4177

Renault M, Pawsey S, Bos MP, Koers EJ et al (2012) Solid-state NMR spectroscopy on cellular preparations enhanced by dynamic nuclear polarization. Angew Chem Int Ed 51(12):2998–3001

Rout AK, Strub M-P, Piszczek G, Tjandra N (2014) Structure of transmembrane domain of lysosome-associated membrane protein type 2a (LAMP-2A) reveals key features for substrate specificity in chaperone-mediated autophagy. J Biol Chem 289(51):35111–35123

Roy E, Gast P, van Gorkom H, de Groot HJ (2007) Photochemically induced dynamic nuclear polarization in the reaction center of the green sulphur bacterium chlorobium tepidum observed by (13)C MAS NMR. Biochim Biophys Acta -Bioenergetics 1767(6):610–615

Sanghera N, Correia BE, Correia JR, Ludwig C et al (2011) Deciphering the molecular details for the binding of the prion protein to main ganglioside GM1 of neuronal membranes. Chem Biol 18(11): 1422–1431

Shahid SA, Bardiaux B, Franks WT, Krabben L, Habeck M et al (2012) Membrane-protein structure determination by solid-state NMR spectroscopy of microcrystals. Nat Methods 9:1212–1217

Simons K, Ikonen E (1997) Functional rafts in cell membranes. Nature 387(6633):569–572

Sobhanifar S, Reckel S, Junge F, Schwarz D, Kai L et al (2010) Cell-free expression and stable isotope labelling strategies for membrane proteins. J Biomol NMR 46(1):33–43

Sounier R, Mas C, Steyaert J, Laeremans T, Manglik A et al (2015) Propagation of conformational changes during [mgr]-opioid receptor activation. Nature 524(7565):375–378

Tang M, Sperling LJ, Berthold DA, Schwieters CD, Nesbitt AE et al (2011) High-resolution membrane protein structure by joint calculations with solid-state NMR and X-ray experimental data. J Biomol NMR 51(3):227–233

Thamarath SS, Alia A, Daviso E, Mance D, Golbeck JH et al (2012) Whole cell nuclear magnetic resonance characterization of two photochemically active states of the photosynthetic reaction center in heliobacteria. Biochemistry 51(29):5763–5773

Van Horn W, Ogilvie M, Flynn P (2008) Use of reverse micelles in membrane protein structural biology. J Biomol NMR 40(3):203–211

Van Horn WD, Kim HJ, Ellis CD, Hadziselimovic A et al (2009) Solution NMR structure of membrane-integral diacylglycerol kinase. Science 324(5935):1726–1729

Vinarov DA, Newman CLL, Markley JL (2006) Wheat germ cell-free platform for eukaryotic protein production. FEBS J 273(18):4160–4169

Vostrikov VV, Mote KR, Verardi R, Veglia G (2013) Structural dynamics and topology of phosphorylated phospholamban homopentamer reveal its role in the regulation of calcium transport. Structure 21(12):2119–2130

Warschawski DE, Arnold AA, Beaugrand M, Gravel A, Chartrand E, Marcotte I (2011) Choosing membrane mimetics for NMR structural studies of transmembrane proteins. Biochim Biophys Acta-Biomembranes 1808(8):1957–1974

Williamson PTF, Verhoeven A, Miller KW, Meier BH, Watts A (2007) The conformation of acetylcholine at its target site in the membrane-embedded nicotinic acetylcholine receptor. Proc Natl Acad Sci U S A 104(46):18031–18036

Wu CH, Ramamoorthy A, Opella SJ (1994) High-resolution heteronuclear dipolar solid state NMR spectroscopy. J Magn Reson A 109(2):270–272

Yamamoto K, Caporini MA, Im S-C, Waskell L, Ramamoorthy A (2015) Cellular solid-state NMR investigation of a membrane protein using dynamic nuclear polarization. Biochim Biophys Acta-Biomembranes 1848(1):342–349

Zech SG, Olejniczak E, Hajduk P, Mack J, McDermot AE (2004) Characterization of protein-ligand interactions by high-resolution solid-state NMR spectroscopy. J Am Chem Soc 126(43):13948–13953

Zhou Y, Cierpicki T, Jimenez RHF, Lukasik SM, Ellena JF et al (2008) NMR solution structure of the integral membrane enzyme DsbB: functional insights into DsbB-catalyzed disulfide bond formation. Mol Cell 31(6):896–908

Characterisation of Conformational and Ligand Binding Properties of Membrane Proteins Using Synchrotron Radiation Circular Dichroism (SRCD)

4

Rohanah Hussain and Giuliano Siligardi

Abstract

Membrane proteins are notoriously difficult to crystallise for use in X-ray crystallographic structural determination, or too complex for NMR structural studies. Circular dichroism (CD) is a fast and relatively easy spectroscopic technique to study protein conformational behaviour in solution. The advantage of synchrotron radiation circular dichroism (SRCD) measured with synchrotron beamlines compared to the CD from benchtop instruments is the extended spectral far-UV region that increases the accuracy of secondary structure estimations, in particular under high ionic strength conditions. Membrane proteins are often available in small quantities, and for this SRCD measured at the Diamond B23 beamline has successfully facilitated molecular recognition studies. This was done by probing the local tertiary structure of aromatic amino acid residues upon addition of chiral or non-chiral ligands using long pathlength cells (1–5 cm) of small volume capacity (70 μl–350 μl). In this chapter we describe the use of SRCD to qualitatively and quantitatively screen ligand binding interactions (exemplified by Sbma, Ace1 and FsrC proteins); to distinguish between functionally similar drugs that exhibit different mechanisms of action towards membrane proteins (exemplified by FsrC); and to identify suitable detergent conditions to observe membrane protein-ligand interactions using stabilised proteins (exemplified by inositol transporters) as well as

The original version of this chapter has been revised. An erratum to this chapter can be found at DOI 10.1007/978-3-319-35072-1_13

R. Hussain (✉) • G. Siligardi
Diamond Light Source, Harwell Science and Innovation Campus, OX11 0DE Didcot, UK
e-mail: rohanah.hussain@diamond.ac.uk; giuliano.siligardi@diamond.ac.uk

I. Moraes (ed.), *The Next Generation in Membrane Protein Structure Determination*, Advances in Experimental Medicine and Biology 922, DOI 10.1007/978-3-319-35072-1_4

the stability of membrane proteins (exemplified by GalP, Ace1). The importance of the in solution characterisation of the conformational behaviour and ligand binding properties of proteins in both far- and near-UV regions and the use of high-throughput CD (HT-CD) using 96- and 384-well multiplates to study the folding effects in various protein crystallisation buffers are also discussed.

Keywords
Circular dichroism (CD) • Synchrotron radiation circular dichroism (SRCD) • Protein secondary structure • Local tertiary structure • High throughput-CD (HTCD) • SRCD UV-denaturation assay • Ligand binding

4.1 Introduction

Circular Dichroism (CD) and Synchrotron Radiation CD (SRCD) has been proven to be highly useful for studies of ligand binding by soluble proteins, particularly as it is a relatively quick and easy spectroscopic measurement, requiring no extensive sample preparation, and it has the potential to be used in high throughput (HT-CD) screening (Mason 1982; Fasman 1996; Berova et al. 2000, 2012; Siligardi and Hussain 1998, 2015; Siligardi et al. 2002; Martin et al. 2011; Johnson 1990; Kelly et al. 2005; Siligardi and Hussain 2010; Jávorfi et al. 2010; Fiedler et al. 2013).

CD spectroscopy has been widely used for the purposes of determining secondary structural content and integrity of membrane proteins (e.g. histidine kinases – Potter et al. 2002; Kim et al. 2010; Keegan et al. 2010; Yeo et al. 2008; membrane transporters – Psakis et al. 2009), protein fragments (e.g. Powl et al. 2010), and protein unfolding (Psakis et al. 2009; Keegan et al. 2010). Sreerama and Woody 2004, found that the CD analysis of secondary structure content of membrane proteins using soluble protein reference sets was slightly inferior to that obtained for soluble proteins. The inclusion of membrane proteins in the soluble protein reference sets – now a common practise – has since improved the CD analysis for both membrane and soluble proteins.

In the last few decades, drug discovery has benefited from information about the three-dimensional structure of proteins at atomic resolution determined by X-ray crystallography (Carvalho et al. 2009) and NMR spectroscopy (Pellecchia et al. 2008). However, the number of solved human protein structures deposited in the Protein Data Bank is still only a small fraction of the approximately 20,000 protein-coding genes in the human genome (Pennisi 2012). As many proteins that cannot be studied by either X-ray crystallography due to their failure to crystallise or by Nuclear magnetic resonance spectroscopy (NMR) due to their size or irregular structure, can be characterized in terms of protein folding by CD, which is the differential absorption between left and right circularly polarized light (Fasman 1996), see Fig. 4.1. CD spectroscopy is sensitive to the absolute configuration and conformation of chiral molecules. With the exception of glycine, amino acids are chiral molecules adopting either L or D stereo-isomers which are non-superimposable mirror-imaged configurations. Natural proteins only consist of L amino acids.

The application of CD spectroscopy for ligand binding studies of membrane proteins has proved more limited, largely due to the technical challenges associated with working with hydrophobic membrane proteins. The use of the B23 beamline (Diamond Light Source synchrotron) for SRCD has overcome the limitations encountered in studying precious membrane proteins often available in very small quantities. The B23 beamline has enabled the use of small volume capacity cuvette cells for the measurements in the far-UV and near-UV spectral regions that could not otherwise be used with bench-top CD instruments. The UV protein denaturation assay, a unique feature of the B23 beamline, is an-

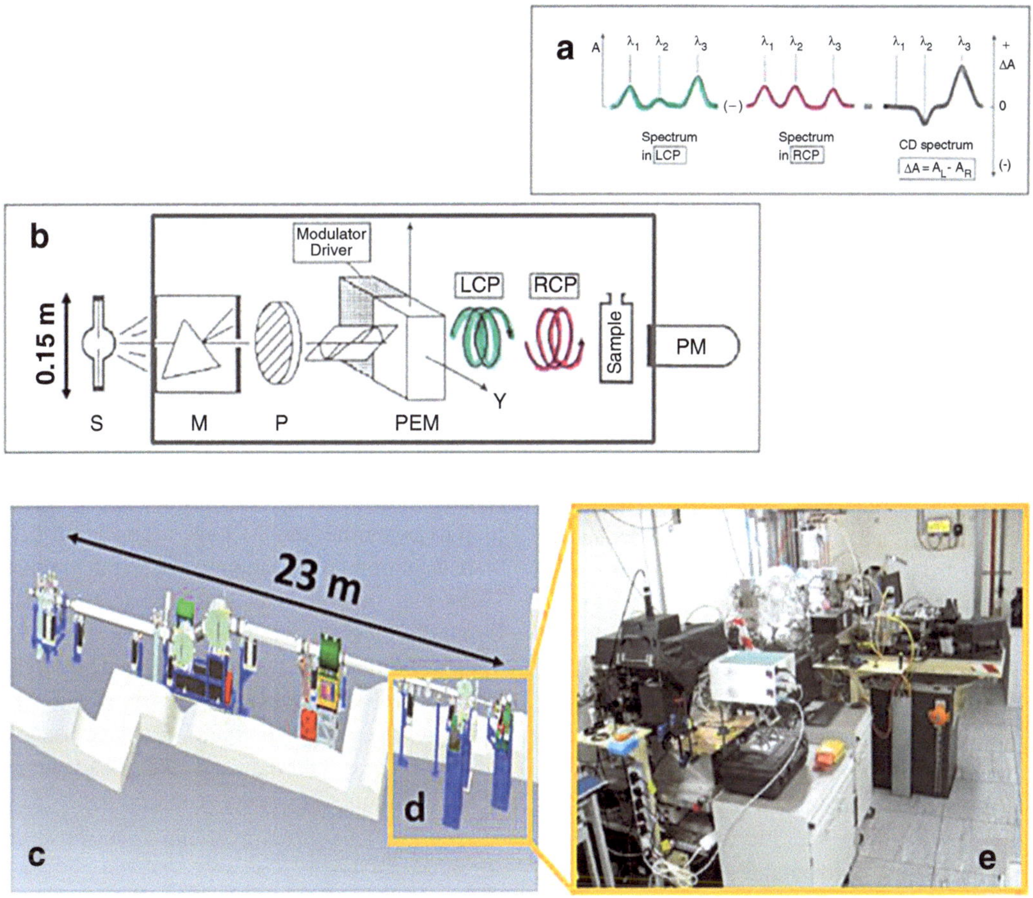

Fig. 4.1 (**a**) Optical layout of a CD spectropolarimeter: *S* light source, *M* monochromator, *P* polariser, photo elastic modulator (PEM), left circularly polarised light (LCP) and right circularly polarised light (RCP), sample and detector. (**b**) CD spectrum is the differential absorption $\Delta A = (A_L - A_R)$ between LCP and RCP light. In the cartoon such a difference is greatly exaggerated. For proteins ΔA is about 10^{-3} to 10^{-4} ($\Delta A \approx$ (measured CD (ellipticity) in millidegrees)/3300). (**c**) The light source is a Xe-arc lamp (length =15 cm) for bench top CD instrument compared to the equivalent 23 m of the bending magnet (BM) of Diamond B23 beamline for synchrotron radiation CD (SRCD). (**d**) and (**e**) B23 has two end-stations, one dedicated for CD imaging, cryogenic to high temperatures and HT-CD, and the other for solution and solid state (dry films) samples

other method developed to assess ligand binding interactions, in particular for achiral fatty acid ligand analogues or small organic molecules devoid of UV chromophores, which are difficult, if not impossible to investigate with bench-top CD instruments. The higher vacuum UV and far-UV photon flux has proved essential for the spectroscopic characterisation of membrane proteins in high salt concentrations to enhance protein stability, which cannot be achieved satisfactorily with bench-top instruments due to absorption cut-off.

In this chapter, several examples of SRCD spectroscopy applied to the study of membrane proteins in solution using the Diamond B23 beamline (Hussain et al. 2012a, b; Jávorfi et al. 2010) are presented. Qualitative and quantitative assessments of ligand binding interactions of the FsrC, SbmA, and Ace1 membrane proteins, identification of functionally similar but mechanistically distinct drug targeting FsrC, and the characterisation of suitable detergent conditions for ligand-membrane proteins interaction studies are reviewed.

4.2 Diamond B23 Beamline for Synchrotron Radiation Circular Dichroism (SRCD)

The B23 beamline is a bending magnet beamline at Diamond Light Source, UK, dedicated to SRCD spectroscopy. The beamline consists of two distinct end-stations: module A in operation since 2011 and module B in operation since 2009, see Fig. 4.1. The beamline operates in the wavelength range of 125–650 nm, delivering a highly collimated beam with a focal spot size at the sample of 0.5 mm^2 and a photon flux in excess of 10^{12} photons/s. The high brilliance of the beamline, coupled with its highly collimated incident micro-beam is enabling a large variety of measurements and experiments to be carried out, ranging from dilute to highly concentrated solutions, to thin solid films. The end-station of module A operates in the wavelength range of 125–500 nm and is primarily dedicated to the study of thin chiral films, whereas module B operating in the 165–650 nm is dedicated to solutions and liquid samples. A novel vertical sample chamber, which is unique among the other SRCD beamlines around the world, has been developed to accommodate a horizontal X-Y motorised stage that can hold either the temperature controlled stage (such as Linkam™ stage), which can operate in cryogenic and high temperature conditions (from −150 °C to 300 °C), or standard microtiter or microwell plates (96- and 384-well Suprasil quartz plates (Siligardi and Hussain 2015) for high throughput CD (HT-CD) screening.

Module B is dedicated to ligand binding interactions that are often carried out as CD titrations using small aperture cuvette cells that require the ability to optimise beam alignment in a fast and reproducible manner using a manual X-Y stage (Borisenko et al. 2015). This allows for measurements of dilute solutions (as low as 5 nM protein concentration) (Laera et al. 2011) using small volume capacity cells such as 0.800 ml for 10 cm pathlength or 0.020 ml for 1 cm pathlength. SRCD stopped-flow can also be carried out in both modules. The higher photon flux of SRCD beamlines has traditionally been considered detrimental as it promotes protein denaturation. Measures have been put in place to eliminate or greatly reduce this effect, however, this phenomenon, has also been exploited at the B23 beamline as UV protein denaturation (Hussain et al. 2012a, b; Jávorfi et al. 2010) can be conveniently used to measure accelerated protein photo stability as a function of protein formulation and qualitatively assessing and screening ligand binding interactions (Hussain et al. 2012a, b; Longo et al. 2014, 2015). For module B, a rotating sample cell holder has also been introduced to eliminate protein denaturation induced by the higher vacuum UV and far-UV photon flux for experiments where the UV denaturation effect is not desirable, such as thermal denaturation. Finally, to support the capabilities and potential of B23 beamline applications, a concerted software development program has been put in place. Users can now reduce and analyse their data through a single software application developed at the beamline – CDApps (Hussain et al. 2015). The software is available for users on-site, as well as via remote access, and allows novice users to quickly process their data and take full advantage of the technique.

4.3 Ligand Binding Studies of the FsrC, Ace1 and SbmA Membrane Proteins

Protein-ligand binding interactions can be assessed by a variety of techniques such as Isothermal Calorimetry (ITC), Surface Plasmon Resonance (SPR), Fluorescence and Circular Dichroism (CD). Each technique has its own strengths and weaknesses. ITC can often be inconclusive when the thermal effects associated with the displacement of the bound solvent molecules by the ligand are rather small. SPR requires one of the components to be immobilized, which can lead to false positive or false negative results when compared to non-immobilized conditions. Fluorescence requires the presence of a fluorophore that does not absorb and emit light at the same wavelength for the ligand and/or protein (Tryptophan and Tyrosine residues). CD relies on detectable spectral differences between the observed spectrum of the protein-ligand mixture at (x:y) molar ratio and the calculated sum of the spectra of the protein at (x) molar ratio and ligand at (y) molar ratio. For achiral ligands, the comparison is facilitated by the fact that the ligand chromophore will acquire an induced CD upon binding of a protein site that is chiral. In this case, the detection of any induced CD, which is any spectral difference from that of the protein alone, is unambiguously indicative of the presence of a bound species, as the free achiral form is devoid of any CD from its UV absorbing chromophore (Siligardi and Hussain 1998; Hussain et al. 2012b; Siligardi et al. 2002; Martin et al. 2011). However, for ligands with weak or with no UV chromophore groups, such as sugars, lipids and irregular peptides, the lack of any CD spectral change upon ligand addition does not necessarily signify that there are no binding interactions. The ligand might bind far from the aromatic side-chains of the protein, without causing any detectable secondary structure conformational change. For these cases, the protein UV denaturation assay using the B23 beamline was developed and is now part of the facilities available at the Diamond B23 beamline (Hussain et al. 2012a, b; Jávorfi et al. 2010). High UV photon flux at B23 was used to qualitatively distinguish the binding of ligands to proteins, such as non-steroidal anti-inflammatory drugs to serum albumin, though not a membrane protein is a transporter (Fig. 4.2) (Hussain et al. 2012a). UV denaturation can also be used to discriminate the effects of different excipients, such as detergents, salts, and buffers on peptide hormones such as vasoactive peptide VIP, which binds to a GPCR (Longo et al. 2015) and antibodies (such as cetuximab which is an inhibitor of EGFR, a transmembrane protein) on the effect of formulating agents (Longo et al. 2015; Siligardi and Hussain 2015). These measurements are easy to carry out, as it only requires repeated scanning of samples and a plot of the denaturation decay, which will provide differential dynamics properties of the system.

Binding studies using the CD titration method are even more challenging for membrane proteins than they are for soluble proteins. A common limitation encountered with membrane proteins is that they are often expressed in rather small quantities. At the B23 beamline, the highly collimated incident micro-beam (0.3×0.5 mm) is enabling the use of a small aperture cuvette cells of low volume capacity (from few μl to 25 μl for 0.01 cm to 1 cm pathlengths) to measure the SRCD of precious and scarce membrane proteins.

FsrC, a membrane protein histidine kinase, showed very little conformational changes in secondary structure upon addition of ligand Gelatinase Biosynthesis-Activating Pheromone (GBAP). However, the local tertiary structure of FsrC showed significant changes when GBAP was added, indicating the involvement of aromatic residues binding to GBAP, see Fig. 4.3. Qualitative studies showed that the affinity of the ligand GBAP to FsrC was 2 μM, monitored at a single wavelength of 277 nm. Further competitive binding studies with another ligand, siamycin I, showed that siamycin I did not compete with GBAP and that aromatic side-chain residues should be involved in the interaction with the ligand (Patching et al. 2012; Phillips-Jones et al. 2013).

SbmA, a bacterial inner membrane protein of Gram-negative bacteria is involved in the transport within the cell of prokaryotic and eukaryotic antimicrobial peptides and glycopeptides, as well as of peptide nucleic acid (PNA) oligomers. A SbmA homolog, BacA, is required for the development of *Sinorhizobium meliloti* bacteroids within plant cells and favours chronic infections with *Brucella abortus* and *Mycobacterium tuberculosis* in mice. A SRCD spectroscopic study provided evidence that SbmA and BacA interact *in vitro* with Bac7 (1–35), a proline rich peptide. Bac7 was titrated at various molar ratios to SbmA and BacA in both the far-UV (180–260 nm) and near-UV (250–330 nm) regions. In the far-UV region, significant changes of secondary structure were observed upon addition of Bac7 to SbmA (see Fig. 4.4a and Table 4.1) and BacA (Fig. 4.4b) indicating binding interactions. In the near-UV region (Runti et al. 2013), characteristic of the local tertiary structure of aromatic side-chain residues (Tryptophan, Tyrosine and Phenylalanine) (Siligardi et al. 1991; Siligardi and Hussain

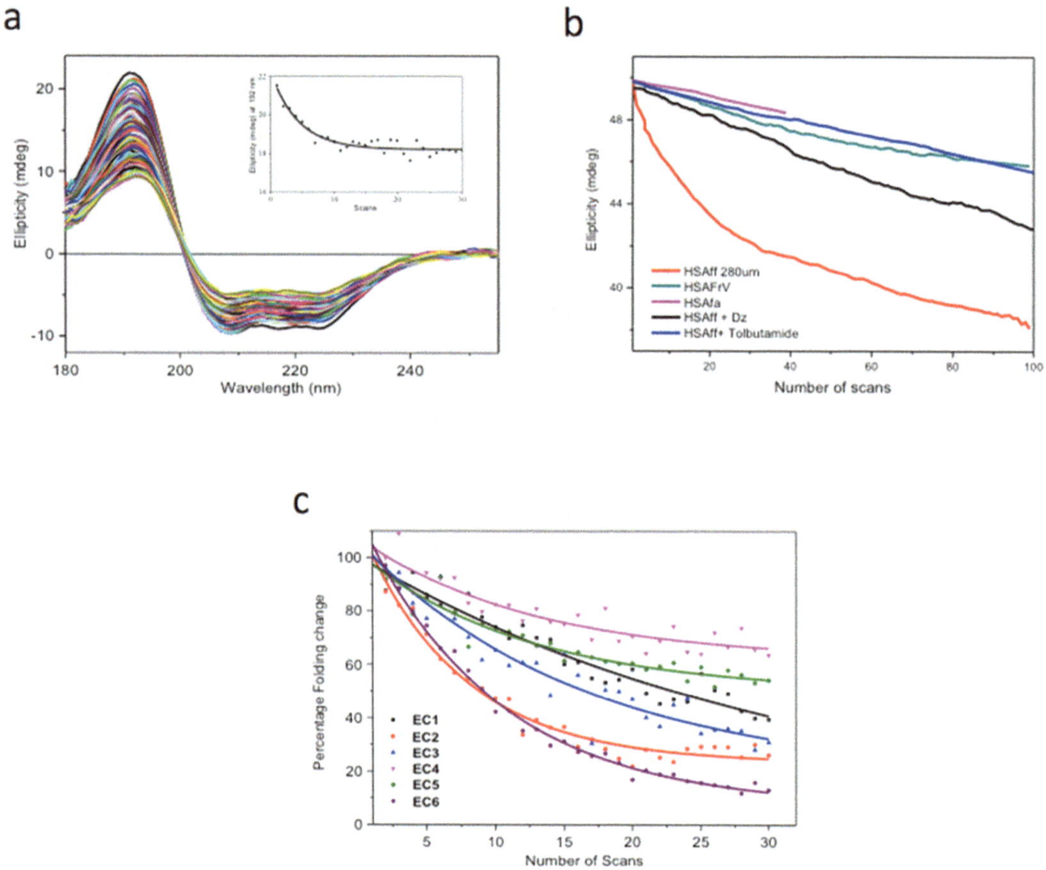

Fig. 4.2 (**a**) Thirty repeated consecutive SRCD spectra of fatty acid and immunoglobulin free human serum albumin (HSAff). The insert is the rate of protein denaturation at 190 nm. (**b**) Rate UV protein denaturation of HSAff in H_2O with and without ligands such as fatty acid (octanoic acid), diazepam and tolbutamide measured at 190 nm for 100 repeated consecutive SRCD spectra. (**c**) Rate of UV protein denaturation of antibody Mab-1 in different formulation agents for 30 repeated consecutive SRCD spectra. The graph is reported in percentage of protein folding damage calculated by dividing the ellipticity at 205 nm of the protein-ligand complex by that of the protein alone and multiplied by 100

1998; Hussain et al. 2012b; Gaspar et al. 2011) no significant changes were observed, which suggests that no aromatic residues were involved at close range (radius of 6 Å) in the interface between ligand and protein. The SRCD spectra were converted to $\Delta A = (A_L - A_R)$ units from millidegree units and the plot of ΔA intensities at 223 nm versus the concentrations of Bac7 used in the titration was analysed with a non-linear regression method (Siligardi et al. 2002) to quantitatively determine the dissociation constant using CDApps software (Hussain et al. 2015). The results showed that the peptide had similarly high binding affinities to SbmA (Kd of 0.26 μM) and BacA (Kd of 0.3 μM).

For the membrane protein Ace1 (actinobacter chlorhexidine efflux transport system 1), no significant conformational changes in the far-UV region (180–250 nm) were observed in the presence of an achiral chlorhexidine antibacterial drug (Hassan et al. 2013). However, in the near-UV region, the SRCD titration unambiguously showed the induced CD of the bound chorhexidine to the AceI membrane protein (Fig. 4.5a). The non-linear regression analysis of the SRCD data revealed adissociation constant

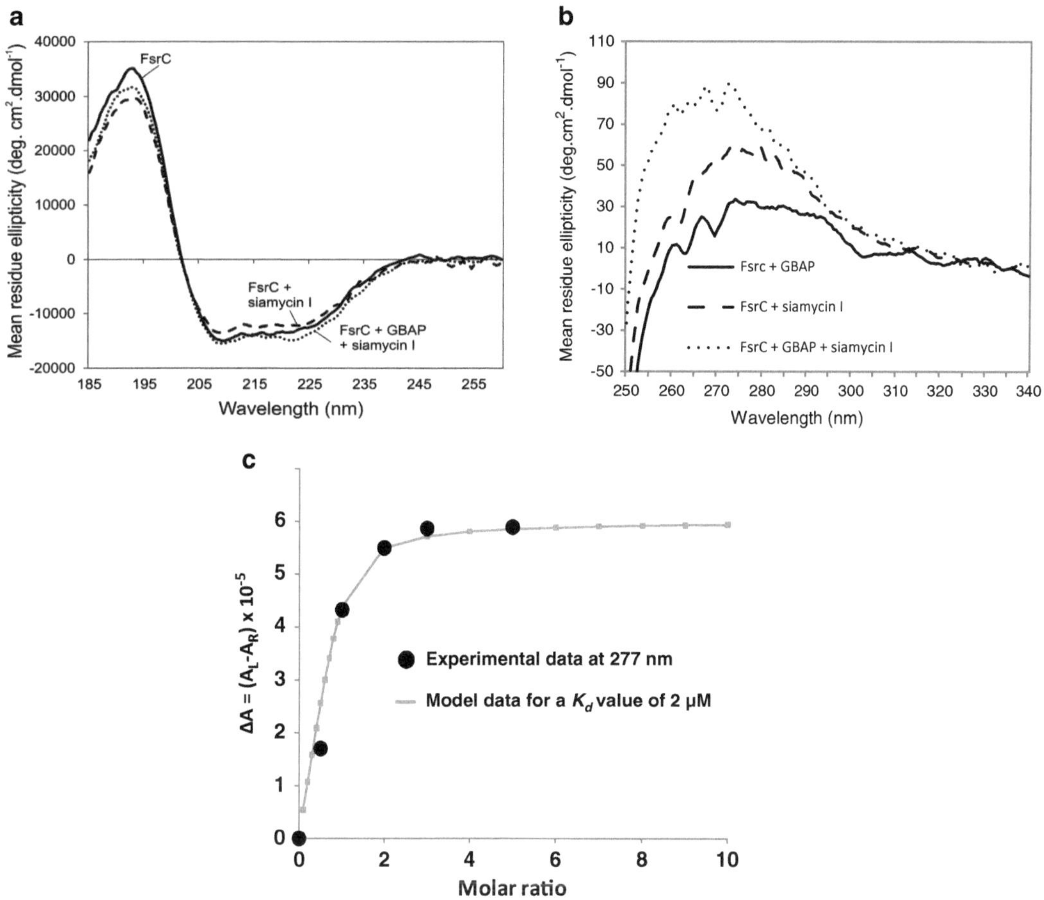

Fig. 4.3 Binding interaction properties of FsrC membrane protein determined using the B23 beamline for synchrotron radiation circular dichroism (SRCD) (a, b, c figures redrawn from Patching et al. 2012 and Phillips-Jones et al. 2013). (**a**) Far-UV SRCD spectra of FsrC membrane protein with (*dashed*) and without (*solid*) GBAP ligand Patching et al. 2012). (**b**) Near-UVSRCD spectra: (*top*) FsrC with (*dashed*) and without (*solid*) GBAP ligand. The insert is the fitting of the GBAP SRCD titration into FsrC protein using a non-linear regression analysis. The Kd calculated from CD data was 2 μM, (*bottom*) FsrC with (*dashed*) and without (*solid*) Siamycin I ligand. (**c**) Near-UV SRCD spectra of (FsrC + GBAP) (*thick black*), (FsrC + Siamycin I) (*dashed*) and (FsrC + GBAP + Siamycin I) (*thin black*). The unique micro-collimated beam of Diamond's B23 beamline enabled the measurements to be carried out using a small volume capacity cell (50–100 μl) of 1 cm pathlength otherwise unattainable with bench-top CD instruments

(K_d) of 6 μM (Fig. 4.5b). In this example the binding was demonstrated by the increased induced CD at about 270 nm, which reach a plateau upon saturation. The ligand binding did not perturb the content of the secondary structure of Ace1, as illustrated by the lack of conformational changes in the far-UV SRCD spectrum.

4.4 Protein Stability

Throughout the course of protein production, different batches of proteins are expressed and different protocols or procedures for purification are used. These variables can have an effect on the integrity of the proteins. This is especially relevant to membrane proteins that have to be

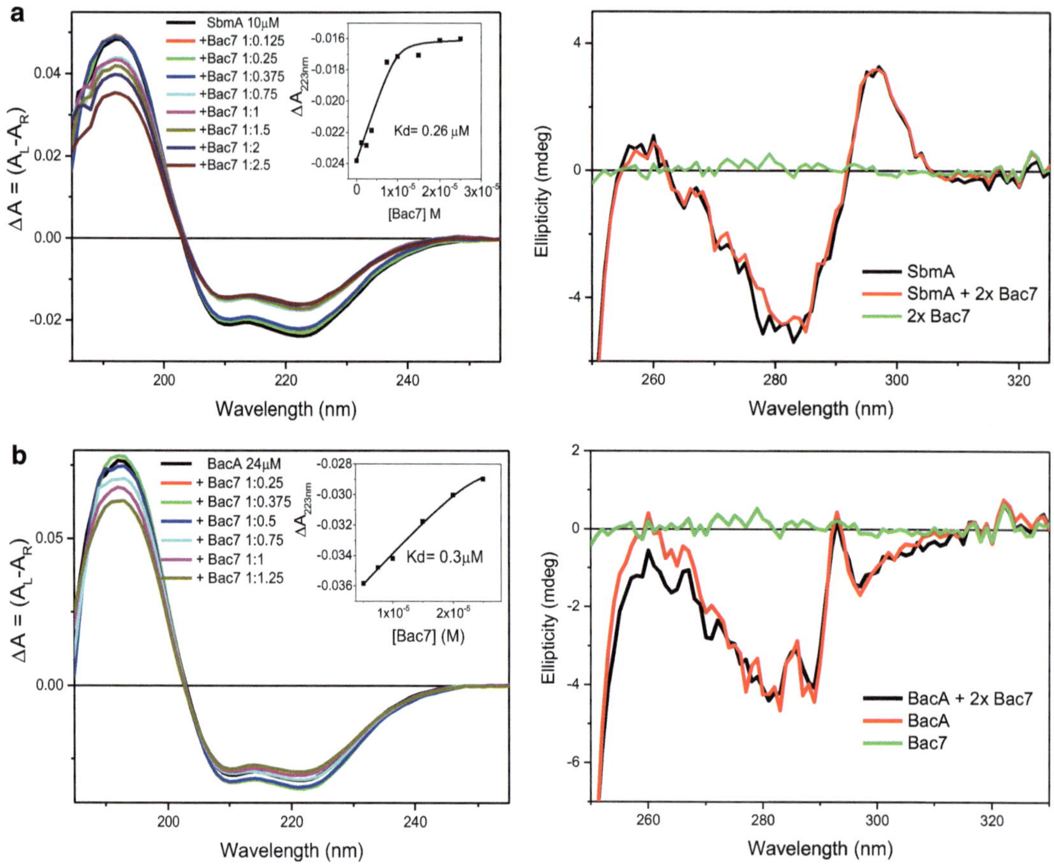

Fig. 4.4 (**a**) SRCD spectra of SbmA with and without Bac7 in the far-UV (*left*) and near-UV (*right*) regions. The insert in the far-UV region shows the determination of the dissociation constant Kd calculated to be 0.28 μM by fitting the CD data at 223 nm versus Bac7 concentration using the non-linear regression analysis (Siligardi et al. 2002) of CDApps (Hussain et al. 2015). (**b**) SRCD spectra of BacA with and without Bac7 in the far-UV (*left*) and near-UV (*right*) regions. The insert in the far-UV region spectra shows the determination of the dissociation constant Kd calculated to be 0.30 μM by fitting the CD data at 223 nm versus Bac7 concentration using the non-linear regression analysis of CDApps

Table 4.1 Protein secondary structure content of SbmA and BacA membrane proteins with and without addition of Bac7 ligand up to 1:35 molar ratios calculated from SRCD data using CONTINLL algorithm (Sreerama and Woody 2004) with SMP50 reference data set of 37 soluble proteins and 13 membrane proteins applied through CDApps beamline software (Hussain et al. 2015)

Protein secondary structure elements (SSE)	SbmA	SbmA + Bac7 [1–35]	BacA	BacA + Bac7 [1–35]
H1, α-helix	0.41	0.32	0.41	0.46
H2, distorted α-helix	0.19	0.17	0.19	0.19
S1, β-strand	0.04	0.09	0.06	0.03
S2, distorted β-strand	0.04	0.04	0.03	0.02
T, turn	0.12	0.10	0.12	0.10
U, unordered	0.21	0.28	0.20	0.20
Spectral fit SD	0.06	0.04	0.06	0.04

Redrawn from Runti et al. 2013

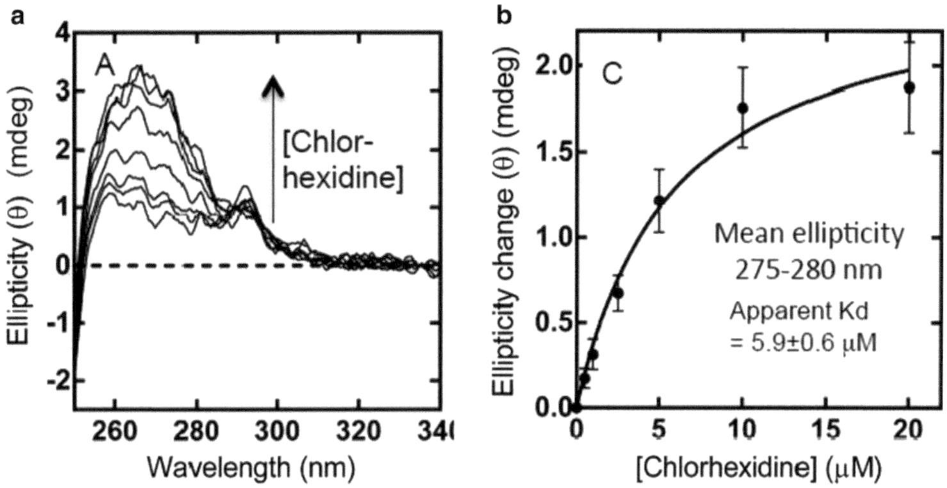

Fig. 4.5 SRCD titration of chlorhexidine into 20 μM purified wild-type AceI protein. (**a**) Near-UVSRCD spectra of AceI protein with and without chlorhexidine. The arrow indicated the spectra with increased concentration of chlorhexidine. (**b**) Plot of ellipticity (θ) intensity versus chlorhexidine concentration. The fitting curve (*solid line*) of the experimental data (*solid black circles*) was calculated using the non-linear regression analysis with a Kd of 5.9 μM (Siligardi et al. 2002)

solubilised in different detergents for stabilisation and for crystallisation with potential ligand molecules.

The example of the FsrC protein (Patching et al. 2012) showed that it required 1.5 h of incubation time for stabilisation, which was determined by measuring consecutive repeated scans of about 3 min each in the 260–180 nm region until the dominating alpha helical spectral feature stopped increasing at about 30 scans, see Fig. 4.6. It was essential to know the specific equilibration time for FsrC, and more generally, the equilibration time for membrane proteins in each specific detergent, when assessing ligand binding interactions, as otherwise it might result in ambiguous results. Interestingly, for FsrC the addition of ligand peptide GBAP appeared to stabilize the protein rather quickly (Fig. 4.6). The sugar transport protein GalP was also monitored for stability over time using the repeated scan method using high UV photon flux (Kalverda et al. 2014) showing that GalP is stable over repeated scans using a smaller bandwidth of 1.1 nm.

4.5 Temperature Denaturation

Temperature denaturation studies on proteins are routinely used to determine the thermodynamic properties of wild type proteins, their mutants, and the effect of ligand binding interactions or that of excipients as stabilisers to withstand long protein storage and transport conditions.

However, for the SbmA protein (described in Sect. 4.4) the thermal behaviour was not affected by the addition of peptide Bac-7 (Fig. 4.7a), even though conformational changes by SRCD were observed in the far UV region (Fig. 4.4) (Runti et al. 2013). The SRCD titration of Bac7 with SbmA indicated that the molecular interaction was accompanied by an 11 % decrease in alpha helical content for either SbmA or Bac7 (Table 4.1). The thermal studies, however, showed no significant differences between the melting temperature ($T_m = 55$ °C) of SbmA and the melting temperature of a mixture of SbmA and Bac7 (Fig 4.7).

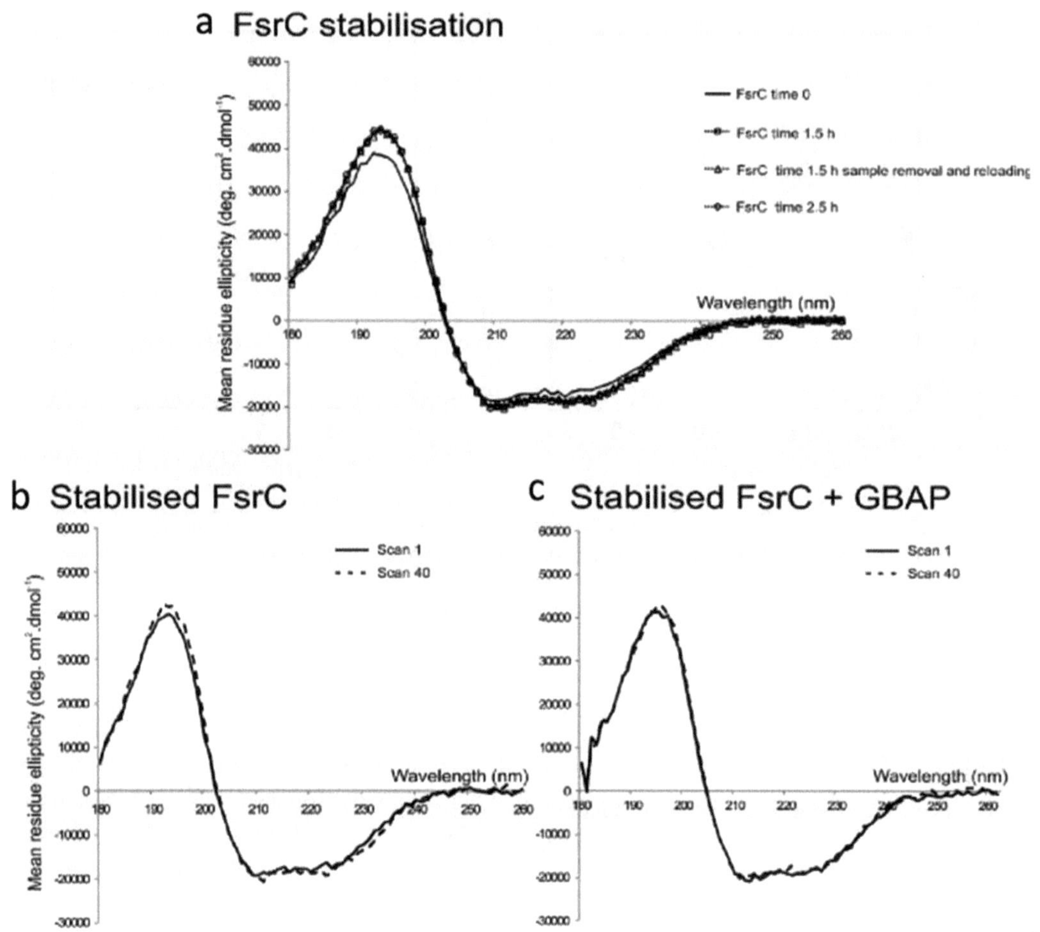

Fig. 4.6 FsrC stability determined by SRCD spectroscopy. (**a**) Far UV SRCD spectra of purified FsrC (6 μM) at 0 h (*solid black line*), 1.5 h (*dashed line, unfilled square*), and 2.5 hr (*dashed line, unfilled circle*) following sample preparation in 10 mM sodium phosphate pH 7.5 containing 0.02 % in n-Dodecyl β-D-maltoside (DDM) at 20 °C. Repeated SRCD spectra after 1.5 h involving sample removal and reloading were also included (*dashed line, unfilled triangle*). SRCD spectra of stabilised FsrC before and after exposure to far UV radiation illustrated in the absence (**b**) and in the presence (**c**) of twofold GBAP ligand. Spectrum 1, immediately after stabilisation (solid line); and spectrum 40, following 100 min exposure to light radiation in the 190–260 nm spectral region during 39 repeated consecutive scans (*dashed line*)

In the temperature denaturation study of the AceI protein (Kalverda et al. 2014) the protein melting temperatures were measured by ramping the temperature from 5 to 90 °C while monitoring the far-UV CD spectrum, particularly at 209 nm or 222 nm. The wild-type protein was observed to be rather stable at temperatures below 60 °C. However, mutant E50Q AceI, known to impair resistance to chlorhexidine, showed thermal denaturation at 40 °C. Chlorhexidine binding greatly increased the thermal stability of the E50Q mutant protein in a dose-dependent manner. Importantly, these experiments yielded evidence, albeit indirect, that the AceI protein was itself involved in binding chlorhexidine as an integral feature of the resistance mechanism.

In a study by Bettaney et al. (2013), the SRCD spectra of three inositol membrane transport proteins (IolF, IolT, and YfiG) measured with the Diamond B23 beamline showed a cut-off below 180 nm, indicated by the voltage of the photo-

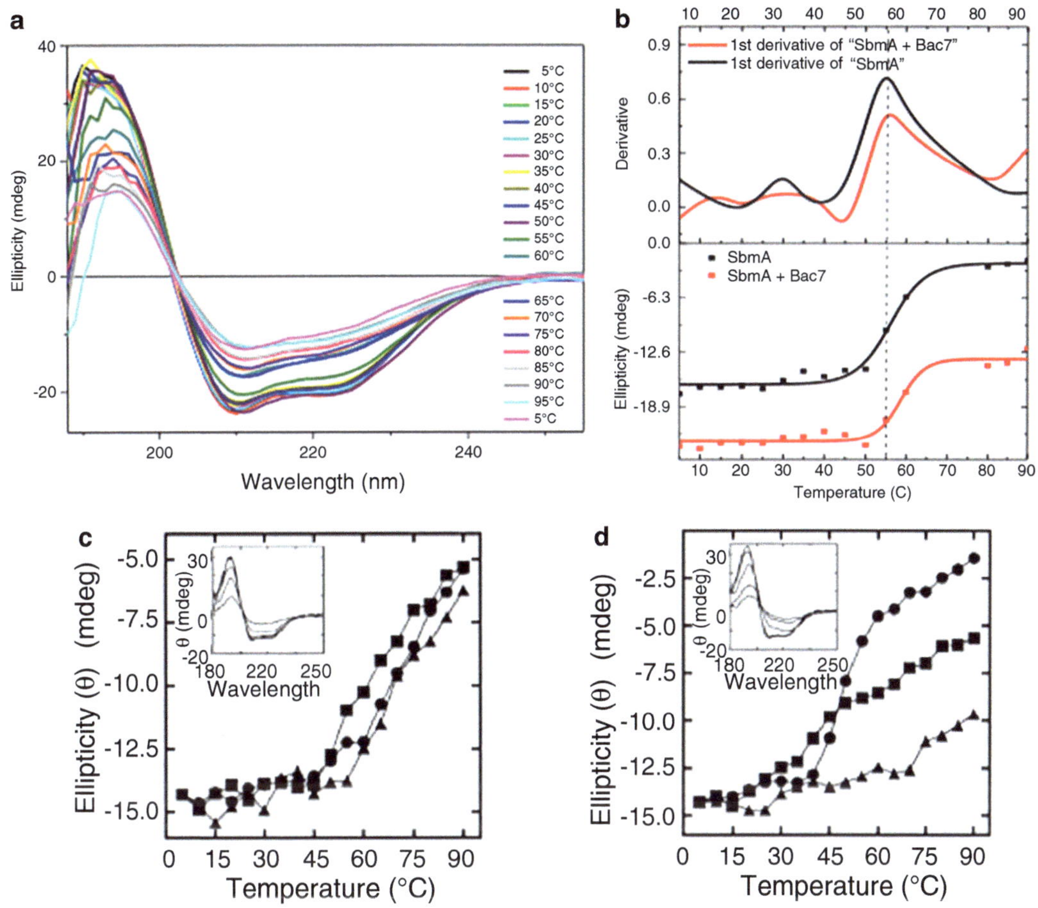

Fig. 4.7 (**a**) SRCD spectra of SbmA + Bac7 [1:5] as a function of temperature (Runti et al. 2013). (**b**) Thermal denaturation of SbmA with and without Bac7 ligand using SRCD spectroscopy. The upper part of the figure is the determination of the melting temperature Tm by 1st derivative method whilst the lower part is by the Boltzman equation. The lower part is also used to represent the thermal denaturation property or the thermal stability of the protein (Runti et al. 2013). (**c**) Thermal stability of wild type Ace1 protein (33 μM) with and without chlorhexidine by SRCD (redrawn from Hassan et al. 2013). (**d**) Thermal stability of mutant Ace1 protein E15Q (33 μM) with and without chlorhexidine by SRCD (redrawn from Hassan et al. 2013). For both (**c**) and (**d**), the ellipticity at 209 nm is shown for protein only (•), protein plus 100 μM chlorhexidine (■; 1:3 molar ratio) and protein plus 500 μM chlorhexidine (▲; 1:15 molar ratio) at increasing temperature. Insets show the characteristic α-helical protein far-UV spectrum of the respective proteins at increased temperature in the absence of chlorhexidine (redrawn from Hassan et al. 2013)

multiplier tube (PMT) detector exceeding 600 V (Fig. 4.8). It is essential for accurate secondary structure estimations that the positive CD bands at about 190–195 nm of the α-helical conformation are measured with the lowest level of noise possible, which can be readily achieved with synchrotron CD beamlines. Repeated scans can also improve the signal-to-noise-ratio, but at an increased overall time (the noise is reduced by the root square of the number of scans). Following successful confirmation of secondary structural integrity and composition, the specificity of IolF, IolT, and YfiG for a large variety of inositols and sugars of D and L configurationn was determined by measuring the cellular uptake of radiolabelled 3H-myo-inositol in the presence of unlabelled competing compounds (Bettaney et al. 2013) as well as differences in thermodynamic properties, which could give indications of any concerns for crystallisation trials.

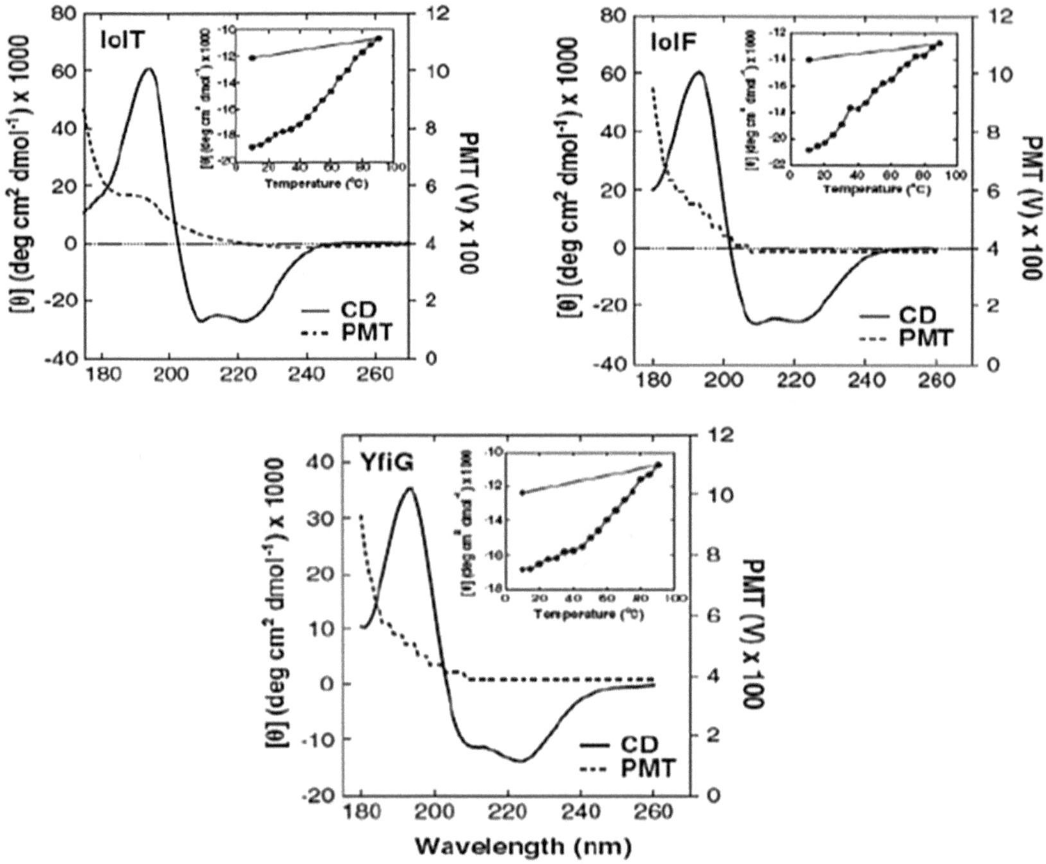

Fig. 4.8 Analysis of secondary structure and thermal denaturation profiles in the purified proteins. Synchrotron radiation circular dichroism (SRCD) spectra at 20 °C in the far-UV region are shown for the purified IolT(His$_6$), IolF(His$_6$) and YfiG(His$_6$) proteins at a concentration of 20 μM in a buffer containing 0.05 % DDM and 10 mM potassium phosphate (pH 7.6) using a 0.2-mm pathlength cell and acquiring spectra with an interval time of 0.5 nm. The spectra are buffer-subtracted and are the average of four scans. The plots show the CD spectrum (CD) and the high tension (HT) voltage values. Inset are thermal stability profiles at a wavelength of 222 nm from CD spectra recorded over a range of temperatures (10–90 °C and then back to 10 °C) for each of the proteins at a concentration of 0.1 mg/ml (redrawn from Bettaney et al. 2013)

4.6 Conformational Analysis and HT-CD Screening of Protein Crystallisation Buffers

The assessment of protein folding in solution is particularly important because conformational changes among a wild type protein and its mutants/constructs are likely to affect the protein activity and stability. Protein folding plays a crucial role in the function of a protein and its characterisation is a prerequisite for the understanding of the mechanism and consequences of sequence amino acid point mutation that could trigger protein misfolding for insulin, α-synuclein, lysozyme, and transthyrethin to name but a few cases with serious health implications (Blancas-Mejía and Ramirez-Alvarado. 2013; Ruzza et al. 2014, 2015; Marchiani et al. 2013).

A recent example regarding the importance of protein folding characterisation (including different constructs of the same protein) by SRCD is the case of the membrane protein GDP-mannose-dependent mannosyltransferase WbdD. In this example, the accurate determination of the α-helical content of the protein was crucial as a molecular ruler to regulate O-antigen chain length in lipopolysaccharide of *E coli* 09A (Hagelueken et al. 2015).

WbdD is a membrane-associated protein and its interaction with WbdA (another GDP-mannose-dependent mannosyltransferase) is essential for the ability of the soluble polymerase to act on the membrane-embedded undecaprenyl lipid–linked acceptor (Clarke et al. 2009). The authors investigated two insertion and two deletion variants of WbdD1–556 to gain some structural insight into the effect of the changes in the coiled-coil region. Multi-techniques such as the crystal structure of WbdD1-556 showing the coiled-coil region, molecular modelling and small angle X-ray scatterings (SAXS) models were coupled to elucidate the 3D structure. The study was further expanded with a bioassay of different constructs, giving various lengths of coupled lipopolysaccharides with CD spectroscopy. Circular dichroism analysis of WbdD1–459 and WbdD1–556 using the CONTINLL algorithm (Sreerama and Woody 2004; Hussain et al. 2015) on spectra collected on B23 at Diamond confirmed a detectable increase in α-helical content, as was expected by the addition of a coiled-coil region (Fig. 4.9). This multi-technique study helps to piece together information gathered and complements the limitations of each of the individual techniques into a final and more complete understanding of the role of the coiled-coil region of the protein as a molecular ruler.

The high photon flux of the B23 beamline has been successfully used to develop a protein UV-denaturation assay that can discriminate the relative stability of different types of protein folding and also to qualitatively determine ligand binding interactions (Hussain et al. 2012a, b; Jávorfi et al. 2010; Longo et al. 2014, 2015). The latter application has been very useful to study ligands with weak or no UV chromophores that would otherwise be difficult or impossible to monitor by conventional benchtop CD spectrometer.

The conformation of peptides and proteins is known to be affected by environmental conditions such as buffer composition, pH, salt concentration, detergents, metal ions and precipitants. The latter is widely used in crystallography to enhance crystallisation. However, the use of CD spectroscopy for such characterisations would be rather laborious and time consuming as this would normally be carried out by measuring the samples in a single cuvette cell, one by one. The unique highly collimated incident micro-

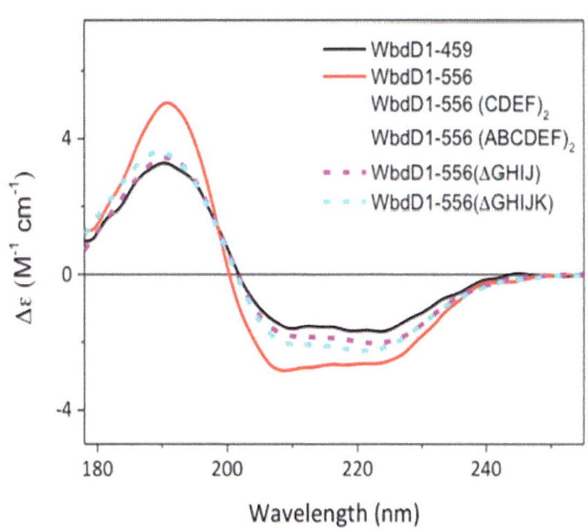

Samples	Experimental α-helical content (%)
WbdD1-459	25
WbdD1-556	29
WbdD1-556 (CDEF)$_2$	33
WbdD1-556 (ABCDEF)$_2$	36
WbdD1-556 (ΔGHIJ)	27
WbdD1-556 (ΔGHIJK)	28

Fig. 4.9 CD spectra of different WbD constructs and their secondary structure content estimation using CONTINLL (Sreerama and Woody 2004) of CD Apps beamline software (Hussain et al. 2015). On the right is shown the percentage of α-helical content calculated from SRCD data using CONTINLL algorithm in CDApps beamline software

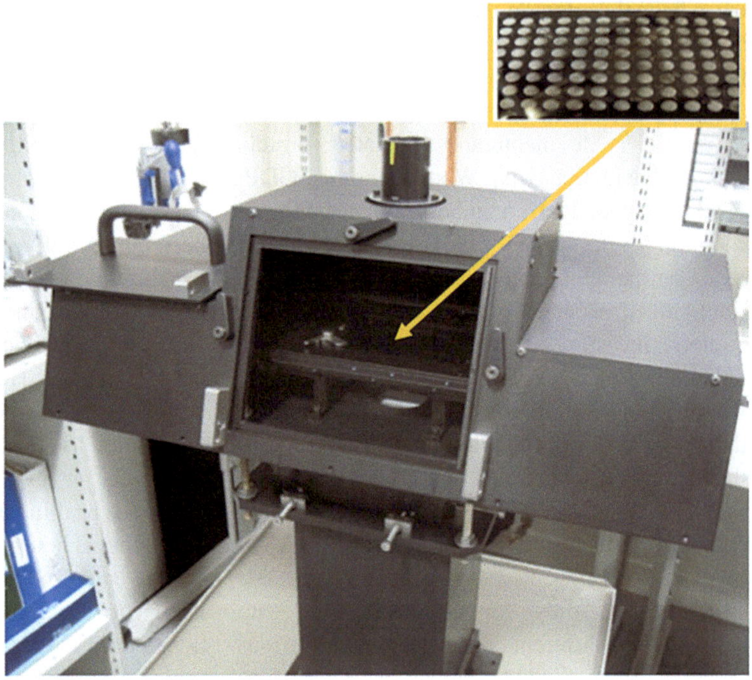

Fig. 4.10 Vertical sample compartment of Diamond B23 beamline module A. The chamber enables the SRCD measurements of horizontally positioned samples at respect to the incident monochromatic light. It has been designed to accommodate the 96- and 384-well multiplates made of fused quartz (Suprasil, Hellma). The central insert shows where the enlarged 96-well multiplate is located inside the chamber (*yellow arrow*)

beam (from 0.3 mm^2 to 1.5 mm^2 cross section) of the B23 has recently been exploited for the use of 96- or 384-well plates, to allow for high throughput CD (HT-CD) (Fig. 4.10) screening to characterise protein folding in crystallisation conditions and in protein-drug binding interactions (Siligardi and Hussain 2015).

Myoglobin, a highly α-helical protein, was investigated by SRCD spectroscopy dissolved in a selection of 48 conditions from the MemGold2 crystallisation screen (Molecular Dimensions) that is widely used for membrane protein crystallization. The corresponding 48 SRCD spectra (Fig. 4.11) showed significant conformational differences that could be attributed to electrolyte concentration and pH. It is important to note that despite myoglobin being a highly α-helical soluble protein its conformation can be perturbed by the membrane protein crystallisation buffer MemGold2. This can be readily illustrated in the pie chart of the secondary structure content determined by SRCD spectroscopy using CONTINLL of the SRCD spectra of the 96-well multiplate (Fig. 4.12).

Most membrane proteins are highly helical and it is not inconceivable that they too might behave similarly to myoglobin under various buffer conditions. This is consistent with the fact that membrane proteins only crystallise in certain buffer conditions (Privé 2007; O'Malley 2011) but not in others. This raises the question whether proteins that do not crystallise do not do so because they have different conformations that do not promote association or that their thermodynamic properties are significantly different, inhibiting crystal formation.

The availability of HT-CD facility at the B23 beamline opens up the possibility of finding cross-correlations between buffer conditions and protein conformation, identifying possible buffer targets that could promote crystallisation for further 3D structural determination by X-ray crystallography.

4.7 Conclusion

The work described in this chapter highlights the use of CD and SRCD spectroscopy for studies of membrane proteins, particularly when a wide spectral range from far-UV to near-UV region (180–350 nm) is achieved for both protein and ligand. This approach enables the determination of the stability and content of the protein sec-

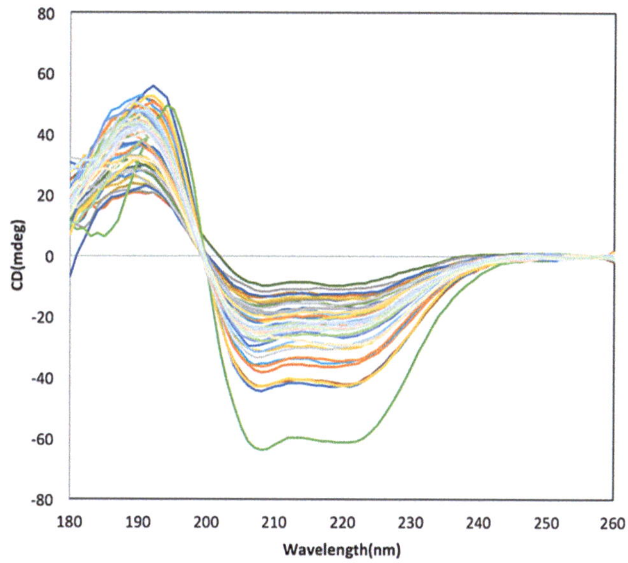

Fig. 4.11 SRCD spectra of myoglobin dissolved in 48 distinct buffers of MemGold2 for membrane protein crystallisation

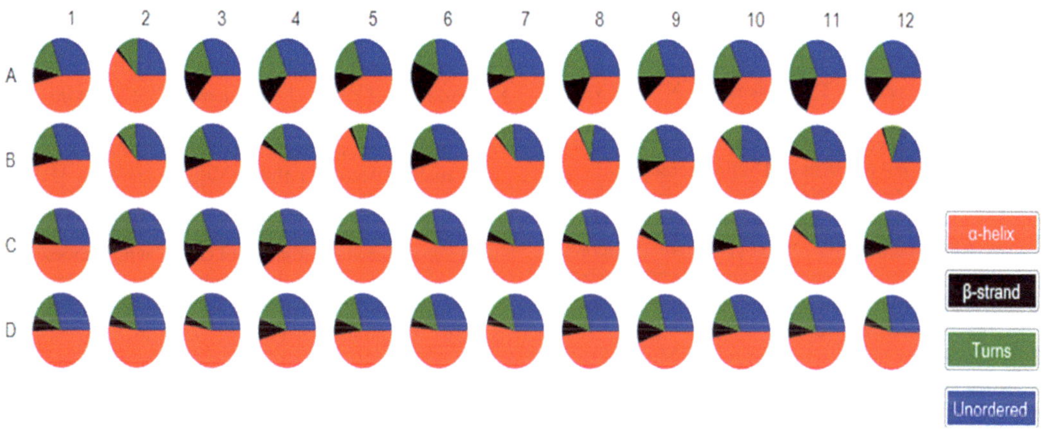

Fig. 4.12 Secondary structure content of myoglobin dissolved in the 48 crystallographic solvent conditions of the MemGold2 multiplate (well coordinates A1-12, B1-12, C1-12 and D1-12). The pie chart was prepared using B23 CDApps software (Hussain et al. 2015). Although the α-helix content is the dominating element of secondary structure, significant different a-helical contents were induced by some of the MemGold2 solvent conditions. At first glance less helical content can be observed in A3, A4, A6, A8 12, C3, C4, D9 (12 out of 48 or 25 %) whilst more helical in B3, B4, B7, B8, B10, B12 and C11 (6 out of 48 or 12.5 %)

ondary structure, the qualitative and quantitative assessment of ligand binding interactions and the characterisation of the solvent (detergent) conditions to optimise protein stability and binding properties. It has been successful in the identification of drugs that despite similar activity revealed different thermodynamics properties when bound to various protein constructs.

The amide bond of the protein backbone structure gives information about the protein secondary structure while the aromatic side-chains of the tryptophan, tyrosine and phenylalanine amino acid residues provide details about their local tertiary structure, as such, these are ideal as molecular probes for ligand binding interactions. The chromophore of the ligand can be used to unambiguously determine ligand binding in both far- and near-UV regions and in particular whether protein conformational changes are occurring upon ligand binding. Further development in high throughput allows the screening of crystallization buffers and the potential screening of ligands in drug discovery with membrane proteins acting as host or target receptor. This broad approach of CD spectroscopy in solution for membrane proteins is advantageous. SRCD in particular provides wavelength extension in the vacuum UV region and higher photon flux, which enables more penetrating measurements with less transparent media (NaCl content up to 500 mM) and in small volume capacity cuvette cells that are unattainable using bench-top CD instruments.

Acknowledgments We gratefully acknowledge the Diamond Light Source Ltd. for the B23 beamtime allocation (www.diamond.ac.uk/Home/ForUsers.html).

References

Berova N, Nakanichi K, Woody RW (2000) Circular dichroism principles and applications, 2nd ed. New York: Wiley-VCH

Berova N, Polavarapu PL, Nakanishi K, Woody RW (2012) Comprehensive chiroptical spectroscopy: applications in sterochemical analysis of synthetic compounds, natural products and biomolecules, vol 2. New Jersey: John Wiley & Sons

Bettaney KE, Sukumar P, Hussain R, Siligardi G, Henderson PJF, Patching SG (2013) A systematic approach to the amplified expression, functional characterization and purification of inositol transporters from Bacillus subtilis. Mol Membr Biol 30:3–14

Blancas-Mejía LM, Ramirez-Alvarado M (2013) Systemic amyloidoses. Annu Rev Biochem 82:745–774

Borisenko KB, Shanmugam J, Williams BAO, Ewart P, Gholipour B et al (2015) Photo-induced optical activity in phase-change memory materials. Sci Rep 5:8770

Carvalho AL, Trincão J, Romão MJ (2009) X-ray crystallography in drug discovery. Methods Mol Biol 572:31–56

Clarke BR, Greenfield LK, Bouwman C, Whitfield C (2009) Coordination of polymerisation, chain termination, and export in assembly of the Escherichia coli lipopolysaccharide 09a antigen in an ATP-binding cassette transporter-dependent pathway. J Biol Chem 284:30662–30672

Fasman GD (1996) Circular dichroism and the conformational analysis of biomolecules. Plenum Press, New York

Fiedler S, Cole L, Keller S (2013) Automated circular dichroism spectroscopy for medium-throughput analysis of protein conformation. Anal Chem 85:868–1872

Gaspar D, Lúcio M, Rocha S, Costa Lima JLF, Reis S (2011) Changes in PLA2 activity after interacting with anti-inflammatory drugs and model membranes: evidence for the involvement of tryptophan residues. Chem Phys Lipids 164:292–299

Hagelueken G, Clarke BR, Huang H, Tuukkanen A, Danciu I et al (2015) A coiled-coil domain acts as a molecular ruler to regulate O-antigen chain length in lipopolysaccharide. Nat Struct Mol Biol 22:50–56

Hassan KA, Jackson S, Penesyan A, Patching S, Tetu SG et al (2013) Transcriptomic and biochemical analyses identify a family of chlorhexidine efflux proteins. Proc Natl Acad Sci U S A 110(50):20254–20259

Hussain R, Jávorfi T, Siligardi G (2012a) Circular dichroism beamline B23 at the Diamond Light Source. J Synchrotron Radiat 19:32–135

Hussain R, Jávorfi T, Siligardi G (2012b) Spectroscopic analysis: synchrotron radiation circular dichroism. Comprehen Chiral 8:438–448

Hussain R, Benning K, Myatt D, Jávorfi T, Longo E et al (2015) CDApps: integrated software for experimental planning and data processing at beamline B23 Diamond Light Source. J Synchrotron Radiat 22:465–468

Jávorfi T, Hussain R, Myatt D, Siligardi G (2010) Measuring circular dichroism in a capillary cell using the B23 synchrotron radiation CD beamline at Diamond Light Source. Chirality 22:149–E153

Johnson WC (1990) Protein secondary structure and circular dichroism: a practical guide. Proteins: Struct Funct Bioinf 7:205–214

Kalverda AP, Gowdy J, Thompson GS, Homans SW, Henderson PJF, Patching S (2014) TROSY NMR with a 52 kDa sugar transport protein and the binding of a small-molecule inhibitor. Mol Membr Biol 31:131–140

Keegan N, Ridley H, Lakey JH (2010) Discovery of biphasic thermal unfolding of OmpC with implications for surface loop stability. Biochemistry 49:9715–9721

Kelly SM, Jess TJ, Price NC (2005) How to study proteins by circular dichroism. Biochim Biophys Acta 1751:119–139

Kim YP, Yeo KJ, Kim MH, KimY-C JYH (2010) Structural characterisation of the intra-membrane histidine kinase YbdK from *Bacillus subtilis* in DPC micelles. Biochem Biophys Res Commun 391:1506–1511

Laera S, Ceccone G, Rossi F, Gilliland D, Hussain R, Siligardi G, Calzolai L (2011) Measuring protein structure and stability of protein-nanoparticle systems with synchrotron radiation circular dichroism. Nano Lett 11:4480–4484

Longo E, De Santis E, Hussain R, van Der Walle C, Casasfinet J, Uddin S, dos Santos A, Siligardi G (2014) The effect of palmitoylation on the conformation and physical stability of a model peptide hormone. Int J Pharm 472:156–164

Longo E, Hussain R, Siligardi G (2015) Application of circular dichroism and magnetic circular dichroism for assessing biopharmaceuticals formulations photostability and small ligands binding properties. Int J Pharm 480:84–91

Marchiani A, Mammi S, Siligardi G, Hussain R, Tessari I, Bubacco L, Delogu G, Fabbri D, Dettori MA, Sanna D, Dedola S, Serra PA, Ruzza P (2013) Small molecules interacting with α-synuclein: antiaggregating and cytoprotective properties. Amino Acids 45:327–338

Martin SR, Schilstra MJ, Siligardi G (2011) Biophysical approaches determining ligand binding to biomolecular targets, detection, measurement and modelling. In: Podjarmy A, Dejaegere A, Kieffer B (eds) Chapter 7: circular dichroism. Cambridge: RSC Publishing, pp 226–246
Mason SF (1982) Molecular optical activity and the chiral discrimination. Cambridge University Press, Cambridge
O'Malley MA, Helgeson ME, Wagner NJ, Robinson AS (2011) Towards rational design of protein detergent complexes: determinants of mixed micelles that are critical for the in vitro stabilisation of a G-protein coupled receptor. Biophys J 101:1938–1948
Patching SG, Edara S, Ma P, Nakayama J, Hussain R, Siligardi G, Phillips-Jones MK (2012) Interactions of the intact FsrC membrane histidine kinase with its pheromone ligand GBAP revealed through synchrotron radiation circular dichroism. Biochim Biophys Acta Biomembr 1818:1595–1602
Pellecchia M, Bertini I, Cowburn D, Dalvit C, Giralt E et al (2008) Perspectives on NMR in drug discovery: a technique comes of age. Nat Rev Drug Discov 7:738–745
Pennisi E (2012) ENCODE project writes eulogy for junk DNA. Science 337:1159–1160
Phillips-Jones MK, Patching SG, Edara S, Nakayama J, Hussain R, Siligardi G (2013) Interactions of the intact FsrC membrane histidine kinase with the tricyclic peptide siamycin I revealed through synchrotron radiation circular dichroism. Phys Chem Chem Phys 15:444–447
Potter CA, Ward A, Lauri C, Williamson MP, Henderson PJF, Phillips-Jones MK (2002) Expression, purification and characterisation of full-length heterologously expressed histidine protein kinase RegB from *Rhodobacter sphaeroides*. J Mol Biol 320: 201–213
Powl AM, O'Reilly AO, Miles AJ, Wallace BA (2010) Synchotron radiation circular dichroism spectroscopy-defined structure of the C-terminal domain of NaChBac and its role in channel assembly. Proc Natl Acad Sci U S A 107:14064–14069
Privé GG (2007) Detergents for the stabilization and crystallization of membrane proteins. Methods 41:388–397
Psakis G, Saidijam M, Shibayama K, Polaczek J, Bettaney KE et al (2009) The sodium-dependent D-glucose transport protein of *Helicobacter pylori*. Mol Microbiol 71:391–403
Runti G, del Carmen LRM, Stoilova T, Hussain R, Jennions M, Choudhury HG, Benincasa M, Gennaro R, Beis K, Scocchi M (2013) Functional characterization of SbmA, a bacterial inner membrane transporter required for importing the antimicrobial peptide Bac7(1-35). J Bacteriol 195:5343–5351
Ruzza P, Siligardi G, Hussain R, Marchiani A, Islami M et al (2014) Ceftriaxone blocks the polymerization of α-synuclein and exerts neuroprotective effects in vitro. ACS Chem Neurosci 5:30–38
Ruzza P, Hussain R, Biondi B, Calderan A, Tessari I, Bubacco L, Siligardi G (2015) Effects of trehalose on thermodynamic properties of alpha-synuclein revealed through synchrotron radiation circular dichroism. Biomolecules 5:724–734
Siligardi G, Hussain R (1998) Biomolecules interactions and competitions by non-immobilised ligand interaction assay by circular dichroism. Enantiomer 3:77–87
Siligardi G, Hussain R (2010) Applications of circular dichroism. In: Lindon J, Tranter G, Koppenaal D (eds) Encyclopedia of spectroscopy and spectrometry, vol 1, 2nd edn. Elsevier, Oxford, pp 9–14
Siligardi G, Hussain R (2015) CD spectroscopy: an essential tool for quality control of protein folding. In: Raymond JO (ed) Structural proteomics: high-throughput methods, methods in molecular biology, vol 1261. New York: Springer; pp 255–276
Siligardi G, Campbell MM, Gibbons WA, Drake AF (1991) Conformational analysis of the melanin concentrating hormone (MCM) by CD spectroscopy: disulphide bridge and aromatic tyrosyl contribution. Eur J Biochem 206:23–29
Siligardi G, Panaretou B, Meyer P, Singh S, Woolfson DN, Piper PW, Pearl LH, Prodromou C (2002) Regulation of Hsp90 ATPase activity by the co-chaperone Cdc37p/p50[cdc37]. J Biol Chem 277:20151–20159
Sreerama N, Woody RW (2004) On the analysis of membrane protein circular dichroism spectra. Protein Sci 13:100–112
Yeo KJ, Kwak S-N, Kim HJ, Cheong C, Kim MH, Jeon YH (2008) Expression and characterisation of the integral membrane domain of bacterial histidine kinase SCO3062 for structural studies. Biochem Biophys Res Commun 376:409–413

Membrane Protein Crystallisation: Current Trends and Future Perspectives

5

Joanne L. Parker and Simon Newstead

Abstract

Alpha helical membrane proteins are the targets for many pharmaceutical drugs and play important roles in physiology and disease processes. In recent years, substantial progress has been made in determining their atomic structure using X-ray crystallography. However, a major bottleneck still remains; the identification of conditions that give crystals that are suitable for structure determination. Over the past 10 years we have been analysing the crystallisation conditions reported for alpha helical membrane proteins with the aim to facilitate a rational approach to the design and implementation of successful crystallisation screens. The result has been the development of MemGold, MemGold2 and the additive screen MemAdvantage. The associated analysis, summarised and updated in this chapter, has revealed a number of surprisingly successfully strategies for crystallisation and detergent selection.

Keywords

Membrane protein • Crystallisation • Screen development • Detergent selection

5.1 Introduction

Protein crystallisation is often described as a 'black box' process, full of mystery and superstition. In fact crystallisation itself is a well documented process following well understood physical chemistry laws and involving the supersaturation of the protein of interest to coax the molecules into a regular three-dimensional crystal (McPherson and Gavira 2014; Chayen and Saridakis 2008). Although not the topic of this chapter, many excellent sources of information are available on how to set up these conditions, using either vapor diffusion (Delmar et al. 2015), microbatch (Chayen 1998) or free-interface diffusion (Segelke 2005). The mystery begins when we try to consider which conditions will coax the proteins to assemble in a regular form and

The original version of this chapter has been revised. An erratum to this chapter can be found at DOI 10.1007/978-3-319-35072-1_13

J.L. Parker • S. Newstead (✉)
Department of Biochemistry, University of Oxford, Oxford OX1 3QU, UK
e-mail: joanne.parker@bioch.ox.ac.uk; simon.newstead@bioch.ox.ac.uk

I. Moraes (ed.), *The Next Generation in Membrane Protein Structure Determination*, Advances in Experimental Medicine and Biology 922, DOI 10.1007/978-3-319-35072-1_5

produce diffraction quality crystals. For many years, the standard experimental set up has involved screening your purified, homogenous protein sample against commercial, sparse-matrix style 'crystallisation screens' (Luft et al. 2014). The idea behind these screens was to sample as much 'crystallisation space' as possible with the minimal protein amount (Newman et al. 2007). Of course, many of these commercial screens were based on the currently available information regarding soluble protein crystallisation. This included the success of large molecular weight (MW) polymers, particularly polyethylene glycol (PEG). High concentrations of salt were also used, as these would naturally help to 'salt out' the protein and hopefully grow protein crystals (Page and Stevens 2004).

Membrane proteins however are different. The requirement to extract these proteins from the membrane using detergents, whilst simultaneously keeping them folded and stable in solution creates a new set of unknown variables (Iwata 2003). Added to this the fact that in a crystallisation experiment involving detergent solubilised membrane proteins, the actual entity being crystallised is the detergent-protein complex and not simply the protein alone (Kunji et al. 2008; Bill et al. 2011). It was against this backdrop that in 2008 a comprehensive analysis of membrane protein crystallization conditions was published (Newstead et al. 2008a). The idea behind this analysis was simple, the number of crystal structures had just reached 121 and our aim was to analyse these conditions and draw conclusions as to which chemicals were successful in growing membrane protein crystals. Could any trends be observed and could this information be used to improve success rates in current projects? The result of this analysis was the release of the first rationally designed sparse matrix style membrane protein crystallisation screen, MemGold (Newstead et al. 2008a).

In the following years the pace of membrane protein structure determination has increased exponentially (White 2007). This increase is due to progress being made in tackling many of the hurdles faced in determining the crystal structure of membrane proteins (Bill et al. 2011; Ghosh et al. 2015). This includes advances in protein production using recombinant systems (Tate et al. 2003; Chen et al. 2013), methods for screening stability (Drew et al. 2008; Kawate and Gouaux 2006; Sonoda et al. 2011) and in X-ray data collection using microfocus beamlines, fast read out detectors and modifications to sample application (Nogly et al. 2005). More recent progress has been made in protein engineering, resulting in either increases in protein stability (Tate and Schertler 2009) or the introduction of additional crystallisation scaffolds, such as T4 lysozyme or BRIL (Chun et al. 2012). However, growing well-ordered three-dimensional crystals still represents a significant hurdle. In 2012, we followed up our first analysis with another review of the current trends in crystallisation, this time based on 254 examples from the Protein Data Bank (PDB) (Parker and Newstead 2012). Our results showed that the initial trends described in 2008 had broadly held, but revealed intriguing new developments such as an increase in the number of cases where additional or mixed detergents had been required and changes in the types of membrane protein being crystallised. The new information enabled the development a sister crystallization screen, MemGold 2, to complement the original MemGold screen released 4 years earlier. In addition to our analysis of crystallisation conditions, an in depth analysis of additives was now possible. The use of additional chemicals to optimise initial crystals to improve diffraction quality is well documented and many commercial kits are available (Chayen and Saridakis 2008; Cudney et al. 1994). An additive screen targeted specifically for membrane proteins however, had so far remained absent from the commercial market. A specific membrane protein additive screen was therefore suggested to facilitate crystal optimization and released along with MemGold 2, called MemAdvantage.

As of August 2015, the number of crystallisation examples in our database is more than 500 and in this chapter we present an updated analysis from these conditions. Here we compare the results of these past analyses with each other and with those focused on soluble proteins (Fazio et al. 2015). The aim of this chapter is to equip the protein crystallographer with the knowledge to design their own screens using information that is up to date and relevant to membrane protein samples.

5.2 Current Trends in the Number and Types of Alpha Helical Membrane Protein Structures

Since 2008, an additional 448 novel alpha helical membrane protein (MP) structures have been added to our original crystallisation database, bringing the total number of entries to 569. We have previously grouped these into eight different families, broadly divided by function (Fig. 5.1). However, we noticed a significant growth in the number of enzymes being reported. Therefore, we have included two additional families for proteases and other enzymes, bringing the total family count up to 10. The 'Other' family now contains examples of either single members of a functional family, such as the tight junction Claudin-15 (PDB:4P79) and BcsA-BcsB cellulose synthease (PDB:4HG6). The data clearly show an increase in the determination of channel and transporter structures, from 29 and 27 to 149 and 157, respectively. This has occurred with a continued decrease in the overall percentage of respiratory complexes, from 24 % to 9 %. Possibly the largest change from 2012 has been in the G protein coupled receptor (GPCR) family, which now makes up 13 % of the database with >70 structures, up from 17 structures in 2012. Significant progress has been made in the structure determination of GPCRs due to a number of technological advancements in protein engineering and lipidic mesophase crystallisation (LCP) (Ghosh et al. 2015); we discuss LCP crystallisation in more detail towards the end of the chapter. The contribution from the photosynthetic and light harvesting complexes (LHCs) has also reduced, falling from 7 % in 2012 to 4 % in 2015. The number of ATPase structures has doubled

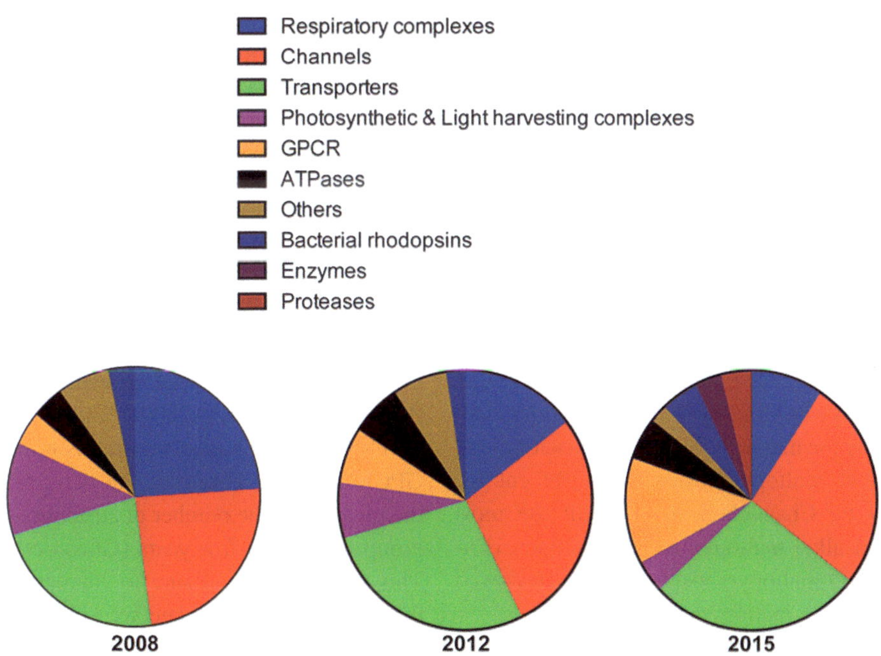

Fig. 5.1 Number and types of alpha helical membrane protein structures reported from 2008 to present (2015). Pie charts showing the change in the proportion of structures belonging to each family group from 2008, 2012 and 2015. Membrane proteins were broken down into the following families: Respiratory complexes (*blue*), GPCRs (*orange*), ATPases (*black*), Bacterial Rhodopsins (*dark blue*), Enzymes (*purple*), Proteases (*dark red*) and where a protein didn't fit in these categories, others (*brown*)

from 16 to 30 since 2012, but their total share of the database has remained the same, at 6 %.

5.3 Detergent Selection

It has been said that when trying to crystallise a membrane protein in detergent what we are actually doing is creating a crystal of detergent contaminated with protein. Membrane proteins by definition contain large surfaces of predominantly hydrophobic residues that would ordinarily reside in the core of the lipid membrane. To purify these proteins, the researcher will need to choose a suitable detergent to solubilise the protein in preparation for purification (Rosenbusch 1990). But which detergent is the right one? This is often the first dilemma faced in the challenge to determine a membrane protein crystal structure. The majority of membrane protein structures deposited in the PDB have been determined using crystals grown from detergent solubilized protein using traditional vapour diffusion experiments. In these experiments the sample being crystallised is a mixture of both protein and associated detergent, making detergent selection a critical parameter for growing well-ordered, well-diffracting crystals. Significant progress has recently been made in the development of novel detergents for use in membrane protein purification and crystallisation (Chae et al. 2010; Tao et al. 2009). However, as in 2012 the alkyl maltopyranosides account for the majority of successfully used detergents accounting for half of all structures in the database (Fig. 5.2), followed by the alkyl Glucopyranosides (23 %), Amine Oxides (7 %) and Polyoxyethylene Glycols (7 %). Transporters still account for the majority for structures determined using alkyl maltopyranosides with 89 entries, followed by Channels with 58. Still the most successful alkyl maltopyranoside detergent is n-dodecyl-β-D-maltopyranoside (DDM), followed by n-decyl-β-D-maltopyranoside (DM).

Although the choice of detergent depends on many different parameters, considerable effort should be made to screen for crystals in shorter chain detergents as these are more likely to diffract to a higher resolution (Sonoda et al. 2010). Analysis of the resolution of reported structures further supports this conclusion (Fig. 5.3), with the alkyl glucopyranoside detergent, n-octyl-β-D-glucopyranoside (OG) having both the highest resolution structure at 0.88 Å, that of a yeast aquaporin, Aqy1 (PDB: 3ZOJ) and highest mean resolution at 2.5 Å. The amine oxides, including n-lauryl dimethylamine n-oxide (LDAO), gave the next most favourable mean resolution of 2.66 Å. There is unlikely to ever be a single panacea detergent that can be applied to all types of membrane proteins. Nevertheless, the data support the continued use of DDM, DM, OG and LDAO as good first choice detergents when screening crystallization conditions. A rational and intelligent approach should always be taken to detergent screening for membrane proteins, which can now be accomplished easily using fluorescent-based methods early on in the structure determination process (Kawate and Gouaux 2006; Newstead et al. 2007). A notable change since 2012 has been the increased success of detergent mixtures. Interestingly, all families, except the bacterial rhodopsin, have had at least one example where > 1 detergent has been reported, suggesting this should be a common approach to adopt early on in the screening and optimisation process. However, as yet no trend exists that may hint at whether certain detergent classes may be paired more successfully.

An important development in membrane protein crystallisation over the past 3 years has been the increased use of the lipidic cubic phase (LCP) as a medium for crystal growth (Caffrey and Porter 2010; Caffrey 2009). This technological development has had an enormous impact on the GPCR field and is one of the main reasons for the increase in the number of structures from this group in the past few years (Ghosh et al. 2015). This methodology is sure to increase in use in the coming years. To date we have recorded 17 structures out of a total of 91, compared with 49 GPCR examples. As highlighted in Fig. 5.3, the mean resolution for structures determined in LCP is 2.5 Å, almost half an ångström lower than

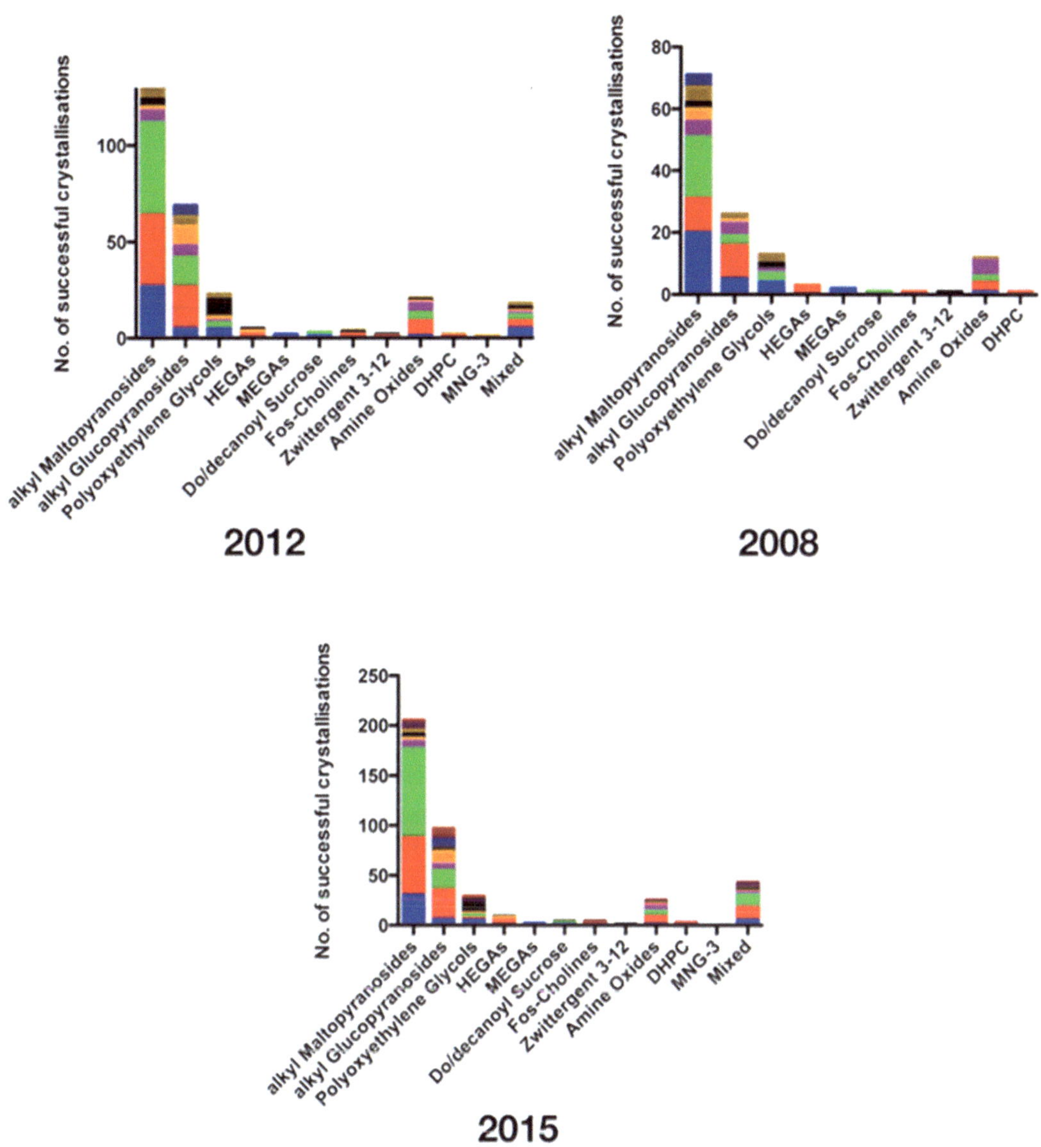

Fig. 5.2 Analysis of successful crystallisationdetergents used for alpha helical membrane proteins. Numbers for each detergent class are shown and the bars are subdivided to represent the different membrane protein families. Respiratory complexes (*blue*), GPCRs (*orange*), ATPases (*black*), Bacterial Rhodopsins (*dark blue*), Enzymes (*purple*), Proteases (*dark red*) and where a protein didn't fit in these categories, others (*brown*)

for the alkyl maltopyranoside detergents and very close to the mean resolution obtained for n-octyl-glucopyranoside (OG), which in many cases is too harsh for alpha helical membrane proteins. This data adds further support to the early adoption of lipidic mesophase crystallisation in any structure project. More information on detergents can be found in Chap. 2 of this book.

5.4 Precipitants – How Do They Differ Between Membrane Proteins and Their Soluble Counterparts?

Our 2008 analysis of precipitants revealed a striking success for small MW PEGs in the crystallisation of channels and transporters, with larger

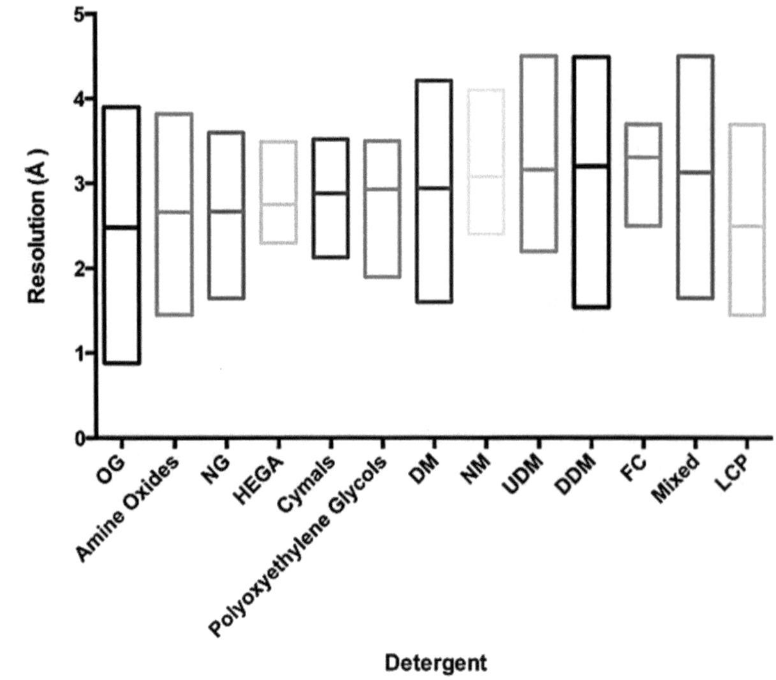

Fig. 5.3 Effect of detergent on mean resolution of reported crystal structures compared to *in meso* crystallisation. *Box plots* showing the lowest, highest and mean resolution (Å) reported for the most commonly used detergents. Abbreviations are OG (octylglucopyranoside), NG (nonlyglucopyranoside), DM (decyl maltoside), NM (nonylmaltoside), UDM (Undecyl maltoside), DDM (dodecyl maltoside), FC (fos-cholines), Mixed refers to a combination of detergents and LCP (lipid cubic phase)

MW PEGs being more successful for respiratory complexes and membrane proteins with large hydrophilic domains (Newstead et al. 2008a). These trends have remained in the updated data set, with the notable appearance of small MW PEGs in the crystallization of the eukaryotic GPCR family. The successful concentration ranges have also been maintained, with small MW PEGs being successful at concentrations between 20 and 40 % v/v, and larger MW PEGs being used at lower concentrations, between 5 and 20 %. The successful use of organic molecules, such as MPD is still low, further confirming their unsuitability in general crystallisation conditions for alpha helical MPs, a situation that is dramatically different for outer membrane proteins where organic molecules are clearly more successful (Newstead et al. 2008b). Of note is the absence of high salt conditions in our database. This contrasts with a recent analysis of crystallisation space reported for the entire PDB in 2014 (Fazio et al. 2015). This analysis clearly demonstrates the most successful crystallization reagents are PEG 3350 and ammonium sulphate, which only make up 4.0 and 3.5 % of our database, respectively. This contrasts with PEG 400, which accounts for 33 % of the reported membrane protein conditions, but doesn't appear in the ten most abundant chemicals reported in a non-redundant analysis of successful crystallisation conditions.

5.5 MemGold and MemGold2 – What's the Difference?

MemGold was the first rationally designed sparse matrix style membrane protein crystallisation screen; the previous screens developed by Jeff Abramson and So Iwata were based on a more systematic screening of PEG 400 and 4000 (Iwata 2003). MemGold was designed based on the then available 121 structures published in the PDB in 2008 and proved to be very successful as a tool for discovering initial crystallisation conditions. However, it was unclear to us at the time whether these conditions were the most optimal, given that 24 % of the conditions were contributed from the respiratory complex family. Respiratory complexes often have much larger extracellular domains that tend to dominate the crystal contacts in the unit cell, which suggested to us that perhaps these proteins tended to favour

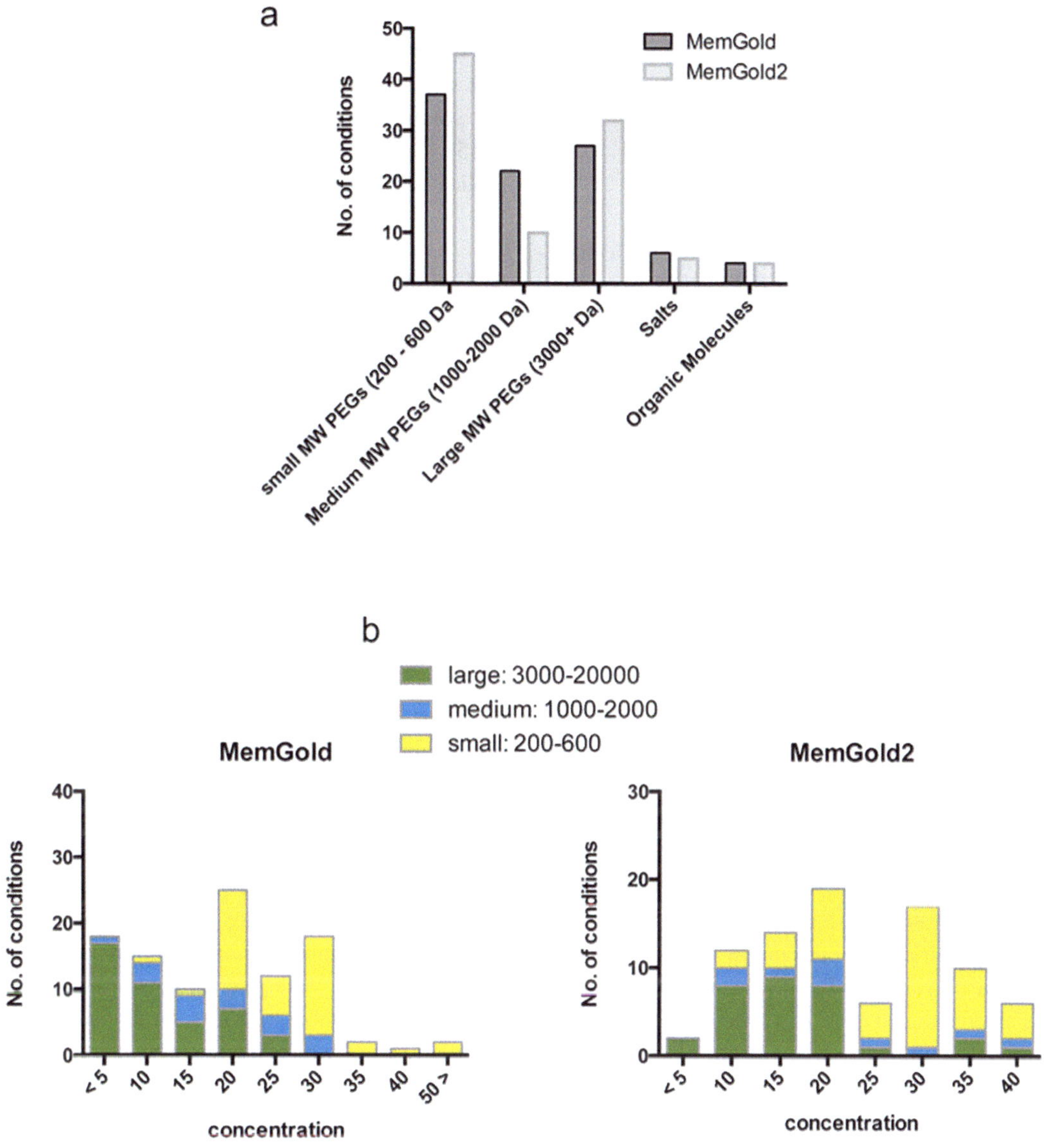

Fig. 5.4 A comparison of MemGold and MemGold2 precipitant conditions. (**a**) Number of conditions that utilize PEGs, salts and organic molecules as the major precipitant solution. (**b**) Analysis of the concentration ranges of PEGs

larger MW PEGs over the more hydrophobic channels and transporters. As the number of respiratory complexes has reduced relative to that of transporters and channels, we designed a new screen based on our 2012 analyses, MemGold2 (Parker and Newstead 2012). As can be seen from Fig. 5.4a, there is a noticeable difference in crystallisation precipitants between MemGold and MemGold2. In particular, we observe an increase in the number of small (200–600 Da) and large MW PEGs (>3000 Da) with a decrease in the medium MW PEGs (1000–2000 Da). The two screensalso differ with respect to the concentration ranges of the precipitants (Fig. 5.4b), reflecting the differences in the make up of the database. In particular, the concentration ranges of all the PEGs have shifted to higher values, with large MW PEGs around 15–20 %

(w/v), medium MW PEGs evenly distributed between 10 and 40 % (w/v) and small MW PEGs clustering around the 30 % (v/v) range.

5.6 Buffers, pH and Salts

Buffering chemicals and salts often have a significant impact on protein crystallisation; in particular polyvalent cations and anions are often essential for crystallisation (Newman 2004; Trakhanov and Quiocho 1995). In MemGold we observed an equal split between pH 7 and pH 8, which were the most successful pH values reported (Fig. 5.5). In MemGold2 however we noticed the number of pH 8 conditions reduced markedly, with pH 6 and 6.5 increasing. This suggests that pH range is an important parameter to optimize and consider when designing membrane protein screens. We also noticed that the spread of pH values from 3.0 to 10.5 appears to be wider for membrane proteins than the recent analysis of the entire PDB, which is fairly narrow between 5 and 9 (Fazio et al. 2015).

We have also observed a significant increase in the number of different polyvalent cations and anions reported. Therefore MemGold2 contains a different set of these chemicals, which can be essential to enable proteins to interact and pack into a crystal ($PepT_{St}$) (Solcan et al. 2012). It is interesting to note that one of the most successful commercial crystallisation screens is the Hampton PEG/Ion screen, which also involves screening many different polyvalent and monovalent salts against the most successful precipitant for soluble proteins, PEG 3350 (Fazio et al. 2015). This is possibly something that should be replicated for membrane proteins.

5.7 MemAdvantage – An Alpha Helical Membrane Protein Additive Screen

For many projects, an initial crystal condition will require optimisation, the addition of small molecules, salts, and specific ligands are well-established methods in this regard (McPherson and Cudney 2014; Luft et al. 2007). Figure 5.6 shows the range of different small molecule and salt additives that have been reported to improve initial crystallisation conditions for alpha helical MPs (Parker and Newstead 2012). As observed previously, mono- and multivalent salts appear prominently in the database. This no doubt due to the role these ions play in mediating intermolecular contacts during crystallisation. A notable difference however is a substantial increase in the number of secondary detergents and non-volatile organic molecules that are now being recorded. Structures of transporters account for much of this increase, suggesting screening secondary detergents for members of this family would be especially worthwhile. Interestingly, the reported

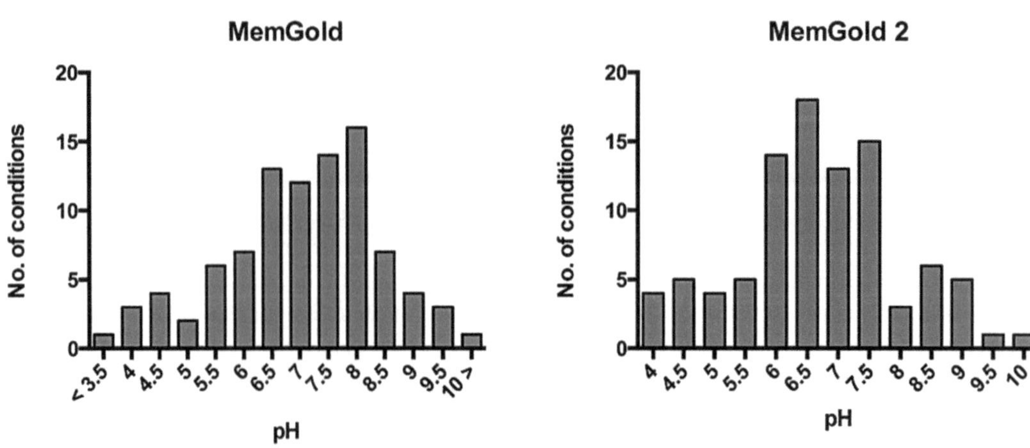

Fig. 5.5 Analysis of the pH ranges screened in MemGold and MemGold2

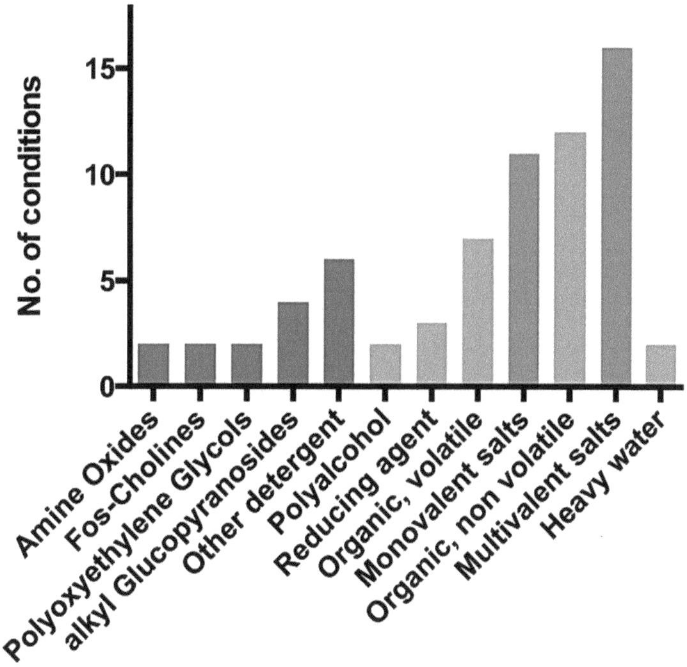

Fig. 5.6 Breakdown of additives used in MemAdvantage screen for crystal optimization

use of additional lipids as additives appears to be mainly isolated to channels, with monovalent salts being more successful for transporters. It is clear that improving the initial crystallisation hits can be achieved using secondary additives. MemAdvantage was designed to facilitate membrane protein crystal optimisation by providing a convenient 96 well format to screen initial crystal conditions against the most successful additives reported in the PDB. However, given the growth of membrane protein structures, further development in this area is likely to continue, especially given the currently small number of lipids, which have shown promise in recent years as a way to improve crystal quality (Gourdon et al. 2011; Malinauskaite et al. 2014).

5.8 MemMeso – A Systematically Designed *in meso* Crystallisation Screen

The recent success of *in meso* crystallisation has prompted the question of whether the current commercial screening kits are suitable for this methodology. To date there are >90 structures of membrane proteins solved using this method (Fig. 5.7a). 53 % of these are GPCRs with the next most successful class of protein being transporters, at 18 % (Fig. 5.7b). However, considering that over half the examples are GPCRs that were crystallised using protein fused to T4 lysozyme or apocytochrome b(562)RIL (BRIL), it would be premature to attempt a rational analysis that could be extended more generally at this time. However, if your area of research is GPCR structural biology the current examples would seem a productive starting point for further screen design. That being said, Molecular Dimensions Ltd. recently released a systematically designed *in meso* crystallization screen, MemMeso, based on work carried out in the laboratory of Osamu Nureki in the University of Tokyo, Japan. This screen comprises only small MW PEGs (200–600), four pH conditions (5, 6, 7 and 8) and 9 different salt conditions.

Our database contains 91 unique examples of membrane proteins crystallised using the *in meso* LCP method. The crystallisation conditions from these examples have been analysed (Fig. 5.7c). As expected, the conditions are dominated by PEG 400. However, this

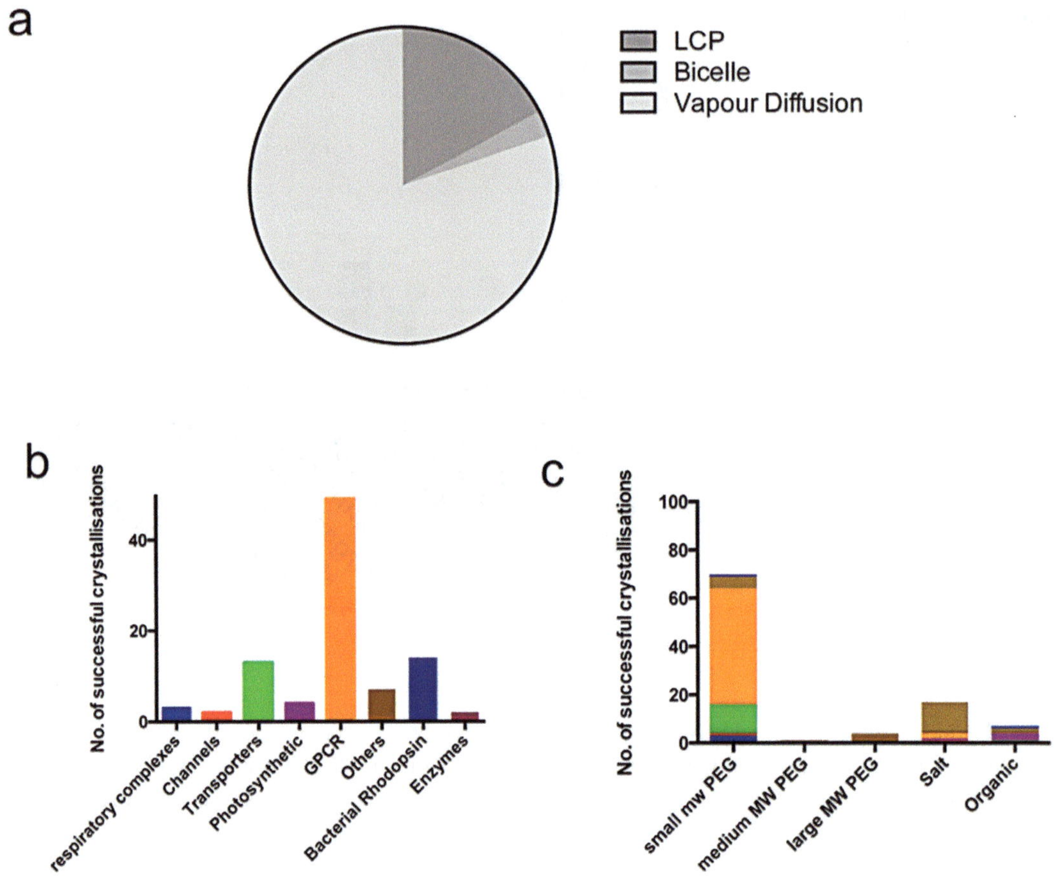

Fig. 5.7 Current trends in *in meso*crystallization of alpha helical membrane proteins. (**a**) Pie chart showing the proportion of structures within our database crystallised using either vapour diffusion, LCP or bicelle methods. (**b**) Breakdown of the LCP crystal structures into different family classes as shown in Fig. 5.1. (**c**) Analysis of types of precipitants used in LCP

result is heavily biased by the GPCR examples. Interestingly we observe a number of conditions using small organic molecules, which we had previously observed were largely unsuccessful for vapour diffusion crystallisation of membrane proteins (Newstead et al. 2008a). For example the Ca^{2+}/H^{+} antiporter (PDB: 4KPP) was crystallised using pentaerythritol propoxylate and sensory rhodopsin I (PDB: 1XIO) was crystallised using MPD. The remaining examples in the organics are Jeffamine-M600, which is similar in chemical composition to polyethylene glycol. Interestingly we also observe a significant number of high salt conditions, contributed by Bacteriorhodopsin, Halorhodosin and sensory rhodopsin II. Although the number of examples in our analysis are small, it suggests that crystallisation in the lipidic cubic phase may be influenced differently to that in solution.

5.9 Conclusions

Membrane proteins represent important pharmaceutical targets and interesting subjects of study with respect to cellular biology and protein biochemistry. However, they still represent challenging targets to crystallise and study. To date our database of 569 unique structures compares to > 110,000 structures in the entire PDB, representing < 1 % of known crystal structures. The field of membrane protein structural biology is still developing at a rapid pace. The introduction of serial injection systems for crystals at synchrotron radiation and free electron sources (Conrad et al. 2015) and the development of *in situ* diffraction data collection methodology (Huang et al. 2015) suggest that what structural biologists need from a crystallisation experiment is likely to change in the coming years. The final chapter on the topic of crystal screen design and optimization is far from being written. As more information is gathered it seems likely that new trends will be discovered and new crystallisation methods invented or traditional methods refined to meet the growing need to understand these important and fascinating proteins at atomic resolution. The information contained in this chapter represents the current snapshot of 'crystallisation space' for alpha helical membrane proteins. It is our wish that this information will encourage the efficient use of the MemGold family of screens but also enable the design of more tailored crystallisation screens for particular projects of interest to you.

Acknowledgments This works was funded through the Wellcome Trust Investigator Award 102890/Z/13/Z to SN.

References

Bill RM, Henderson PJ, Iwata S, Kunji ER, Michel H et al (2011) Overcoming barriers to membrane protein structure determination. Nat Biotechnol 29:335–340

Caffrey M (2009) Crystallizing membrane proteins for structure determination: use of lipidic mesophases. Annu Rev Biophys 38:29–51

Caffrey M, Porter C (2010) Crystallizing membrane proteins for structure determination using lipidic mesophases. J Vis Exp 45

Chae PS, Rasmussen SG, Rana RR, Gotfryd K et al (2010) Maltose–neopentyl glycol (MNG) amphiphiles for solubilization, stabilization and crystallization of membrane proteins. Nat Methods 7:1003–1008

Chayen NE (1998) Comparative studies of protein crystallization by vapour-diffusion and microbatch techniques. Acta Crystallogr D Biol Crystallogr 54(1): 8–15

Chayen NE, Saridakis E (2008) Protein crystallization: from purified protein to diffraction-quality crystal. Nat Methods 5:147–153

Chen H, Shaffer PL, Huang X, Rose PE (2013) Rapid screening of membrane protein expression in transiently transfected insect cells. Protein Expr Purif 88:134–142

Chun E, Thompson AA, Liu W, Roth CB, Griffith MT et al (2012) Fusion partner toolchest for the stabilization and crystallization of G protein-coupled receptors. Structure 20:967–976

Conrad CE, Basu S, James D, Wang D, Schaffer A et al (2015) A novel inert crystal delivery medium for serial femtosecond crystallography. IUCrJ 2:421–430

Cudney R, Patel S, Weisgraber K, Newhouse Y, McPherson A (1994) Screening and optimization strategies for macromolecular crystal growth. Acta Crystallogr D Biol Crystallogr 50(4):414–423

Delmar JA, Bolla JR, Su CC, Yu EW (2015) Crystallization of membrane proteins by vapor diffusion. Methods Enzymol 557:363–392

Drew D, Newstead S, Sonoda Y, Kim H, von Heijne G, Iwata S (2008) GFP-based optimization scheme for the overexpression and purification of eukaryotic membrane proteins in Saccharomyces cerevisiae. Nat Protoc 3:784–798

Fazio VJ, Peat TS, Newman J (2015) Crystallization: digging into the past to learn lessons for the future. Methods Mol Biol 1261:141–156

Ghosh E, Kumari P, Jaiman D, Shukla AK (2015) Methodological advances: the unsung heroes of the GPCR structural revolution. Nat Rev Mol Cell Biol 16:69–81

Gourdon P, Andersen JL, Hein KL, Bublitz M, Pedersen BP et al (2011) HiLiDe-systematic approach to membrane protein crystallization in lipid and detergent. Cryst Growth Des 11:2098–2106

Huang CY, Olieric V, Ma P, Panepucci E, Diederichs K et al (2015) In meso in situ serial X-ray crystallography of soluble and membrane proteins. Acta Crystallogr D Biol Crystallogr 71(6):238–1256

Iwata S (ed) (2003) Methods and results in crystallization of membrane proteins, vol 4. Internat'l University. Linc La Jolla, USA

Kawate T, Gouaux E (2006) Fluorescence-detection size-exclusion chromatography for precrystallization screening of integral membrane proteins. Structure 14(4):673–681

Kunji ER, Harding M, Butler PJ, Akamine P (2008) Determination of the molecular mass and dimensions of membrane proteins by size exclusion chromatography. Methods 46:62–72

Luft JR, Wolfley JR, Said MI, Nagel RM, Lauricella AM et al (2007) Efficient optimization of crystallization conditions by manipulation of drop volume ratio and temperature. Protein Sci 16:715–722

Luft JR, Newman J, Snell EH (2014) Crystallization screening: the influence of history on current practice. Acta Crystallogr F Struct Biol Commun 70:835–853

Malinauskaite L, Quick M, Reinhard L, Lyons JA, Yano H et al (2014) A mechanism for intracellular release of Na + by neurotransmitter/sodium symporters. Nat Struct Mol Biol 21(11):1006–1012

McPherson A, Cudney B (2014) Optimization of crystallization conditions for biological macromolecules.

Acta Crystallogr F Struct Biol Commun 70:1445–1467
McPherson A, Gavira JA (2014) Introduction to protein crystallization. Acta Crystallogr F Struct Biol Commun 70:2–20
Newman J (2004) Novel buffer systems for macromolecular crystallization. Acta Crystallogr D Biol Crystallogr 60(3):610–612
Newman J, Egan D, Walter TS, Meged R, Berry I et al (2007) Towards rationalization of crystallization screening for small- to medium-sized academic laboratories: the PACT/JCSG+ strategy. Acta Crystallogr D Biol Crystallogr 61(10):1426–1431
Newstead S, Kim H, Von Heijne G, Iwata S, Drew D (2007) High-throughput fluorescent-based optimization of eukaryotic membrane protein overexpression and purification in Saccharomyces cerevisiae. Proc Natl Acad Sci U S A 104:13936–13941
Newstead S, Ferrandon S, Iwata S (2008a) Rationalizing alpha-helical membrane protein crystallization. Protein Sci 17(3):466–472
Newstead S, Hobbs J, Jordan D, Carpenter EP, Iwata S (2008b) Insights into outer membrane protein crystallization. Mol Membr Biol 25(8):631–638
Nogly P, James D, Wang D, White TA, Zatsepin N et al (2005) Lipidic cubic phase serial millisecond crystallography using synchrotron radiation. IUCrJ 2:168–176
Page R, Stevens RC (2004) Crystallization data mining in structural genomics: using positive and negative results to optimize protein crystallization screens. Methods 34:373–389
Parker JL, Newstead S (2012) Current trends in α-helical membrane protein crystallization: an update. Protein Sci 21(9):1358–1365
Rosenbusch JP (1990) The critical role of detergents in the crystallization of membrane proteins. J Struct Biol 104:134–138
Segelke B (2005) Macromolecular crystallization with microfluidic free-interface diffusion. Expert Rev Proteomics 2:165–172
Solcan N, Kwok J, Fowler PW, Cameron AD, Drew D et al (2012) Alternating access mechanism in the POT family of oligopeptide transporters. The EMBO J 31:3411–3421
Sonoda Y, Cameron A, Newstead S, Omote H, Moriyama Y et al (2010) Tricks of the trade used to accelerate high-resolution structure determination of membrane proteins. FEBS Lett 584:2539–2547
Sonoda Y, Newstead S, Hu NJ, Alguel Y, Nji E et al (2011) Benchmarking membrane protein detergent stability for improving throughput of high-resolution X-ray structures. Structure 19:17–25
Tao Z, Zhang YW, Agyiri A, Rudnick G (2009) Ligand effects on cross-linking support a conformational mechanism for serotonin transport. J Biol Chem 284:33807–33814
Tate CG, Schertler GF (2009) Engineering G protein-coupled receptors to facilitate their structure determination. Curr Opin Struct Biol 19:386–395
Tate CG, Haase J, Baker C, Boorsma M, Magnani F et al (2003) Comparison of seven different heterologous protein expression systems for the production of the serotonin transporter. Biochim Biophys Acta 1610:141–153
Trakhanov S, Quiocho FA (1995) Influence of divalent cations in protein crystallization. Protein Sci 4(9):1914–1919
White S (2007) Membrane proteins of known 3D structure. http://blanco.biomol.uci.edu/mpstruc/

Crystal Dehydration in Membrane Protein Crystallography

6

Juan Sanchez-Weatherby and Isabel Moraes

Abstract

Crystal dehydration has been successfully implemented to facilitate the structural solution of a number of soluble and membrane protein structures over the years. This chapter will present the currently available tools to undertake controlled crystal dehydration, focusing on some successful membrane protein cases. Also discussed here will be some practical considerations regarding membrane protein crystals and the relationship between different techniques in order to help researchers to select the most suitable technique for their projects.

Keywords

Crystal dehydration • HC1 • Free mounting system • *In situ* • Membrane proteins • Relative humidity

6.1 Introduction

The effect of dehydration in crystal diffraction has been known since the beginnings of protein crystallography (Perutz 1946). It can often be detrimental to the crystal but also be very beneficial, dramatically improving diffraction properties. Partly exacerbated by the initial unavailability of cryo-cooling techniques, and tight control of water content in crystals prior or during data collection, dehydration has always been of paramount importance. In the early days, when most data were collected at room temperature, it was necessary to maintain the stability of samples for hours or even days (Pickford et al. 1993). This meant that researchers were more aware of the benefits and pitfalls of hydration control and accustomed to accounting for the role it played in their observed results. Dehydration has been reported to be beneficial on the diffraction quality of macromolecular crystals (Heras and Martin 2005; Newman 2006; Russo et al. 2012). In most cases the discovery has occurred as a result of casual observations (such cracked capillaries,

The original version of this chapter has been revised. An erratum to this chapter can be found at DOI 10.1007/978-3-319-35072-1_13

J. Sanchez-Weatherby (✉)
Diamond Light Source, Harwell Science and innovation Campus, Didcot OX11 0DE, UK
e-mail: juan.sanchez-weatherby@diamond.ac.uk

I. Moraes
Membrane Protein Laboratory, Diamond Light Source/Imperial College London, Harwell Campus, Didcot, Oxfordshire, UK
e-mail: isabel.de-moraes@diamond.ac.uk

I. Moraes (ed.), *The Next Generation in Membrane Protein Structure Determination*,
Advances in Experimental Medicine and Biology 922, DOI 10.1007/978-3-319-35072-1_6

drops left to dry, badly sealed trays and others) (Esnouf et al. 1998; Abergel 2004).

The changes caused by dehydration are of such importance that several past and present studies have focused solely in the understanding of alterations caused by dehydration and its implications on the functional interpretation of protein and DNA structures. (Biswal and Vijayan 2002; Kaushal et al. 2008; Saraswathi et al. 2002; Nagendra et al. 1998; Sukumar et al. 1999; Kuo et al. 2003; Bowler et al. 2006; Perutz 1946; Amunts et al. 2007; Dobrianov et al. 2001).

As X-ray sources became stronger, cryo-cooling rapidly became the norm (Hope 1988; Mitchell and Garman 1994; Garman 1999; Juers and Matthews 2004; Alcorn and Juers 2010). Once crystal samples are cooled, most of the alterations are prevented, so the need for careful control of humidity is diminished and shifted from the data collection stage to the pre-cooling sample preparation step (Farley et al. 2014; Pflugrath 2015). Crystal handling can have a significant effect on the hydration state of samples and, as it has been highlighted recently, it can have potential consequences in the structural solution (Farley et al. 2014). Therefore, crystallographers have a number of techniques and "tricks" in order to prevent these effects with greater or lesser success. Chemically, the cryo-protection process also causes alterations in osmotic pressure in the crystal samples, similar to a hydration/dehydration process, thus it can change the structure and packing, often leading to changes in diffraction quality (Heras and Martin 2005).

Despite documented benefits of controlled dehydration in macromolecular crystallography, dehydration is often not pursued due to a lack of well-established protocols and equipment. To address this, in the last few years a number of novel techniques have been made available, which focus on trying to make the experiments easier to carry out, reproducible and where possible, with X-ray feedback. These techniques range from simple tools to aid capillary mounting (Basavappa et al. 2003; Kalinin et al. 2005; Mac Sweeney and D'Arcy 2003; Yadav et al. 2005), to bespoke equipment designed to maintain a humid environment around the samples while mounted in the beam path of a synchrotron beamline (Sanchez-Weatherby et al. 2009; Russi et al. 2011; Bowler et al. 2015; Kiefersauer et al. 2000a; Kiefersauer et al. 2000b; Kiefersauer et al. 2002; Kiefersauer et al. 2014).

The move to more complex crystallographic challenges and the availability of new techniques has meant more researchers are trying these experiments. As more experiments are being carried out and successes have been coming through, interest in crystal dehydration has been recently reignited (Moraes et al. 2014; Bowler et al. 2015; Kiefersauer et al. 2014).

This chapter will briefly review the current status of the field and the available tools, it will discuss the basic aspects of a dehydration experiment and give some guidance when attempting membrane protein crystal dehydration.

6.2 The Dehydration Method

As noted in the past (Newman 2006), dehydration can be a bit of a confusing term. Here we consider crystal dehydration as any process able to remove available water molecules from the crystals lattice. Dehydration is achieved by either altering the vapour (Sect. 6.2.1) or the chemical (Sect. 6.2.2) equilibrium of the crystal. These two methods of dehydration are different but also interrelated as the chemical alterations can be used to induce a change in the vapour environment and, the changes in the vapour surrounding the sample can induce alterations the chemistry. The relative amount of water present in a volume of air, Relative Humidity (RH), is directly linked to the temperature and chemical nature of a solution in equilibrium with the air. For this reason, salt solutions are universal calibrating standards for any equipment designed to measure air RH (Greenspan 1977).

When dehydrating crystals, it is advisable to try methods based on both approaches. In some cases, both methods can lead to very similar results but in others (particularly membrane protein crystals with detergents) the outcome of the process can be dependent on the particular technique used.

Crystallographers working in closed systems, like crystallisation plates or capillaries, can guide their dehydration experiments based on the well-established number of standard salts. Furthermore, the empirical equilibrium relative humidity when using the humidity control device (HC1) (Bowler et al. 2015; Russi et al. 2011; Sanchez-Weatherby et al. 2009) or the free mounting system (FMS) (Kiefersauer et al. 2002; Kiefersauer et al. 2000a, b) can be tabulated and tools exist that allow the researcher to try to estimate the theoretical alterations caused by chemical modification or the formulation of solutions required to induce a particular change in RH (Bowler et al. 2015; Wheeler et al. 2012).

6.2.1 Dehydration by Modifying the Vapour Equilibrium

This method is based on altering the water vapour pressure of the air surrounding the samples. This makes the water in the surrounding crystallisation buffer and crystal solvent channels equilibrate with the surrounding air making it saturate the air and lowering its availability in solution. Methods to carry this type of dehydration range from simple exposure of the samples to ambient dry air, through controlling the air environment using chemicals (Hellert et al. 2014; Kalinin et al. 2005; Kalinin and Thorne 2005), to the use of bespoke equipment designed to expose the samples to controlled dehydration (Bowler et al. 2015; Russi et al. 2011; Sanchez-Weatherby et al. 2009; Kiefersauer et al. 2002; Kiefersauer et al. 2000a, b).

6.2.2 Dehydration by Modifying the Chemical Equilibrium

Chemical alteration is achieved by the addition of a compound that will directly bond to the water molecules of the crystallisation buffer that surrounds the crystals. This lowers the number of interactions that proteins can establish with water, forcing it to interact with other molecules, thus inducing conformational changes. Chemicals used can include salts, precipitants and alcohols. From a practical point of view, the researcher either replaces the solution surrounding the crystals samples via careful pipetting or dialysis, or physically transfers the crystals into a new solution. This is the most frequent way of undertaking dehydration and it is often done in combination with cryo-protection or ligand binding. The key parameters in this process are the chemicals used, their concentration increment (if necessary) and the time samples are allowed to equilibrate in the new solution prior to harvesting (Shi et al. 2008; Adachi et al. 2009).

6.2.3 Relative Humidity and Its Relationship with Cryo-Protection

Key in dehydration experiments is the concept of Relative Humidity (RH). Relative Humidity is defined as the relative amount of water vapour in a given volume of air. It is expressed as a percentage of the saturation humidity and is pressure- and temperature-dependent (Winston and Bates 1960). As the chemical composition of a solution alters the saturation vapour pressure, it can be used to cross calibrate both the vapour and chemical dehydration experiments. For example at 20 °C a saturated solution of LiCl is tabulated to be around 11 % RH, NaCl 75 % RH, KCl 86 % RH, and K_2SO_4 97 % RH (see Table 6.1

Table 6.1 Examples of equilibrium RH for a number of saturated salt solutions (Greenspan 1977)

Saturated salt	RH at 20 °C
Potassium hydroxide	9.32
Lithium chloride	11.31
Potassium acetate	23.11
Magnesium chloride	33.07
Sodium iodide	39.65
Potassium carbonate	43.16
Magnesium nitrate	54.38
Sodium bromide	59.14
Potassium iodide	69.90
Sodium chloride	75.47
Ammonium chloride	79.23
Potassium bromide	81.67
Ammonium sulfate	81.34
Potassium chloride	85.11
Potassium nitrate	94.62
Potassium sulfate	97.59

Table 6.2 Examples of empirical equilibrium RH determined for the HC1 for a number non saturated salt solutions from (Bowler et al. 2015)

Concentration (M)	Sodium chloride	Sodium acetate	Sodium malonate	Ammonium sulfate
0.5	99.8	99.9	99.9	99.9
1.0	98.9	98.9	99.3	99.1
1.5	97.2	97.1	97.0	97.7
2.0	95.3	94.8	95.2	95.9
2.5	94.2	93.2	92.5	94.3
3.0	92.8	91.2	89.9	92.8

Table 6.3 Examples of empirical equilibrium RH determined for the HC1for a number standard precipitants from (Bowler et al. 2015)

% (w/w)	PEG10000	PEG6000	PEG4000	PEG1500	PEG400	PEG200	Glycerol	Ethylene glycol
10	99.9	99.9	99.9	99.9	99.9	99.9	99.5	98.5
20	99.9	99.9	99.9	99.9	99.9	99.5	96.5	94.0
30	99.9	99.9	99.9	99.9	98.8	97.3	92.0	89.5
40	99.3	99.9	99.2	98.3	96.3	94.9	88.5	85.0

for a more comprehensive list). It is important to note that any standard calibration table applies to closed systems, therefore in devices like the HC1 and FMS that operate using an open airflow, the empirical equilibrium RH values are greater than those presented on Table 6.1 (see also Tables 6.2 and 6.3). For this reason the European Synchrotron Radiation Facility/the European Molecular Biology Laboratory – Grenoble Outstation (ESRF/EMBL) created a web tool (www.embl.fr/CrystalDehydrationCollaboration/RH.html) that can be used to estimate the RH of solutions that includes theoretical and empirical values for better reference (Bowler et al. 2015; Wheeler et al. 2012). It is also important to remember that an individual crystallisation droplet will be in equilibrium with the well solution, but during the course of the crystallisation experiment the well might have dried up over time, thus it is much better to measure the empirical value of each tray before commencing an experiment.

In general, macromolecules tend to crystallise in highly hydrated solutions that require high RH to be stable but lack "cryo-protective" properties. Cryo solutions prevent ice formation by hydrogen bonding to the surrounding water molecules. These bonds also reduce the available water surrounding the crystal structure leading to a desiccation process. The empirical test of several precipitants commonly used in protein crystallisation clearly shows a correlation between the ability to cryo-protect and lower Relative Humidity (Wheeler et al. 2012). This highlights the intimate link between both effects (dehydration and cryo-protection) on crystal diffraction quality, however, it is difficult to understand the contribution from each of them when assessing diffraction quality collected at cryo-temperatures. Despite the key role dehydration may play in structure determination, researchers often do not carry out systematic studies on the effect it could have on the data quality, thus overlooking the role it might have played in their success or failure.

6.2.4 Crystal Changes Induced by Dehydration

The beneficial effects of dehydration are firstly, dependent on the original imperfections of crystal structure and then on the possibility of altering them by removing water molecules.

Macromolecular crystals are far from being "perfect", always presenting a certain degree of

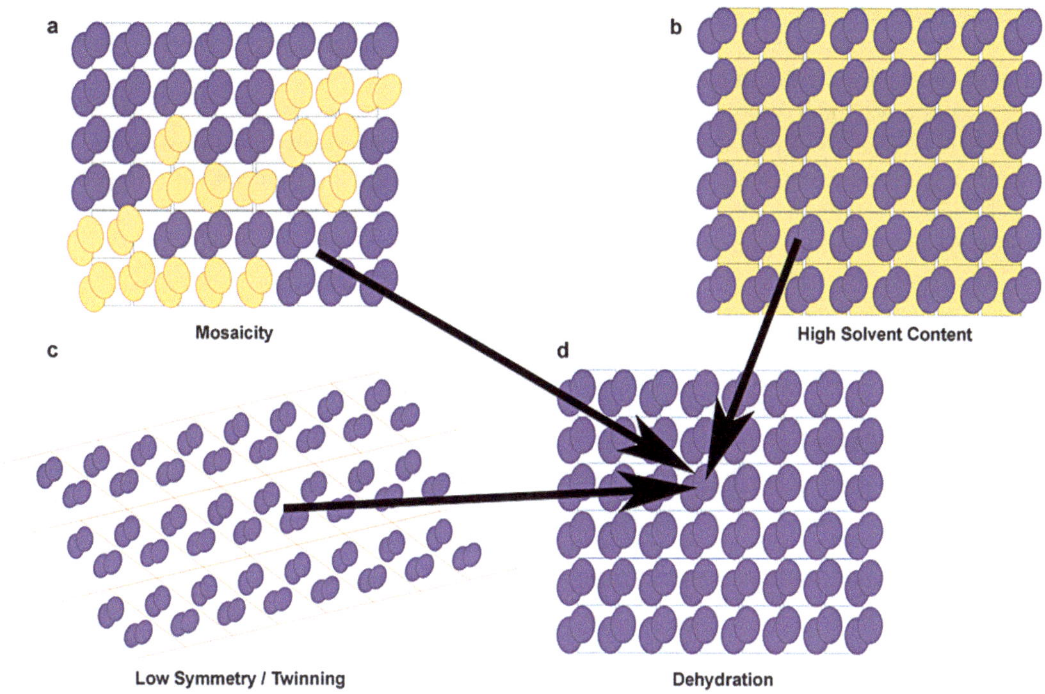

Fig. 6.1 Representation of possible lattice changes induced by dehydration. Most of these changes would lead to increased order or improved protein/solvent ration leading to improved diffraction properties. (**a–d**) Represents the reduction of crystal mosaicity. (**b–d**) Represents an increased or altered symmetry or the case of reduced twinning. (**c–d**) Shows the effect of reduced solvent bulk content leading to improved order or increased protein/solvent ratio

mosaicity (meaning they lack of internal periodicity in certain areas called mosaic blocks), see Fig. 6.1. High mosaic spread in crystals it is mainly caused by the cryo-cooling process and a major contributor to the diffraction intensity reduction. Dehydration can reorganise these mosaic blocks, re-align them, and in principle leading to a crystal with better diffraction properties (Fig. 6.1a).

In other crystals, high solvent content also makes them scatter with less power. If by dehydration, one can lower their global solvent content maintaining the protein molecules order (Fig. 6.1b) (higher protein to solvent ratio) could lead to an improved diffracting power. Changes in solvent content need to be accompanied by small local rearrangements in the crystal contacts and/or flexible areas, therefore diffraction can improve as local motions are reduced and crystal structure changes. In these cases, the degree, direction and the energetics of these changes will condition whether changes are for the better. The actual physical change in the lattice may be subtle, but due to the diffraction phenomenon the results can be very dramatic.

The most dramatic cases are those where the alterations cause enormous difference in the lattice packing triggering a space group change. Often, as the molecules turn and twist, screw axes are generated or removed and alternated indexing solutions start emerging. These alterations might be positive if they induce greater symmetry.

Finally, as it was mentioned early in this section, poor diffraction often is associated with the physical alterations of the crystals when cooled to 100 K. Nevertheless, the cryo-cooling process is very beneficial as it prevents radiation damage, lowers thermal motions of the molecules, reduces background and in certain cases also improves diffraction resolution. Dehydration prior to cryo-cooling in many cases helps to overcome the issues mentioned above as the pre-contraction induced by the dehydration process can yield a better lattice re-arrangement for the cryo-cooling. In other cases, dehydration is also beneficial as it can act as a way of cryo-protection (Pellegrini et al. 2011). Either by dry mounting, preventing ice forming around the crystals, or by concentrating the solutes within the solvent channels thus preventing hexagonal ice from nucleating in the solvent channels.

6.2.5 Effect of Dehydration on the Membrane-Protein Crystal Lattice

Membrane proteins can crystallise in two main forms depending on how they are grown. Crystals grown by the Lipid Cubic Phase (LCP) method (Caffrey 2003) tend to be formed by proteins organised in planar layers through protein–detergent–lipid hydrophobic interactions stacked on top of each other by polar interactions. These crystals are extremely small and fragile two dimensional plates known as type I 3D crystals. These crystals are difficult to harvest, difficult to see and most of their properties are intimately linked to the biochemical properties of the LCP. As the LCP structure is temperature and humidity dependent, this type of crystal is very unlikely to be useful for controlled dehydrations studies. On the other hand, membrane protein crystals can also grow using more "standard" crystallisation methods such vapour diffusion leading to crystals organised in a more three dimensional fashion (type II 3D crystals) similar to soluble proteins and more suited for dehydration experiments. Despite their similarity with soluble crystals, these membrane protein crystals tend to have lower protein content (10–25 % rather than 45–60 %) and higher solvent content (mainly water in soluble proteins) made up of detergents/lipids free micelles.

Integrity and dynamics of membrane proteins are closely related to the properties of the surrounding phospholipids in the native membrane. To avoid protein aggregation once removed from its natural environment, detergent is used in the protein solution in order to mimic the lipid membrane by surrounding the hydrophobic region of the protein generating a water-soluble protein–detergent complex (PDC). Therefore, membrane protein crystals formation and stability are strongly linked to the protein overall molecular structure (hydrophilic and hydrophobic exposed regions) and to the size, concentration and critical micelle concentration (CMC) of the detergent. While some membrane protein crystals are able to tolerate an increase or decrease of detergent once they have been formed others are more sensitive and lose their order if the equilibrium (bound detergent/free detergent) is changed. This will have a strong impact on the crystal dehydration ability. More information in the use of detergents in membrane protein structure determination can be found in Chap. 2 of this book.

In soluble protein crystals, dehydration tends to disturb the whole solvent volume and thus induce variation across the entire structure. Here, dehydration can affect solvent exposed areas, flexible loops, crystal contacts and internal hydrogen bonding networks. The idea behind dehydrating these crystals is to induce structural changes that will modify the packing (normally by altering crystal contacts or stabilisation of flexible areas) leading to better diffraction without negatively affecting areas that are already well ordered.

In membrane proteins, most of the interactions that keep the global lattice stable (via the detergent, lipid or interatomic interactions) are partially fluid and malleable. These regions are mainly hydrophobic and thus not severely affected by dehydration. The few hydrophilic interactions existent act as small anchors that stabilise the lattice. Humidity will affect these small soluble portions and induce changes that can propagate to the whole crystal via rearrangements on the global lattice structure, hopefully inducing a new ordered lattice that will promote better diffraction without affecting the integrity of the protein structure in study.

The magnitude of dehydration to induce a notable change on a crystal structure is, broadly speaking, different between membrane and soluble proteins. In soluble protein crystals, a small change in the global solvent content will cause a huge change in the whole crystal structure therefore these crystals usually require less dehydration. In membrane protein crystals, most of the alterations are targeted at the small solvent rich areas scattered around the lattice, and in order to remove enough water to successfully induce a change on the whole structure, the extent of dehydration needed is often much greater than for soluble proteins. The less amount of water in the lattice, the stronger the process needs to be to extract it. A good example of this are the crystals of the bile acid sodium symporter ASBT (Hu et al. 2011) that have required dehydration from an initial RH of 95 % down to 45 % in order to improve diffraction limit from around 8 Å at room temperature to around 2 Å after cryo-cooling.

Dehydration undertaken on membrane protein crystals without mother liquid in its surroundings (dry mounted samples – see Sect. 6.3.2) only induces changes in water content within the solvent channels. Note that although these channels have great amounts of detergent and other solutes there is very little water available for the dehydration process. On the contrary, dehydration methods that are performed on wet samples surrounded by their mother liquor (see Sect. 6.3.1) affect the whole drop. As water content is higher, dehydration will induce an increased concentration of solutes and detergent surrounding the crystals that can in turn alter the equilibrium with the crystal and disturb its order. For these reasons it is worth trying both wet-mounting and dry-mounting dehydration methods.

6.3 Tools and Techniques for Crystal Dehydration

Practical considerations and timing, determine what most researchers can do with their samples. Crystal availability and beam time do not always match and researchers will always struggle to get their best crystals ready in time for a dehydration study at a synchrotron or home source. In this section, the available dehydration methods and tools have been separated by the ability to execute them in a standard laboratory or at an X-ray source.

6.3.1 Dehydration in the Laboratory

Doing dehydration at the home laboratory has its advantages. The process can be performed using standard tools available within any molecular biology or biochemistry laboratory and with total control of the researcher. This also allows the researcher to test a great number of crystals (if available) without the usual time constrains during a synchrotron visit. In general, when dehydrating crystal samples in the home laboratory, the dehydration procedure is the most basic one where samples are just transferred from one solution to the other or exposed to air for a few seconds before proceeding to the cryo-protection process. This way of dehydrating is normally undertaken casually, not regarded important and often not well described in the methods sections of publications.

The most common dehydration techniques used in a standard laboratory are described below, see also Fig. 6.2.

6.3.1.1 Simple Air Exposure

This is the simplest dehydration method consisting in opening the crystallisation plate allowing the drops to dry for a certain amount of time prior to cryo-cooling (Fig. 6.2a). It is, of course, dependent on the ambient RH, temperature, size of the drops and chemical composition of the crystallisation solution. It is mainly responsible for the "I am getting better at harvesting crystals" syndrome, as with time, crystallographers either start harvesting faster (allowing for less dehydration in delicate samples) or they harvest numerous samples from an individual drop (allowing for the last ones to dehydrate more). It can also be induced during the final harvesting prior cryo-cooling by allowing the sample to dehydrate on the loop before the final plunge in liquid nitrogen. Uncontrolled dehydration is difficult to avoid but careful note taking and repetition is the key to getting to control or understand its effects. As

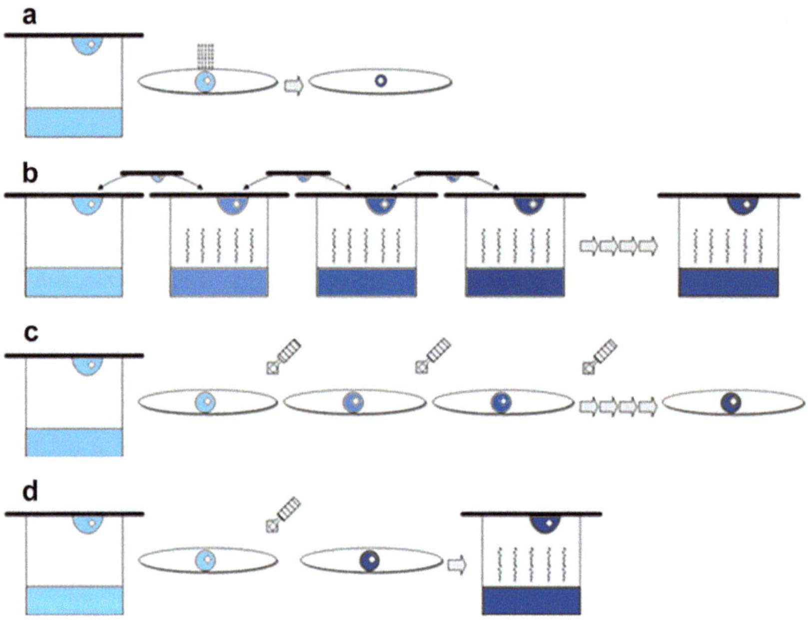

Fig. 6.2 Diagram showing alternate options for laboratory dehydration. (**a**) Simple air exposure, (**b**) vapour control by exchanging well solution, (**c**) chemical exchange by serial soaking and (**d**) chemical exchange by soaking and slow equilibration

crystals become more fragile, smaller and grown in smaller volumes the effect of dehydration by air exposure becomes more prevalent.

6.3.1.2 Vapour Control by Exchanging Well Solution

This is probably the most controllable way of undertaking dehydration in the laboratory. It is, for obvious reasons, only possible in vapour diffusion crystallisation samples and relies on using the same crystallisation drop for the actual experiment. The idea is to replace the well solution with increased amounts of a desiccating solution (Fig. 6.2b). The new solution will lower the water vapour pressure and in return "pull" water out of the crystallisation drop inducing a change in the crystals. This is normally undertaken by using some of the standard saturated calibration salt solutions but can also be done using crystallisation precipitants. Depending on the volumes available, dehydration can be achieved by simple addition of salts to the pre-existing well buffer or by replacing the well with a totally new solution. Dehydration is a slow and gradual process most of the time, which depends on the drop size and well volume. It can be undertaken in one single step or in progressive steps and normally, after a period of equilibration (normally between 8 and 24 h), samples are harvested and stored for data collection. The most attractive feature of this dehydration method is its simplicity and easy implementation. It can be undertaken whenever samples are available and once the optimal conditions are found it is very easy to reproduce. It can be achieved at any temperature and in general the presence of highly volatile compounds in the crystallisation solutions is not a concern.

6.3.1.3 Dehydration by Chemical Exchange of Crystallisation Droplet

As discussed earlier, this process is potentially happening during the standard cryo-cooling or soaking procedures (Fig. 6.2c,d). The main difference between this and the previous method described (Sect. 6.3.1.2) is that here it is possible to change the chemical components of the crystallisation drop and not only the concentration of the pre-existing ones. This procedure opens the possibility that the chemical used for dehydration alters the structure and/or the resulting diffraction by chemical modification rather than dehydration. In addition, the rate of exchange is more difficult to control and normally carried out by slow diffusion of buffers with increased concentration. If crystallisation droplets are very small, it is easy to cause a shock on the crystals, therefore pre-mixing solutions to the target concentration can sometimes help. The simplest way of chemical dehydrating is by dialysis. This only requires replacing the solutions and, due to the dialysis membrane, it is very gentle and controllable.

6.3.2 Modern Dehydration Methods with X-ray Feedback

6.3.2.1 Capillary Dehydration

Dehydration performed in home laboratories, despite being the simple and cost effective, it is a partially blind process by the fact that is decoupled from the X-ray data collection. In addition, the cryo-cooling process undertaken after the dehydration procedure can muddle results.

This was noted early on in the use of macromolecular crystallography and initial attempts were made by developing tools that would couple salt-based vapour dehydration to the, then available, data collection using the capillary (glass or quartz) mounting method (Huxley and Kendrew 1953; Pickford et al.1993; Kiefersauer et al. 2000a, b). For a time, this proved to be useful (Rockland 1960; Dobrianov et al. 2001) but fell into disuse due to the high X-ray scatter background produced compared to loops. The technique was re-introduced a few years later (Mac Sweeney and D'Arcy 2003; Basavappa et al. 2003) by adapting new lower background capillaries (made of a transparent polymer) to the then standard pin-based mounts (Kalinin and Thorne 2005; Kalinin et al. 2005; Hellert et al. 2014) (Fig. 6.3a,d). The later method is cheap, easy to implement and very popular amongst researchers attempting to dehydrate whilst collecting X-ray data *in-house* and for simple cryo-cooling (Yao et al. 2004; Kim et al. 2007). As the mounts (MicroRT system – MiTeGen) are standard, the technique is also available at any synchrotron beamline where manual mounting is possible. It is most frequently used as a quick dehydration method and as an easy way of keeping samples stable during room temperature data collections.

The process involves pre-loading (using a gel-loading tip) a transparent polymer capillary that is sealed at one end with a small volume of stabilising or dehydrating solution. Crystals are harvested with a loop or mesh mounted onto a pin and then covered with the capillary through the open end such that it seals. The sample is maintained at a close distance from the liquid. The pin can then be mounted and data collected through the capillary whilst the crystal is kept stable or whilst being dehydrated. The dehydration rate is dependent on the solution composition, the sample volume, the solution volume and the distance between the sample and solution. Cryo-cooling is possible at any point by quickly removing the capillary and plunging in liquid nitrogen (Kalinin and Thorne 2005; Warkentin et al. 2008).

6.3.2.2 Dehydration Using the FMS

The Free Mounting System (FMS) was the first device able to automate the process of dehydration. It operates based on the alteration of RH by the adjustment of the dew point and is able to regulate humidity in a crystal from 98 % to 50 % at temperatures ranging from 15 °C to 25 °C. It has been used in some academic laboratories and private companies yielding interesting results (Kiefersauer et al. 2000a, b).

The key feature of the device is a holding block (Fig. 6.3b,e) that is able to keep the temperature of the sample and the surrounding airflow stable. The samples are mounted onto meshes or loops and placed inside the block. The block is then placed over a standard goniometer mount to allow users to centre the samples onto the X-ray beam path and collect data whilst dehydrating samples.

The major advantage of this device over classic methods is that complex protocols such

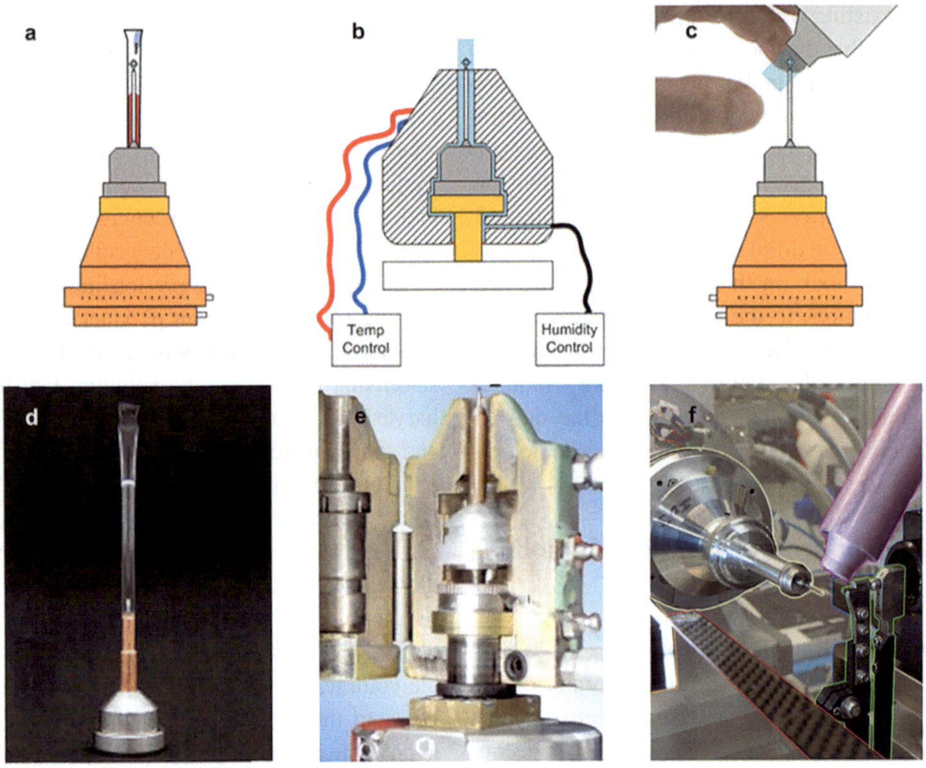

Fig. 6.3 Methods of dehydration with X-ray feedback. (**a**) and (**d**) Capillary dehydration. (**b**) and (**e**) Free Mounting System. (**c**) and (**f**) The humidity control dehydration device (HC1)

gradient dehydration, rehydration and reverse hydration can be achieved with a high level of reproducibility while at the same time allowing diffraction feedback. The only disadvantage compared to the classic method is the fact that only one or two crystals can be analysed at the same time.

The FMS has been used in a number of successful cases, most notably the improvement in diffraction of the F1-ATPase protein (Bowler et al. 2006). It also has allowed for the determination of the extremely interesting case of carbon monoxide oxidase (CODH) (Kiefersauer et al. 2000a, b) that required a complex dehydration-rehydration process only possible with the use of the FMS.

Despite its success, the complexity of sample handling and bulky design have prevented FMS from being widely used in laboratories worldwide and also making synchrotrons sources shy away from installing it on macromolecular crystallography (MX) beamlines. Lately the developers of the device have also proved it can be used in combination with a heating laser. This gives rise to new alternate options to automate dehydration and new cryo-cooling methods, but no new crystal structures have as yet resulted from this sort of work (Kiefersauer et al. 2014).

6.3.2.3 Dehydration Using the HC1 System

The humidity control dehydration device (HC1) was developed a few years after the FMS in order to address the factors limiting the wider use of this method by the community (Sanchez-Weatherby et al. 2009; Russi et al. 2011). The HC1 key difference is that, despite also being based on dew point, the delivery of humid airflow is achieved via a nozzle rather than a cooling block (Fig. 6.3c,f). The HC1 nozzle makes it unable to alter the temperature of the air around the sample but makes it very easy to install and use it in a standard academic laboratory with a home X-ray source or any MX synchrotron beamline. The low complexity of the device reduced the cost and simplified its maintenance making it better received by many academic laboratories and a large number of synchrotron beamlines around the world. To date nine devices are available at several world synchrotrons. The ESRF/EMBL/ILL operate two devices available across the MX beamlines. The Helmholtz-Zentrum Berlin (Mueller et al. 2012) and Max IV Laboratory (Ursby et al. 2013) operate the device installed in one of their MX branches. In the US, both APS (NE-CAT 24-ID-E) and the ALS (Berkeley Center for Structural Biology beamlines) have one each, the Canadian Light Source (Fodje et al. 2014) has one and Diamond Light Source in the UK has another two that can be both installed in any of the MX beamlines and also in a laboratory as an off-line setup.

Considerable efforts have been made at several of the synchrotrons in order to incorporate the HC1 within their hardware and software. For example, workflows within MxCuBE (Gabadinho et al. 2010) have been adapted to undertake dehydration experiments (Brockhauser et al. 2012) and the Diamond beamline software (GDA) now includes a bespoke data collection perspective solely dedicated at undertaking these sorts of experiments. This allows synchrotron users to be able to undertake a range of dehydration experiments as part of their normal beamtime allocation. In some synchrotron facilities (BM14 at the ESRF and I02, I03 and I04 at Diamond), installations also include rapid swapping between the use of the HC1 and the standard cryo-temperature allowing constant routine use without the local beamline support. Others (like MaxLab and BESSY) have opted for a more permanent installation (Mueller et al. 2012; Ursby et al. 2013).

Either in the laboratory or at the synchrotron beamline, any HC1 experiment requires the mounting of samples in the airflow at their starting Relative Humidity (RHi). The relative humidity is empirically determined by observing the behaviour of a droplet solution (crystallisation solution) under the humid airstream while varying the RH. Once the optimum humidity has been determined, crystal samples can be mounted using micromeshes (MiTeGen) so that any excess buffer can be removed by wicking. This makes the experiments more reproducible and ensures samples don't move during the process.

Dehydration protocols can then be applied to each sample by starting at the RHi and lowering humidity to desired target points. At these points samples can either be directly analysed with X-rays or cryo-cooled for later analysis. The key to any experiment is to try to induce changes in the crystal lattice that will yield a better diffracting sample. Once a potential dehydration protocol has been determined then a number of crystals can be dehydrated and cryo-protected to find the elusive structure. If available, this final harvest can be undertaken in a laboratory in order to save precious beam time and improve throughput.

When using the HC1 in a beamline, as it is a room temperature experiment, it is important to try to separate the effects from dehydration from the expected effects of radiation damage. A burn test may be appropriate to define a baseline. The best way of following an on-line dehydration experiment is to collect weak, wide oscillation images (1–2°) that will allow manual indexing to determine lattice dimensions and mosaicity. Following the changes in the lattice, the researcher can determine any transition points of interest that could suggest points of improved diffraction.

If undertaking the experiment in the laboratory it is important to also harvest some control samples under known conditions and try to ensure samples have had enough time to undergo any transition (10–15 min at least) prior to cooling. Harvesting a number of repeats helps to assess the true final outcomes of the process giving an indication of the likeliness of the method in study.

6.3.2.4 *In situ* Dehydration

In most of the MX beamlines across the world, sample environments (also called experimental hutches) have been made more automation-friendly. Consequently, the space available for additional and special setups became more restrictive. Despite this, due to the developments of photon counting pixel array detectors and their potential to help outrun radiation damage

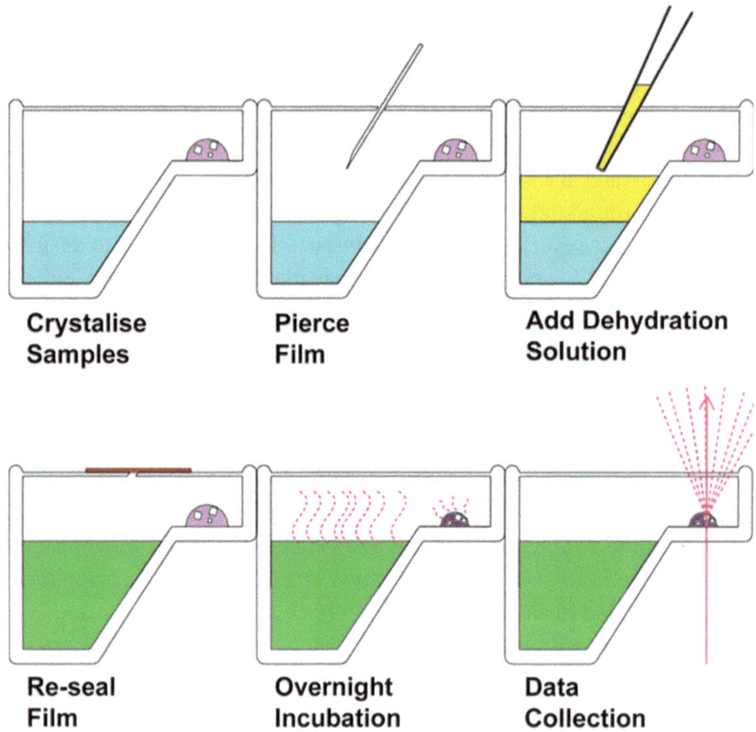

Fig. 6.4 Diagram showing the different stages required for an *in situ* dehydration experiment

(Kraft et al. 2009; Aishima et al. 2010; Rajendran et al. 2011) there have been a number of technical upgrades to allow data collection of room temperature samples within these crowded spaces. These include new crystallisation plates (with lower profile and better geometry), upgraded goniometry, dedicated robotics and imaging systems, make room temperature data collection possible. The option of collecting data directly from crystallisation plates without the need of any physical manipulation of the crystals brings new links between the experiment in the laboratory and the X-ray source. Essentially, in this method, crystals from the same crystallisation condition are grown in a full 96-well crystallisation plate (always advisable to use crystallisation plates with flat bottoms to avoid visual artefacts when mounted at the beamline). Once crystals are grown, gradients of different dehydration solutions are added to the plate wells through small pierces (without touching the crystal drops) and re-sealed (see Fig. 6.4). The dehydration is allowed to proceed for a period of 8–12 h before the plate be taken to an X-ray source with capabilities for *in situ* plate screening (Douangamath et al. 2013; Moraes and Archer 2015; Axford et al. 2015). This method is essentially a vapour control, by exchanging well solutions, which allows multiple samples to be dehydrated in a number of ways in one single procedure. With high throughput tools it is possible, with a reasonable amount of beam time, to analyse these complex dehydration results without the blurring effect of cryo-cooling. In summary, the fundamental advantages of using *in situ* is that there is no need for direct crystal handling and therefore small or delicate samples are not a problem, crystals can grow at 4 °C with the procedure carrying on at the same temperature and there is no need for cryo-cooling.

In these sorts of experiments, the key is to analyse large populations of crystals and observe general trends of change amongst them. These experiments are analysed by studying changes in lattice dimensions, mosaicity, resolution and diffraction properties as a function of the chemical composition of wells. The nature of the initial cohort of crystals can be well defined; therefore their susceptibility to change, any transition points, any regions of stability and a minimum lattice size should all be possible using limited beamtime. Once points of interest are identified, a sub-population of test samples can then be harvested and cryo-cooled in order to confirm any beneficial effects. These cryo-cooling tests should help define a best protocol for dehydration and cryo-protection.

6.3.2.5 Analysing Results and Making Decisions

As it has been described throughout this chapter, there are a number of valid options when carrying out dehydration. The researcher's choice will depend on a number of constraints such sample, resources and time availability (see Table 6.4). Also depending on the method used the outcomes will vary. In certain occasions, observations will be immeasurable due to crystal cracking, shrinking or dissolving, in others observations will be easy to measure (lattice dimensions, mosaicity and resolution limits) and unfortunately, in others negative results will be the outcome (lack of diffraction, presence of ice rings or non indexable lattices). For these reasons it is advisable to correlate all the different information available and try to make sense of partial data. All the different types of dehydration experiments can be linked via their relation to Relative Humidity. By using available tabulated information (salt dehydration) or measuring empirical humidity (HC1/FMS) researchers can correlate results from different experiments. The idea is to try to extract as much information and try new experiments that may lead to a positive outcome with the available tools. For example, prior to any dehydration attempt, most researchers have already tried standard cryo-cooling, soaks and handled their crystals in a number of ways. Researchers are familiar with crystals stability under the microscope and their methods for soaking in cryo-protectant are useful when planning dehydration protocols either using the HC1/FMS or a vapour diffusion experiment. Similarly, measuring or calculating the equilibrium RH of these solutions can give to the crystallographer clues of why some of the solutions are better than others and suggest new cooling protocols. Often, cryo diffraction data is also available and so a quick comparison with room temperature data collection can possibly yield fruitful clues. Variable indexing in cryo lattices can be an indication of a varied initial crystal population suggesting a variable response to cryo making crystals interesting candidates for dehydration. A review of old crystal forms observed during past experiments (despite not diffracting well) can also give useful clues to plan dehydration experiments. During dehydration, the observation of lattice transitions, minimum lattices and points of loss of diffraction can be used later when defining starting points for cryo-cooling methods. Even dehydration itself can be used as a simple cryo-protecting strategy useful in preventing adverse effects of penetrative cryo-protectant (Pellegrini et al. 2011). Using Relative Humidity as a way of cross-referencing results from the different experiments allows decisions to be made on quantitative measurements and permits experiment planning that will yield easier to compare results.

As it was mentioned earlier, when attempting dehydrating membrane protein crystals it is worth keeping in mind the particular packing of these crystals and how this might affect the results. Detergent micelle size/concentration and limited hydrophilic contacts between the molecules in the crystal lattice should be the main points to take into account during interpretation of results. In addition, as these crystals have larger solvent channels, often require greater dehydration than required for soluble proteins. Therefore, one should include both vapour diffusion and chemical modification methods as they might yield markedly different results. Finally, given the fragile nature these crystals, they are likely to benefit from handling free techniques such as in situ dehydration and dialysis.

6.4 Conclusions: Success in Dehydration

Unfortunately due to a multitude of constraints in most research projects, controlled dehydration is often not attempted or just coarsely tested. Researchers frequently associate the positive

Table 6.4 Comparative of the different properties of the possible dehydration methods

Dehydration method	Reproducible	Gentle de-hydration	Fast dehy-dration	Temperature control	Automated	Reversible	Volatiles	High throughput	Direct X-ray feedback	No handling required	Low cost/ accessible
Simple air exposure	×	×	✓	✓	×	×	×	×	×	×	✓
Vapour control by exchanging well solution	✓	✓	×	✓	×	✓	✓	✓	×	×	✓
Chemical exchange of crystallisation droplet	×	✓	✓	✓	×	✓	✓	✓	×	×	✓
Capillary dehydration	✓	✓	×	×	×	✓	✓	×	✓	×	✓
Free mounting system	✓	✓	✓	✓	✓	✓	×	×	✓	×	×
HC1 dehydration	✓	✓	✓	×	✓	✓	×	×	✓	×	×
In situ dehydration	✓	✓	×	✓	×	✓	✓	✓	✓	✓	✓

outcomes of dehydration with the time required for the experiment. Later, by asking colleagues and consulting the available literature often conclude that it is unlikely to be useful for them. This makes dehydration very much a last resort technique, in particular when more standard approaches have failed and the project is important enough to pursue alternatives. In addition, it is known that many of dehydration success cases are not reported in methods sections of scientific papers and it is considered of less importance.

With regard to membrane proteins, a recent review (Bowler et al. 2015; Russo et al. 2012) on dehydration showed that despite the large number of soluble protein structures solved compared with membrane proteins, the number of reported membrane protein targets facilitated by dehydration seem to be very similar to soluble proteins, suggesting that success is inclined towards membrane proteins. It is difficult to be sure but three potential explanations for this phenomenon are:

- Membrane proteins are more costly (in time and money) to produce and investigators may be willing to try every available technique possible before moving onto other projects
- The nature of the membrane protein crystals themselves could be the blame. Membrane protein crystals have smaller water content but overall higher solvent content facilitating motions across the solvent channels making them more susceptible to improvement by dehydration
- It seems that a membrane protein structure which is facilitated by a dehydration method always has a reference within the results section of the scientific paper as it is perceived to be important for the research field

Whatever the reason, it is clear that undertaking controlled dehydration in membrane protein crystals is well worth the effort in most cases.

Acknowledgments We gratefully acknowledge the Diamond Light Source Ltd and the Membrane Protein Laboratory (Wellcome Trust (grant 099165/Z/12/Z).

References

Abergel C (2004) Spectacular improvement of X-ray diffraction through fast desiccation of protein crystals. Acta Crystallogr D Biol Crystallogr 60(8):1413–1416

Adachi H, Umena Y, Enami I, Henmi T, Kamiya N, Shen JR (2009) Towards structural elucidation of eukaryotic photosystem II: purification, crystallization and preliminary X-ray diffraction analysis of photosystem II from a red alga. Biochim Biophys Acta 1787(2): 121–128

Aishima J, Owen RL, Axford D, Shepherd E, Winter G et al (2010) High-speed crystal detection and characterization using a fast-readout detector. Acta Crystallogr D Biol Crystallogr 66(9):1032–1035

Alcorn T, Juers DH (2010) Progress in rational methods of cryoprotection in macromolecular crystallography. Acta Crystallogr D Biol Crystallogr 66:366–373

Amunts A, Drory O, Nelson N (2007) The structure of a plant photosystem I supercomplex at 3.4 A resolution. Nature 447(7140):58–63

Axford D, Foadi J, Hu NJ, Choudhury HG, Iwata S et al (2015) Structure determination of an integral membrane protein at room temperature from crystals in situ. Acta Crystallogr D Biol Crystallogr 71(6):0

Basavappa R, Petri ET, Tolbert BS (2003) A quick and gentle method for mounting crystals in capillaries. J Appl Crystallogr 36(5):1297–1298

Biswal BK, Vijayan M (2002) Structures of human oxy- and deoxyhaemoglobin at different levels of humidity: variability in the T state. Acta Crystallogr D Biol Crystallogr 58(7):1155–1161

Bowler MW, Montgomery MG, Leslie AG, Walker JE (2006) Reproducible improvements in order and diffraction limit of crystals of bovine mitochondrial F(1)-ATPase by controlled dehydration. Acta Crystallogr D Biol Crystallogr 62(9):991–995

Bowler MW, Mueller U, Weiss MS, Sanchez-Weatherby J, Sorensen TLM et al (2015) Automation and experience of controlled crystal dehydration: results from the European synchrotron HC1 collaboration. Cryst Growth Des 15(3):1043–1054

Brockhauser S, Svensson O, Bowler MW, Nanao M, Gordon E, Leal RM, Popov A, Gerring M, McCarthy AA, Gotz A (2012) The use of workflows in the design and implementation of complex experiments in macromolecular crystallography. Acta Crystallogr D Biol Crystallogr 68(8):975–984

Caffrey M (2003) Membrane protein crystallization. J Struct Biol 142(1):108–132

Dobrianov I, Kriminski S, Caylor CL, Lemay SG, Kimmer C, Kisselev A, Finkelstein KD, Thorne RE (2001) Dynamic response of tetragonal lysozyme crystals to changes in relative humidity: implications for post-growth crystal treatments. Acta Crystallogr D Biol Crystallogr 57(1):61–68

Douangamath A, Aller P, Lukacik P, Sanchez-Weatherby J et al (2013) Using high-throughput in situ plate screening to evaluate the effect of dehydration on protein crystals. Acta Crystallogr D Biol Crystallogr 69(5):920–923

Esnouf RM, Ren J, Garman EF, Somers DO, Ross CK, Jones EY, Stammers DK, Stuart DI (1998) Continuous and discontinuous changes in the unit cell of HIV-1 reverse transcriptase crystals on dehydration. Acta Crystallogr D Biol Crystallogr 54(5): 938–953

Farley C, Burks G, Siegert T, Juers DH (2014) Improved reproducibility of unit-cell parameters in macromolecular cryocrystallography by limiting dehydration during crystal mounting. Acta Crystallogr D Biol Crystallogr 70(8):2111–2124

Fodje M, Grochulski P, Janzen K, Labiuk S, Gorin J et al (2014) 08B1-1: an automated beamline for macromolecular crystallography experiments at the Canadian Light Source. J Synchrotron Rad 21(3):633–663

Gabadinho J, Beteva A, Guijarro M, Rey-Bakaikoa V, Spruce D, Bowler MW, Brockhauser S, Flot D, Gordon EJ, Hall DR, Lavault B, McCarthy AA, McCarthy J, Mitchell E, Monaco S, Mueller-Dieckmann C, Nurizzo D, Ravelli RB, Thibault X, Walsh MA, Leonard GA, McSweeney SM (2010) MxCuBE: a synchrotron beamline control environment customized for macromolecular crystallography experiments. J Synchrotron Radiat 17(5):700–707

Garman E (1999) Cool data: quantity AND quality. Acta Crystallogr D Biol Crystallogr 55(10):1641–1653

Greenspan L (1977) Humidity fixed points of binary saturated aqueous solutions. J Res Natl Inst Stand Technol 81A(1):89–96

Hellert J, Krausze J, Schulz TF, Luhrs T (2014) Crystallization, room-temperature X-ray diffraction and preliminary analysis of Kaposi's sarcoma herpesvirus LANA bound to DNA. Acta Crystallogr F 70(11):1570–1574

Heras B, Martin JL (2005) Post-crystallization treatments for improving diffraction quality of protein crystals. Acta Crystallogr D Biol Crystallogr 61(9):1173–1180

Hope H (1988) Cryocrystallography of biological macromolecules: a generally applicable method. Acta Crystallogr B 44(1):22–26

Hu NJ, Iwata S, Cameron AD, Drew D (2011) Crystal structure of a bacterial homologue of the bile acid sodium symporter ASBT. Nature 478(7369):408–411

Huxley HE, Kendrew JC (1953) Discontinuous lattice changes in haemoglobin crystals. Acta Crystallogr 6(1):76–80

Juers DH, Matthews BW (2004) Cryo-cooling in macromolecular crystallography: advantages, disadvantages and optimization. Q Rev Biophys 37(2):105–119

Kalinin Y, Thorne R (2005) Crystal growth in X-ray-transparent plastic tubing: an alternative for high-throughput applications. Acta Crystallogr D Biol Crystallogr 61(11):1528–1532

Kalinin Y, Kmetko J, Bartnik A, Stewart A, Gillilan R, Lobkovsky E, Thorne R (2005) A new sample mounting technique for room-temperature macromolecular crystallography. J Appl Crystallogr 38(2):333–339

Kaushal PS, Sankaranarayanan R, Vijayan M (2008) Water-mediated variability in the structure of relaxed-state haemoglobin. Acta Crystallogr Sect F Struct Biol Cryst Commun 64(6):463–469

Kiefersauer R, Than ME, Dobbek H, Gremer L, Melero M et al (2000a) A novel free-mounting system for protein crystals: transformation and improvement of diffraction power by accurately controlled humidity changes. J Appl Crystallogr 33(5):1223–1230

Kiefersauer R, Dobbek H, Grazulis S, Than M, Huber R (2000b) A novel free-mounting system for protein crystals: transformation and improvement of diffraction power by accurately controlled humidity changes. Acta Crystallogr Sect A 56(s1):s441

Kiefersauer R, Dobbek H, Than ME, Huber R (2002) A novel free-mounting system for protein crystals: improvement of diffraction power by accurately controlled humidity changes. Acta Crystallogr Sect A 58(s1):c150

Kiefersauer R, Grandl B, Krapp S, Huber R (2014) IR laser-induced protein crystal transformation. Acta Crystallogr D Biol Crystallogr 70(5): 1224–1232

Kim CU, Hao Q, Gruner SM (2007) High-pressure cryocooling for capillary sample cryoprotection and diffraction phasing at long wavelengths. Acta Crystallogr D Biol Crystallogr 63(5):653–659

Kraft P, Bergamaschi A, Broennimann C, Dinapoli R, Eikenberry EF et al (2009) Performance of single-photon-counting PILATUS detector modules. J Synchrotron Rad 16(3):368–375

Kuo A, Bowler MW, Zimmer J, Antcliff JF, Doyle DA (2003) Increasing the diffraction limit and internal order of a membrane protein crystal by dehydration. J Struct Biol 141(2):97–102

Mac Sweeney A, D'Arcy A (2003) A simple and rapid method for mounting protein crystals at room temperature. J Appl Crystallogr 36(1):165–166

Mitchell EP, Garman EF (1994) Flash freezing of protein crystals: investigation of mosaic spread and diffraction limit with variation of cryoprotectant concentration. J Appl Crystallogr 27(6):1070–1074

Moraes I, Archer M (2015) Methods for the successful crystallization of membrane proteins. In: Structural proteomics. Springer, New York, pp 211–230

Mueller U, Darowski N, Fuchs MR, Forster R, Hellmig M, Paithankar KS, Puhringer S, Steffien M, Zocher G, Weiss MS (2012) Facilities for macromolecular crystallography at the Helmholtz-Zentrum Berlin. J Synchrotron Radiat 19(3):442–449

Nagendra HG, Sukumar N, Vijayan M (1998) Role of water in plasticity, stability, and action of proteins: the crystal structures of lysozyme at very low levels of hydration. Proteins 32(2):229–240

Newman J (2006) A review of techniques for maximizing diffraction from a protein crystal in stilla. Acta Crystallogr D Biol Crystallogr 62(1):27–31

Pellegrini E, Piano D, Bowler MW (2011) Direct cryocooling of naked crystals: are cryoprotection agents always necessary? Acta Crystallogr D Biol Crystallogr 67(10):902–906

Perutz MF (1946) The composition and swelling properties of haemoglobin crystals. Trans Faraday Soc 42:B187–B195
Pflugrath J (2015) Practical macromolecular cryocrystallography. Acta Crystallogr Sect F Struct Biol Cryst Commun 71(6):622–642
Pickford MG, Garman EF, Jones EY, Stuart DI (1993) A design of crystal mounting cell that allows the controlled variation of humidity at the protein crystal during X-ray diffraction. J Appl Crystallogr 26(3):465–466
Rajendran C, Dworkowski FS, Wang M, Schulze-Briese C (2011) Radiation damage in room-temperature data acquisition with the PILATUS 6M pixel detector. J Synchrotron Radiat 18(3):318–328
Rockland LB (1960) Saturated salt solutions for static control of relative humidity between 5 and 40 °C. Anal Chem 32(10):1375–1376
Russi S, Juers DH, Sanchez-Weatherby J, Pellegrini E, Mossou E et al (2011) Inducing phase changes in crystals of macromolecules: status and perspectives for controlled crystal dehydration. J Struct Biol 175(2):236–243
Russo KI, Sica F, Mattia CA, Merlino A (2012) Increasing the X-ray diffraction power of protein crystals by dehydration: the case of bovine serum albumin and a survey of literature data. Int J Mol Sci 13(3):3782–3800
Sanchez-Weatherby J, Bowler MW, Huet J, Gobbo A, Felisaz F et al (2009) Improving diffraction by humidity control: a novel device compatible with X-ray beamlines. Acta Crystallogr D Biol Crystallogr 65(12):1237–1246
Saraswathi NT, Sankaranarayanan R, Vijayan M (2002) Effect of stabilizing additives on the structure and hydration of proteins: a study involving monoclinic lysozyme. Acta Crystallogr D Biol Crystallogr 58(7):1162–1167
Shi J, Sivaraman J, Song J (2008) Mechanism for controlling the dimer-monomer switch and coupling dimerization to catalysis of the severe acute respiratory syndrome coronavirus 3C-like protease. J Virol 82(9):4620–4629
Sukumar N, Biswal BK, Vijayan M (1999) Structures of orthorhombic lysozyme grown at basic pH and its low-humidity variant. Acta Crystallogr D Biol Crystallogr 55(4):934–937
Ursby T, Unge J, Appio R, Logan DT, Fredslund F, Svensson C, Larsson K, Labrador A, Thunnissen MM (2013) The macromolecular crystallography beamline I911-3 at the MAX IV laboratory. J Synchrotron Radiat 20(4):648–653
Warkentin M, Stanislavskaia V, Hammes K, Thorne RE (2008) Cryocrystallography in capillaries: critical glycerol concentrations and cooling rates. J Appl Crystallogr 41(4):791–797
Wheeler MJ, Russi S, Bowler MG, Bowler MW (2012) Measurement of the equilibrium relative humidity for common precipitant concentrations: facilitating controlled dehydration experiments. Acta Crystallogr Sect F Struct Biol Cryst Commun 68(1): 111–114
Winston PW, Bates DH (1960) Saturated solutions for the control of humidity in biological research. Ecology 1:232–237
Yadav MK, Gerdts CJ, Sanishvili R, Smith WW, Roach LS, Ismagilov RF, Kuhn P, Stevens RC (2005) In situ data collection and structure refinement from microcapillary protein crystallization. J Appl Crystallogr 38(6):900–905
Yao M, Yasutake Y, Tanaka I (2004) Flash-cooling of macromolecular crystals in a capillary to overcome increased mosaicity. Acta Crystallogr D Biol Crystallogr 60(1):39–45

Nonlinear Optical Characterization of Membrane Protein Microcrystals and Nanocrystals

7

Justin A. Newman and Garth J. Simpson

Abstract

Nonlinear optical methods such as second harmonic generation (SHG) and two-photon excited UV fluorescence (TPE-UVF) imaging are promising approaches to address bottlenecks in the membrane protein structure determination pipeline. The general principles of SHG and TPE-UVF are discussed here along with instrument design considerations. Comparisons to conventional methods in high throughput crystallization condition screening and crystal quality assessment prior to X-ray diffraction are also discussed.

Keywords

Nonlinear optical microscopy • Second harmonic generation • Two-photon excited UV fluorescence • Microscopy • High-throughput screening • Crystal detection • Crystal positioning • Crystal optimisation

7.1 Introduction

Due to the amount of time required to solve a protein structure, much effort has been placed into the optimization of the crystal structure determination pipeline. Most commonly, structure determination by crystallography begins by focusing on the selection of a suite of constructs and/or homologous targets for expression and characterization. From those, the most readily expressible, stable, biochemically active, and monodisperse candidates are prepared for crystallization screening, which typically consists of initial "sparse matrix" screens followed by refinement of conditions around the most promising positive hits. Once diffraction-quality crystals are produced, individual crystals are typically extracted into loops and cryo-cooled, followed by single-crystal X-ray diffraction at a synchrotron facility and structure determination from the resulting data. Synchrotron radiation is unique in delivering high flux with wavelength tunability, such that the large majority of high resolution protein structures generated from diffraction are now based on data acquired at synchrotron facilities (Anderson 2014; Redecke et al. 2013).

The time between the selection of an initial set of targets and the final determination of the

J.A. Newman • G.J. Simpson (✉)
Department of Chemistry, Purdue University, West Lafayette, IN 47907, USA
e-mail: newman21@purdue.edu; gsimpson@purdue.edu

I. Moraes (ed.), *The Next Generation in Membrane Protein Structure Determination*,
Advances in Experimental Medicine and Biology 922, DOI 10.1007/978-3-319-35072-1_7

structure of the protein or protein complex can still be quite lengthy. Obviously, the overall duration hinges quite directly on the complexity and availability of the protein target (Skarina et al. 2014). However, there are several other bottlenecks in the current pipeline that significantly limit the throughput for obtaining a crystal structure, including screening of conditions leading to crystallization, assessment of crystal quality, and detection and location of crystals at the synchrotron (Kissick et al. 2013). The trends toward higher-throughput platforms and analysis of ever-smaller crystals is rapidly placing a greater burden on reliable, automated crystal detection.

Several relatively recent trends have aided in helping to reduce the timeframe for crystallization (Bogan 2013). Robotic and microfluidic platforms for the preparation of crystallization trials have enabled massively parallel analysis of many crystallization conditions with smaller per-trial volumes of mother liquor and less protein consumption (Zhu et al. 2014; Brodersen et al. 2013). However, highly parallel preparation substantially increases the number of wells required for inspection of crystal formation and decreases the sizes of the crystals within each lower-volume well. Automated scoring of crystallization trials based on bright field, birefringence, and ultraviolet fluorescence imaging have aided in identifying the most obvious cases of relatively large crystal formation in conventional 96 well-plate architectures, but routinely exhibit relatively high false negative rates, particularly for smaller <5–10 μm crystals. Additionally, crystallization screening can typically require several weeks to learn whether or not a particular target under a particular set of conditions will lead to protein crystal formation. The need is quickly growing for fast and reliable automated assessment of crystal formation.

However, crystal formation alone is a necessary but not always sufficient condition for structure determination. Assuming crystals can be formed, assessment of the diffraction quality of the crystals is often only meaningfully measured after extraction, cryo-cooling, and diffraction at a synchrotron facility, which can routinely require several months per iteration due to the high value of synchrotron beam time. Efforts to increase the throughput of synchrotron facilities have included fully automated systems including robotic crystal positioning and automated diffraction analyses. Similarly, reductions in the beam size have been implemented to target serial crystallography of pooled data acquired from many smaller crystals rather than one single crystal (Sanishvili et al. 2011; Wasserman et al. 2012; Mader et al. 2011). However, in both of these efforts, challenges for automated analysis rest in the need to rapidly identify crystal positions prior to data collection.

Single-shot, "diffract and destroy" methods using X-ray free electron lasers (XFEL) provides another alternative option for suspensions of many protein nanocrystals $< \sim 1$ μm in diameter (Chapman et al. 2011; Cherezov 2011), but come with even longer lead times between access to diffraction facilities, as well as challenges regarding the optimization of crystal quality for these small dimension crystals. A need for fast analysis of presence and quality of nanocrystals prior to analysis with the XFEL is quickly becoming evident as the XFEL becomes a more widely used tool for protein crystal diffraction. Based on the previously discussed trends in protein crystallography, the need is clearly growing for methods that can quickly close the feedback loops in crystal production, optimization, and X-ray data collection.

Nonlinear optical methods such as, second harmonic generation (SHG, also known as second order nonlinear optical imaging of chiral crystals, or SONICC) and two-photon excited ultra-violet fluorescence (TPE-UVF) have recently emerged as new contrast mechanisms for protein crystal detection (Kissick et al. 2009; Madden et al. 2011; Haupert et al. 2012) and are now routinely being used in the protein crystallization pipeline for high-throughput crystallization condition screening, crystal quality assessment (DeWalt et al. 2014; DeWalt et al. 2013), and detection and location of crystals in synchrotron beams (Kissick et al. 2013; Madden et al. 2013; Newman et al. 2016). The following chapter will discuss the advantages and limitations of SHG and TPE-UVF relative to other conventional imaging techniques and describe their current role in minimizing overall timeframe for protein structure determination.

7.2 Nature of Contrast from SHG

Recently, SHG has been used as a high contrast, label-free, alternative imaging method for the detection of protein crystals (Kissick et al. 2009). SHG requires the use of a sufficiently intense light, as found in the focus of a laser, to achieve reasonably efficient nonlinear processes. As molecules approach isotropic ordering, the coherent SHG generated from the ensemble approaches zero due to internal cancellation (Boyd 2008). However, as the molecules become more ordered the SHG generated from the individual proteins constructively adds, resulting in detectable signals. SHG is strictly forbidden in centrosymmetric space groups due to the inversion symmetry cancelling out the coherent addition of SHG within the crystal. Fortuitously, the chiral nature of proteins results in crystals necessarily adopting non-centrosymmetric space groups (Haupert et al. 2012). While proteins crystallize in noncentrosymmetric space groups it is important to note that the **432** crystal class is SHG forbidden due to the high symmetry elements present in the crystal class. However, protein crystals of the **432** crystal class represent <1 % of the current PDB entries. Due to these symmetry requirements no detectable SHG is generated from amorphous aggregates that can potentially form in the crystallization of proteins. Similarly, many common salts used in the crystallization mother liquor crystallize in centrosymmetric space groups and will thus yield no signal if the salts happen to crystallize during the protein crystallization trials (Closser et al. 2013). As a result of the symmetry requirements discussed above, SHG microscopy has become a useful tool for visualizing protein crystals with many advantages over other more widely used techniques (Kissick et al. 2009).

Bright field images are sufficient for detecting large crystals in transparent media (Fig. 7.1a row 1) but suffer in translucent and turbid media (Fig. 7.1b row 1) and when the crystals are small (Fig. 7.1c row 1). Birefringence is another common imaging method used in crystal detection; however, when the sample medium is highly birefringent (Fig. 7.1b row 2) it cannot be reliably used to locate crystals. Another common technique is ultraviolet fluorescence (UVF) which uses the native autofluorescence as contrast (Vernede et al. 2006). However, UVF suffers from high background noise due to the presence of protein in solution, from protein aggregates, and from stray fluorescence from other optical elements. The resulting background can at times be sufficiently high resulting in the signal from crystals being buried in the noise. As seen in the line trace from Fig. 7.1b row 3, UVF can only detect one crystal above the noise floor, whereas the line-trace from the SHG image is able to detect multiple crystals from the same line trace with a high signal to noise ratio (SNR). The high SNR of SHG microscopy also allowed for the detection of nanocrystalline showers (Fig. 7.1c row 5) that were undetected by the other imaging techniques (Fig. 7.1c rows 1–4).

7.3 TPE-UVF vs. UVF

Fluorescent techniques have been widely used for imaging protein crystals. Conventional methods utilize a ~280 nm light source to excite tryptophan and other aromatic residues commonly found in most proteins (Judge et al. 2005). Detection of the red-shifted fluorescence allows for the localization of protein crystals within droplets. However, there are several drawbacks to imaging protein crystals by single photon UV fluorescence. The first drawback to using single photon UV fluorescence is the contribution to the background from out-of-plane signal generators such as aggregates, protein in solution, and autofluorescence from optics and the crystallization plates (Madden et al. 2011). The second drawback is the use of UV-C light. Absorption of UV-C wavelengths have been shown to break disulphide bonds, which will negatively affect the accuracy of the crystal structure that can

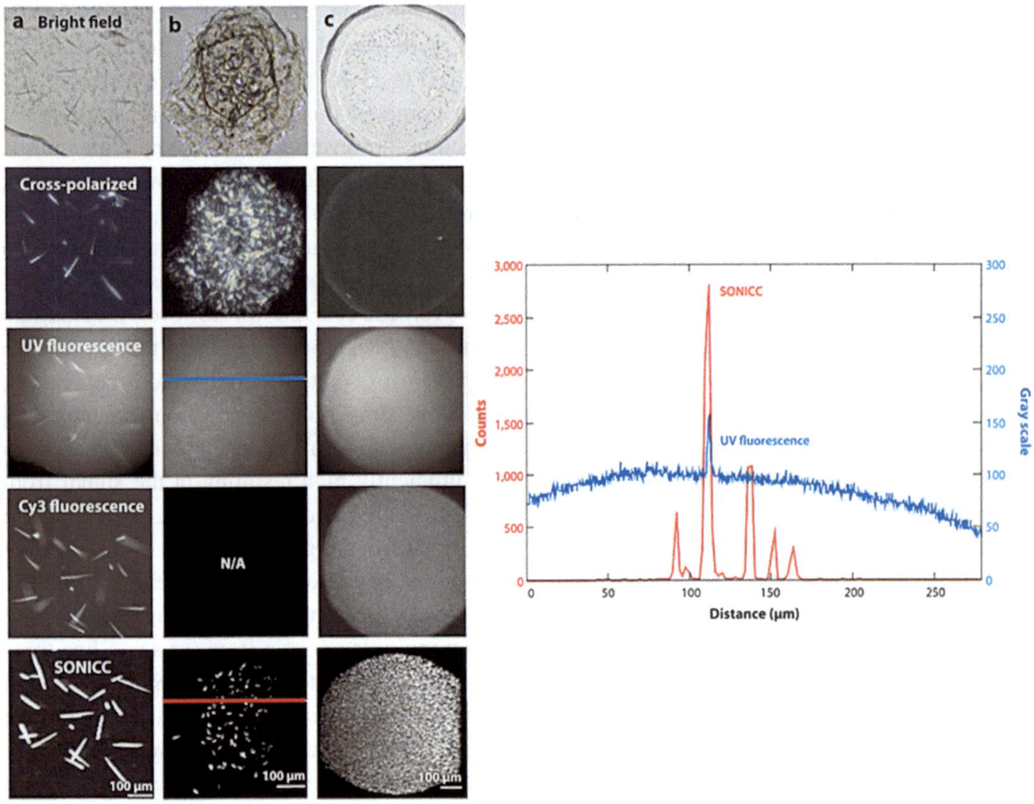

Fig. 7.1 A comparison of conventional imaging techniques along with SHG microscope images for three different samples of protein crystals grown in lipidic cubic phase (columns **a**–**c**). SHG was the only method that was able to detect the presence of crystals with high signal to noise. Line traces for fluorescence imaging and SONICC imaging shows that the significant improvement in signal to noise allows for the detection of crystals that would have otherwise gone undetected by UV fluorescence

ultimately be solved (Wien et al. 2005). Damage from these wavelengths is not limited to the focal plane, as absorption can arise with comparable efficiency above and below the focal plane as well. The operating costs for using this high energy excitation wavelength are substantially higher than those for visible wavelengths due to the need for UV transparent optics and UV transparent 96-well plates, but still relatively inexpensive.

TPE-UVF is an alternative UV fluorescence imaging method. TPE-UVF works by combining two photons of 532 nm light to excite the 266 one-photon absorption band (Madden et al. 2011). Because 2 photons of the 532 laser are required to achieve the energy required for the tryptophan residue absorption band only the most intense portion of the focus will generate the fluorescence signal. As such, this technique does not require the use of a UV light source, effectively eliminating the potential for out-of-plane sample damage and contributions to the background. This is more clearly illustrated in Fig. 7.2 comparing glucose isomerase crystals imaged by conventional single photon UV fluorescence (Fig. 7.2a) and TPE-UVF (Fig. 7.2b). Because only in-plane fluorescence is generated by TPE-UVF the image contrast is much higher for Fig. 7.2b compared to the single photon image (Fig. 7.2a) where the background has significant fluorescence contributions from out-of-plane sample.

Other nonlinear fluorescent methods have been used to image protein crystals. Native

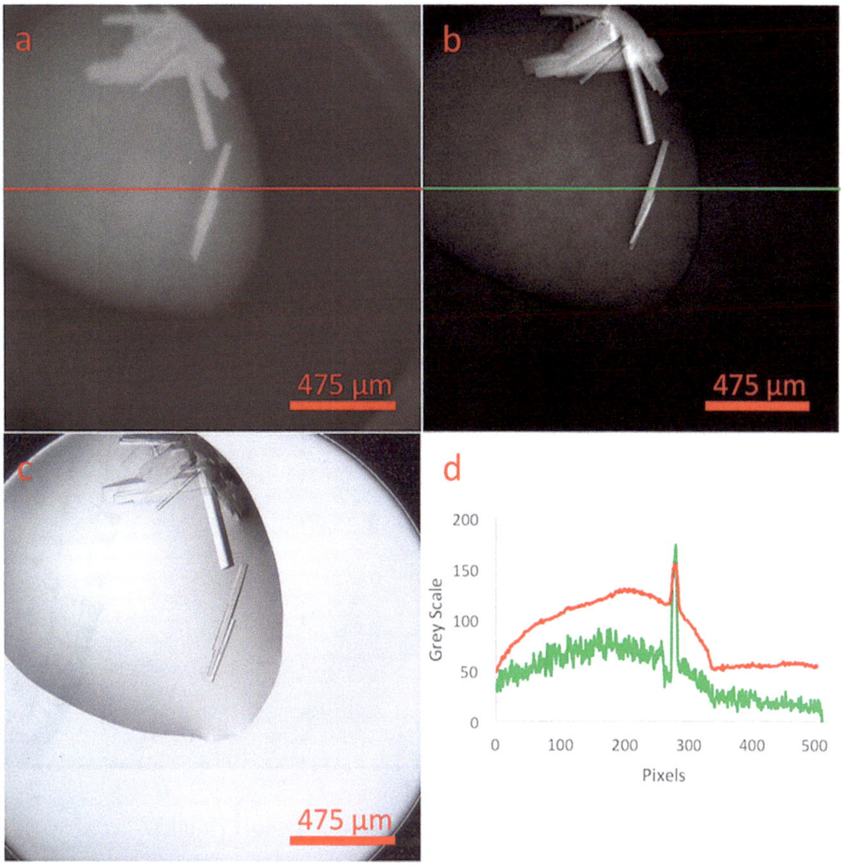

Fig. 7.2 Glucose isomerase crystals imaged by single photon UV fluorescence (**a**), two-photon excited UV fluorescence (**b**) and bright field imaging (**c**) to illustrating the improved signal to noise from the use of a two-photon process arising from the reduction in contributions to the background from out-of-plane signal generation from single photon imaging methods. A line scan is given in (**d**) to demonstrate the reduced out-of-plane signal contributions and increased SNR by using TPE UVF

two-photon fluorescence using an excitation laser wavelength in the near IR range has been used (Padayatti et al. 2012). This method was able to easily detect protein crystals with native chromophores with absorption profiles in the 400–600 nm range. TPEF was also able to detect fluorescence from the oxidation of aromatic residues, with increases in TPEF autofluoresence signals from older proteins (>3 months). Fluorescent dyes have been used to noncovalently attach to protein crystals to increase their TPEF response (Groves et al. 2007). By using a two-photon imaging method the fluorescent background from uncrystallised protein and solvated dye was significantly reduced as compared to single photon fluorescent imaging methods.

7.4 Protein Crystal Coverage and False Negatives

7.4.1 Native SHG and TPE-UVF Protein Crystal Coverage

While SHG and TPE-UVF are capable of imaging protein crystals it is important to discuss the potential of false-positives and false-negatives. As discussed previously, SHG requires that the crystals be noncentrosymmetric to be

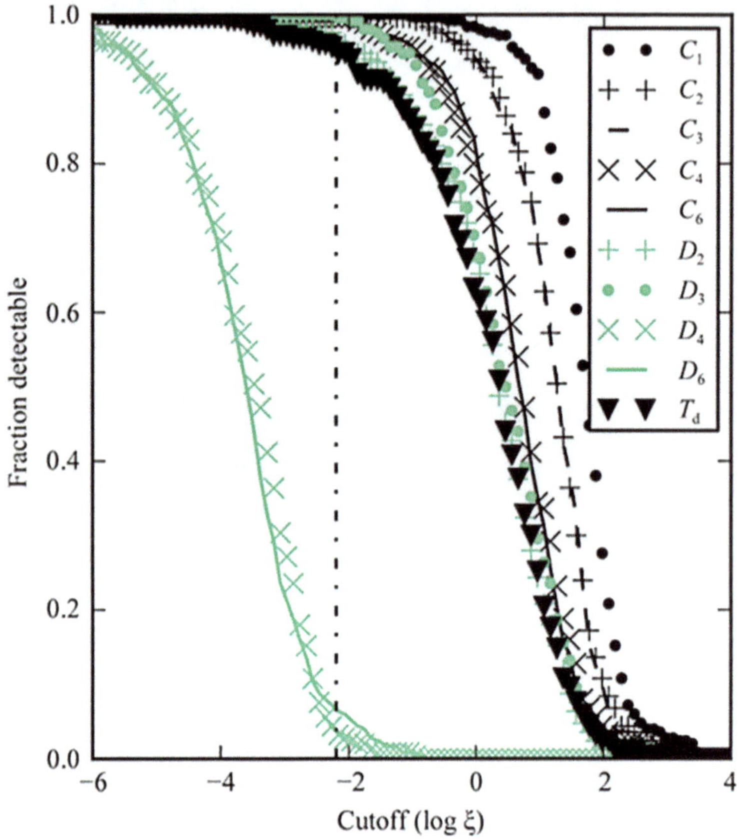

Fig. 7.3 Fraction of protein crystals detectable as a function of instrument sensitivity for different protein crystal classes. ξ is the squared magnitude of the SHG activity normalized by unit cell volume. The vertical *dashed line* represents an instrument that is sensitive enough to detect the tetragonal lysozyme polymorph with low signal to noise (ξ = 0.0064)

detectable. As a result, false-positives can arise from other noncentrosymmetric crystals that can form during a crystallization trial. While rare, a few salts commonly used in protein crystallization can generate strong SHG signals that can interfere with the SHG generated by protein crystals (Closser et al. 2013). Similarly crystals of other small molecules such as ligands can also form, leading to an increase in false-positives. Coupling SHG with other NLO techniques such as TPE-UVF reduces false-positives as most interfering salts do not exhibit fluorescence (Madden et al. 2011; Closser et al. 2013).

SHG intensity varies significantly due to crystal symmetry, with lower symmetry crystals generally exhibiting strong SHG responses compared to more symmetrical crystal lattices. Because of the variability in crystal symmetry the SHG response, from protein to protein, can span several orders of magnitude with ~84 % of the known protein data bank (PDB) entries being detectable by SHG microscopy performed with relatively long acquisition times (Haupert et al. 2012). Figure 7.3 shows the detectability of ten protein crystal classes based on the sensitivity of the instrument. The vertical dashed line represents the ability to detect tetragonal lysozyme with weak signal to noise and represents a practical lower limit for detectability. Based on this threshold greater than 90 % of all crystals in each of the eight out of ten crystal classes can be detected, however less than 10 % of protein crystals in both D_4 and D_6 crystal classes are detectable. As the measurement time is reduced, the fraction of detectable crystals will similarly decrease.

The tetragonal crystal polymorph of lysozyme is an excellent example of a crystal that is weak

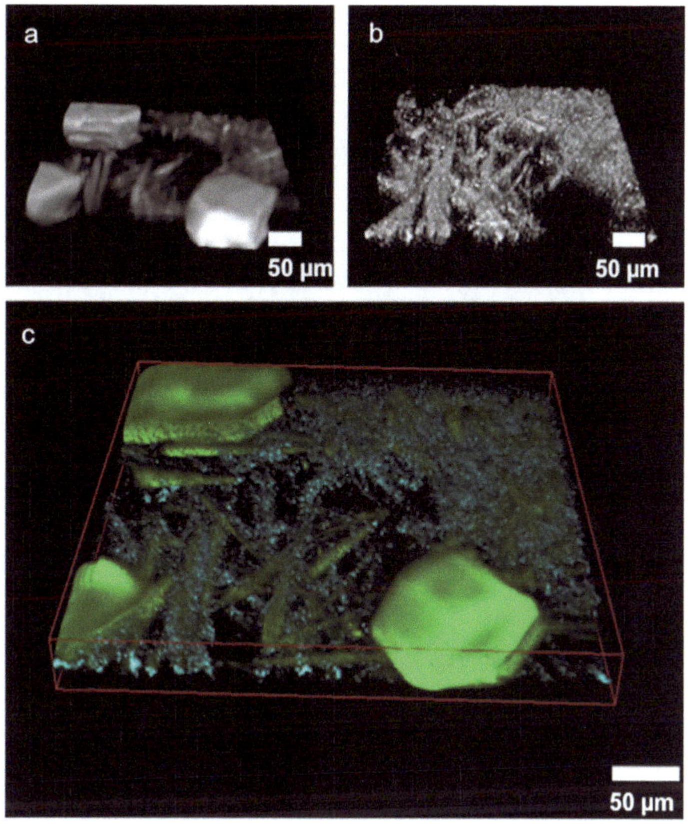

Fig. 7.4 TPE-UVF was used in panel (**a**) to detect tetragonal lysozyme polymorphs that are SHG inactive along with the SHG active monoclinic polymorphs. SHG was used in panel (**b**) to detect the SHG active monoclinic lysozyme polymorph but was unable to detect the tetragonal polymorph due to the high symmetry of the crystal lattice. The images are combined in panel (**c**)

for SHG microscopy, as seen in Fig. 7.4b. As a result of the fourfold symmetry of the $P4_32_12_1$ crystal lattice, large voids are seen in the SHG image where the crystals are located. However, SHG can readily detect the $P2_1$ lysozyme polymorph (Fig. 7.4b) (Haupert et al. 2012). Combining TPE-UVF and SHG imaging modalities onto a single imaging platform can provide complementary imaging techniques that extend the detection coverage. Because aromatic residues are commonly found in most proteins TPE-UVF is a useful method for visualizing protein crystals (Madden et al. 2011). However, like SHG, it too does not have 100 % coverage of protein crystals. There are several instances where either the lack of aromatic residues, such as in insulin, or internal fluorescence quenching, such as in catalase, limit the protein coverage for TPE-UVF (Madden et al. 2011). However, due to the presence of tryptophan in lysozyme, TPE-UVF can readily image the $P4_32_12_1$ Lysozyme polymorph, as well as the $P2_1$ polymorph (Fig. 7.4a). TPE-UVF can also be used to distinguish protein crystals from salt crystals in the rare case that the salt crystal is SHG active (Closser et al. 2013). By combining SHG and TPE-UVF in a single platform, both polymorphs of lysozyme can be detected as shown in Fig. 7.4c, and more generally an increase in protein coverage is observed.

7.4.2 Methods to Increase SHG Coverage

In an effort to increase the SHG coverage of protein crystals, a method has been developed to incorporate SHG active dyes into the crystal lattice to generate brighter SHG responses from the protein crystals (Newman et al. 2015). Unlike trace-labelling methods, which can potentially

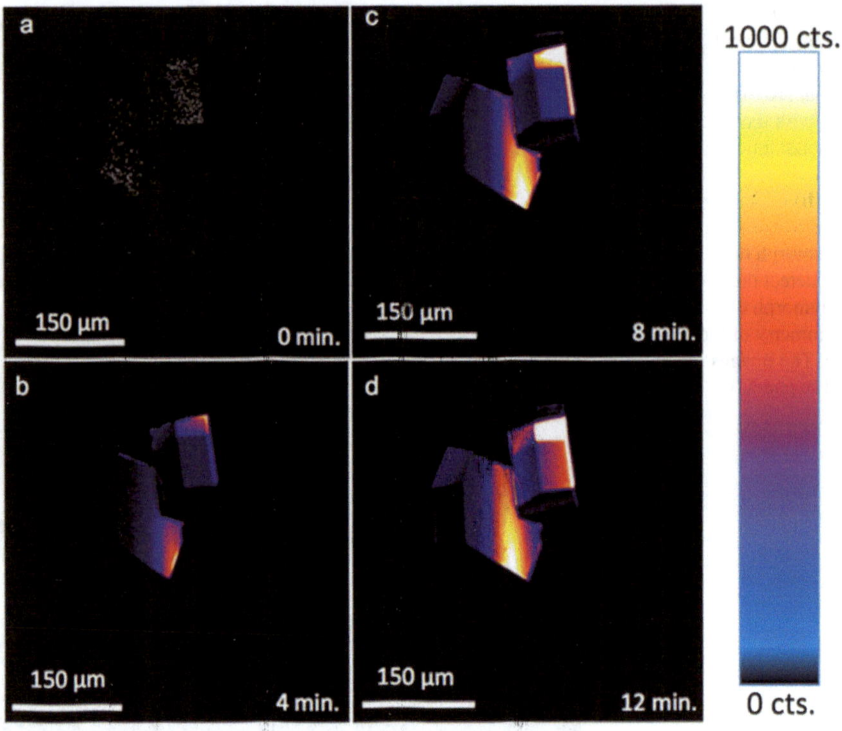

Fig. 7.5 Intercalation of 500 μM solution of malachite green into lysozyme crystals. At time zero (**a**) there is no detectable SHG. The scale was adjusted to show low counts. Within minutes detectable signal was observed (**b**–**d**) with an approximate 700-fold overall SHG enhancement after 12 min of dye intercalation

affect the protein folding (Forsythe et al. 2006), the SHG active dyes are non-covalently bound to the protein. In this method the dye can be introduced to the protein in one of two ways. First, the dye can be mixed into the mother liquor prior to crystallization. While this approach has the distinct advantage of simplicity, there is a potential risk of the dyes binding at the intermolecular contact points in the lattice, which can potentially affect crystallization. The second method is to soak the already formed crystal in a dye solution. The dye molecules intercalate into the crystal and are templated by the crystalline lattice to orient in an ordered alignment, allowing signal generation for SHG microscopy. Figure 7.5 shows the uptake of SHG active malachite green into a $P4_32_12_1$ Lysozyme crystal. After only a few minutes the SHG intensity of nominally weakly SHG-active tetragonal lysozyme (Fig. 7.5a) is increased by several orders of magnitude (Fig. 7.5b–d). By intercalating SHG active dyes into protein crystals the SHG coverage can be extended beyond the predicted 84 % coverage by native SHG. More importantly the stronger SHG signals afforded by dye intercalation allow for shorter acquisition times for improved compatibility with high-throughput crystal screening and for routine detection of nanocrystals prior to serial crystallography. However, high concentrations of dye can potentially negatively affect crystallisability (Newman et al. 2015).

7.5 Considerations for Instrument Design

7.5.1 Why Ultrafast Lasers?

The nonlinear optical methods discussed in this chapter require high optical peak powers to access efficient nonlinear optical signal generation.

However, in beam scanning microscopy, low average powers are preferred to minimize sample damage due to local heating. These dual competing criteria can be met by using an ultrafast pulsed laser ($\sim$100 fs pulses), resulting in the high peak power needed for efficient nonlinear optical generation (Haupert and Simpson 2011). As such, a 100 mW, 80 MHz laser producing 100 fs pulses generates a peak power of 12,000 W per pulse, but only over a timeframe of less than a trillionth of a second.

Several ultrafast laser sources have been used for SHG detection of protein crystals. High-powered Ti: sapphire lasers offer considerable wavelength tunability (700–1050 nm) and high average power (up to 3 W). Relatively recently, ultrafast fiber lasers have made considerable strides towards serving as viable alternatives to solid state lasers. Fiber lasers incorporate the gain medium directly into the fiber, with the fiber acting as the optical cavity of the laser. A wide variety of dopants can be used offering fixed wavelengths ranging from 1030 nm to 1550 nm. Similar to solid state lasers, average powers can reach $\sim$1 W with 100–200 fs pulse durations in an amplified system, and several hundred mW directly from systems requiring just an oscillator. Fiber lasers, in principle, have the potential to be comparatively low cost, stable, and able to be operated hands-free (Fermann and Hartl 2009). However, the technology is still maturing and considerable variability in laser stability still exists across different systems.

7.5.2 A Beam-Scanning Approach

Nonlinear optical methods, such as SHG or TPE-UVF, most commonly produce images by focusing the laser on one pixel at a time and collecting the signal on a fast, large area photodetector, such as a photomultiplier tube (PMT). By scanning the laser across the sample, the intensity of each pixel is encoded in the time trace of the PMT response. The images can then be reconstructed by knowing which pixels correspond to which times.

There are several different methods for scanning the laser beam across the sample. Beam-scanning can either be performed by using two galvanometer-driven mirror pairs to translate the laser across the sample, or by using a slow scan galvanometer-driven mirror paired with a fast scan mirror (e.g. resonant scanner or rotating polygon). By incorporating a resonant mirror the laser can be translated rapidly (4–15 kHz) across one axis of the sample significantly reducing the pixel dwell time (Haupert and Simpson 2011). Reducing the single-pass dwell time significantly reduces potential perturbations to the sample due to local heating, allowing for higher incident powers (Fu et al. 2006). Given the nonlinear dependence of signal with peak power, the advantages of fast scanning can be quite significant. The reduction in signal due to a shorter per-pixel dwell time can be overcome by performing multiple sweeps over the same pixel to recover comparable signals generated from one inspection of a pixel with a longer dwell time.

The imaging timeframes for SHG and TPE-UVF are sufficiently short to be compatible with platforms for high throughput screening of protein crystals in 96-well plates. Using an 8 kHz resonant mirror, imaging frame rates as high as 15 frames/s (fps) can be easily realized in 512×512 imaging (Haupert and Simpson 2011), or 30 fps for bidirectional scanning. However, in high-throughput applications, a trade-off emerges between crystal detection limits and measurement time. Higher analysis speeds correspond to fewer recovered photons per crystalline sample, potentially complicating both manual and automated crystal scoring. Fortunately, the low inherent backgrounds associated with photon counting detectors allow reasonable detection of protein crystals with only a modest total number of detected photons in most instances. To achieve images with high signal-to-noise sufficient for the analysis described in the next section, longer integration times may be necessary, particularly for crystals generating weak SHG signals.

7.6 Polarization Resolved SHG for Crystal Quality Assessment

High-resolution structures determined by X-ray crystallography require diffraction from high quality protein crystals. However, determining in advance which protein crystals are of high enough quality to yield high resolution diffraction remains a difficult task. Twinned, multi-domain, or highly mosaic crystals can significantly complicate structure determination from X-ray diffraction. Twinning and multidomain crystal formation are often difficult to identify with conventional imaging techniques, since the linear optical properties are often identical for different twinned domains (Yeates and Fam 1999). *In situ* diffraction has shown some promise, but the current X-ray fluxes used do not allow for assessment of high resolution features, analysis of crystals $< \sim 10$ μm in length, or localized analysis within individual domains in larger crystals (Bingel-Erlenmeyer et al. 2011). Currently the best assessment of crystal quality comes after the crystals are extracted, cryo-cooled, and exposed to X-rays at a synchrotron source. This requirement significantly increases sample screening times at synchrotron facilities. As such, there is clearly a need for a high-throughput, non-perturbative method for assessing protein crystal quality prior to sample preparations for a synchrotron run.

The polarization-dependence of SHG provides a potential optical "handle" for initial assessments of quality for crystals still within the mother liquor. Owing to its coherent nature, the polarization state of the detected SHG signal generated from the sample is highly dependent on both the polarization state of the incident light as well as the crystal symmetry and orientation. SHG has up to 18 unique polarization-dependent tensor elements defining its orientation and polarization-dependent response, compared with just three for linear optics (Begue et al. 2009). As a result, a significantly greater amount of information is available from polarization resolved NLO measurements and analysis compared to analogous linear measurements such as birefringence.

In studies of multi-domain crystal conglomerates, polarization dependent SHG images were used to identify unique crystalline domains, confirmed by subsequent X-ray diffraction (DeWalt et al. 2013). Polarization-dependent SHG was measured by rotating waveplates to modulate the input polarization, and the polarization response was analysed by principle component analysis (PCA), shown in Fig. 7.6. The white and black portions indicate regions where the PCA determined different polarization responses, consistent with the presence of multiple domains within the crystal. X-ray diffraction measurements acquired at localized regions were consistent with a multi-domain crystal, with non-overlapping reflections arising depending on the location of diffraction.

While PCA provides some level of contrast, there are significant advantages to relating the measurements directly back to analytical models for the polarization-dependence. Nonlinear optical Stokes ellipsometry (NOSE) measurements enable the recovery of the full set of polarization-dependent free parameters available in a given sample orientation (DeWalt et al. 2014). In the case of SHG, 10 unique observables are recovered in a single measurement (5 for each detector, for vertical and horizontal detection). Once the set of free parameters is determined, the amplitudes of each coefficient can be conveniently represented as an RGBCM colour map, such that a single image per detector contains all the information from the polarization-dependent analysis. This representation also has the distinct advantage of enabling detection of multi-domain crystals based on false-colour contrast in the RGBCM maps. An obvious example is shown in Fig. 7.7b, which clearly indicates two distinct crystalline domains (Fig. 7.7a).

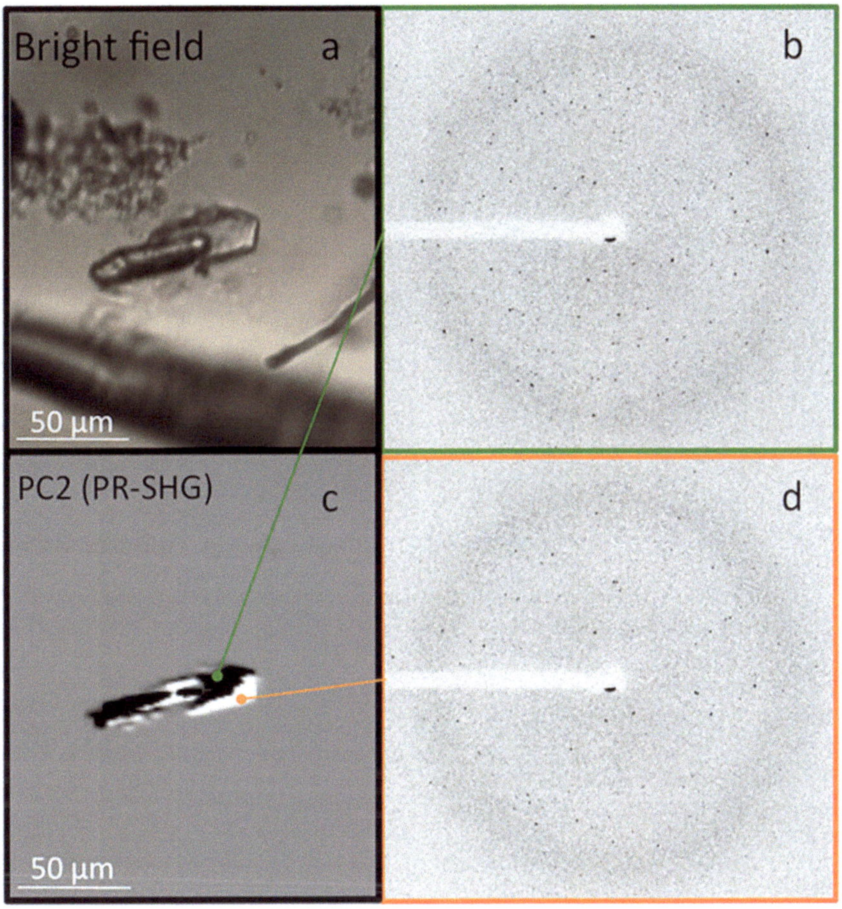

Fig. 7.6 Bright field image of a multi-domain crystal (**a**) along with the PCA image for the polarization dependent SHG signal (**c**). The white and black regions indicate locations identified by the PCA that have a different polarization response indicative of a multi-domain crystal. Diffraction from both the black (**b**) and white (**d**) region confirms the presence of multiple crystal domains

The use of an EOM allows for rapid polarization modulation (8 MHz) of the input laser resulting in the acquisition of the 5 coefficients while still maintaining video frame rates. Because of rapid data acquisition rates, the incorporation of polarization resolved SHG measurements into routine screening of crystallization trays is consistent with the timeframes typically required for high throughput SHG screening of protein crystals. Early identification of multi-domain crystals would significantly reduce the amount time spent on crystallization preparation, speed optimization, and help identify crystals likely to produce quality X-ray diffraction at synchrotron sources.

Acknowledgments The authors wish to acknowledge support from the National Institutes of Health (NIH) grants NIH-R01GM103401 and NIH-R01GM103910. We would also like to thank Ellen Gualtieri from Formulatrix for supplying the glucose isomerase crystal images in Fig. 7.2.

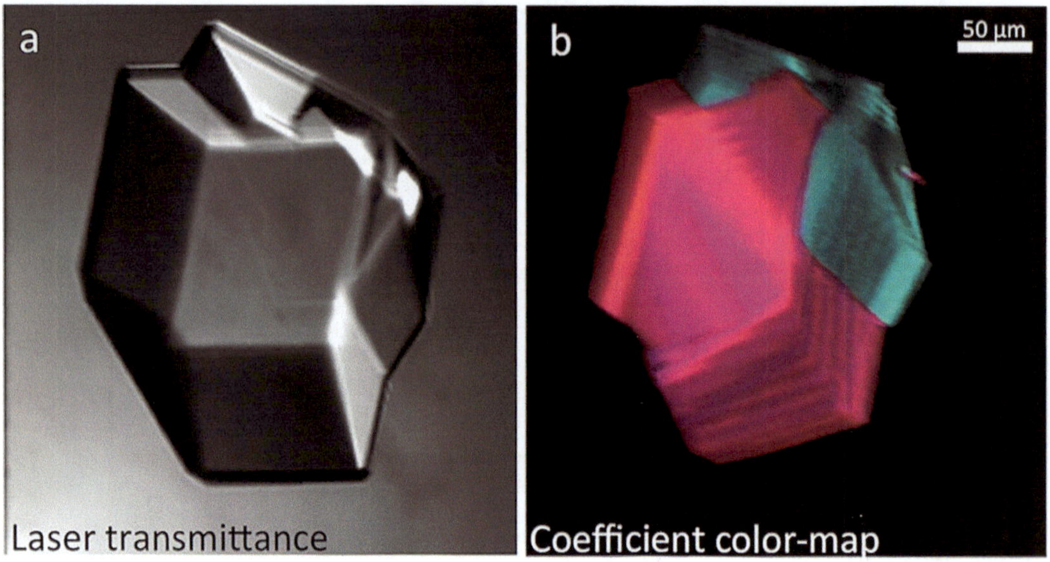

Fig. 7.7 Laser transmittance (**a**) and the coefficient colour-map (**b**) resulting from polarization resolved SHG analysis of a multi-domain glucose isomerase crystal

References

Anderson WF (2014) Structural genomics and drug discovery: methods and protocols, vol 1140. Humana Press Inc., Totowa

Begue NJ, Moad AJ, Simpson GJ (2009) Nonlinear optical stokes ellipsometry. 1. Theoretical framework. J Phys Chem C 113(23):10158–10165

Bingel-Erlenmeyer R, Olieric V, Grimshaw JPA, Gabadinho J, Wang X et al (2011) SLS crystallization platform at beamline X06DA—a fully automated pipeline enabling in situ X-ray diffraction screening. Cryst Growth Des 11(4):916–923

Bogan MJ (2013) X-ray free electron lasers motivate bioanalytical characterization of protein nanocrystals: serial femtosecond crystallography. Anal Chem 85(7):3464–3471

Boyd RW (2008) Nonlinear optics, 3rd edn. Elsevier Inc, Cambridge

Brodersen DE, Andersen GR, Andersen CBF (2013) Mimer: an automated spreadsheet-based crystallization screening system. Acta Crystallogr F Struct Biol Commun Acta 69(7):815–820

Chapman HNF, Fromme P, Barty A, White TA, Kirian RA et al (2011) Femtosecond X-ray protein nanocrystallography. Nature 470:73–77

Cherezov V (2011) Lipidic cubic phase technologies for membrane protein structural studies. Curr Opin Struct Biol 21:559–566

Closser RG, Gualtieri EJ, Newman JA, Simpson GJ (2013) Characterization of salt interferences in second-harmonic generation detection of protein crystals. J Appl Crystallogr 46(6):1903–1906

DeWalt EL, Begue VJ, Ronau JA, Sullivan SZ, Das C, Simpson GJ (2013) Polarization-resolved second-harmonic generation microscopy as a method to visualize protein-crystal domains. Acta Crystallogr D Biol Crystallogr 69(1):74–81

DeWalt EL, Sullivan SZ, Schmitt PD, Muir RD, Simpson GJ (2014) Polarization-modulated second harmonic generation ellipsometric microscopy at video rate. Anal Chem 86(16):8448–8456

Fermann ME, Hartl I (2009) Ultrafast fiber laser technology. IEEE J Sel Top Quantum Electron 15(1):191–206

Forsythe E, Achari A, Pusey ML (2006) Trace fluorescent labeling for high-throughput crystallography. Acta Crystallogr D Biol Crystallogr 62(3): 339–346

Fu Y, Wang H, Shi R, Cheng J-X (2006) Characterization of photodamage in coherent anti-Stokes Raman scattering microscopy. Opt Express 14(9):3942–3951

Groves MR, Muller IB, Kreplin X, Muller-Dieckmann J (2007) A method for the general identification of protein crystals in crystallization experiments using a noncovalent fluorescent dye. Acta Crystallogr D Biol Crystallogr 63(4):526–535

Haupert LM, Simpson GJ (2011) Screening of protein crystallization trials by second order nonlinear optical imaging of chiral crystals (SONICC). Methods 55(4):379–386

Haupert LM, DeWalt EL, Simpson GJ (2012) Modeling the SHG activities of diverse protein crystals. Acta Crystallogr D Biol Crystallogr 68(11):1513–1521

Judge RA, Swift K, Gonzalez C (2005) An ultraviolet fluorescence-based method for identifying and distinguishing protein crystals. Acta Crystallogr D Biol Crystallogr 61(1):60–66

Kissick DJ, Gualtieri EJ, Simpson GJ, Cherezov V (2009) Nonlinear optical imaging of integral membrane protein crystals in lipidic mesophases. Anal Chem 82(2):491–497

Kissick DJ, Dettmar CM, Becker M, Mulichak AM, Cherezov V et al (2013) Towards protein-crystal centering using second-harmonic generation (SHG) microscopy. Acta Crystallogr D Biol Crystallogr D 69(5):843–851

Madden JT, DeWalt EL, Simpson GJ (2011) Two-photon excited UV fluorescence for protein crystal detection. Acta Crystallogr D Biol Crystallogr 67(10): 839–846

Madden JT, Toth SJ, Dettmar CM, Newman JA, Oglesbee RA, Hedderich HG, Everly RM, Becker M, Ronau JA, Buchanan SK et al (2013) Integrated nonlinear optical imaging microscope for on-axis crystal detection and centering at a synchrotron beamline. J Synchrotron Radiat 20(4):531–540

Mader K, Marone F, Hintermuller C, Mikuljan G, Isenegger A, Stampanoni M (2011) High-throughput full-automatic synchrotron-based tomographic microscopy. J Synchrotron Radiat 18(2):117–124

Newman JA, Scarborough NM, Pogranichniy NR, Shrestha RK, Closser RG, Das C, Simpson GJ (2015) Intercalating dyes or enhanced contrast in second harmonic generation imaging of protein crystals. Acta Crystallogr D Biol Crystallogr 71(7): 1471–1477

Newman JA, Zhang S, Sullivan SZ, Dow XY, Becker M, Sheedlo MJ, Stepanov S, Carlsen MS, Everly RM, Das C et al (2016) Guiding synchrotron Xray diffraction by multimodal video-rate protein crystal imaging. J Synchrotron Radiat 23(4)

Padayatti P, Palczewska G, Sun W, Palczewski K, Salom D (2012) Imaging of protein crystals with two-photon microscopy. Biochemistry 51(8):1625–1637

Redecke L, Nass K, DePonte DP, White TA, Rehders D et al (2013) Natively inhibited Trypanosoma brucei Cathepsin B structure determined by using an X-ray laser. Science 339:227–230

Sanishvili R, Yoder DW, Pothineni SB, Rosenbaum G, Xu S, Vogt S, Stepanov S, Makarov OA, Corcoran S, Benn R et al (2011) Radiation damage in protein crystals is reduced with a micron-sized X-ray beam. Proc Natl Acad Sci U S A 108(15):6127–6132

Skarina T, Xu X, Evdokimova E, Savchenko A (2014) High-throughput crystallization screening. Methods Mol Bio 1140:159–168

Vernede X, Lavault B, Ohana J, Nurizzo D, Joly J et al (2006) UV laser-excited fluorescence as a tool for the visualization of protein crystals mounted in loops. Acta Crystallogr D Biol Crystallogr 62(3): 253–261

Wasserman SR, Koss JW, Sojitra ST, Morisco LL, Burley SK (2012) Rapid-access, high-throughput synchrotron crystallography for drug discovery. Trends Pharmacol Sci 33(5):261–267

Wien F, Miles AJ, Lees JG, Vronning Hoffmann S, Wallace BA (2005) VUV irradiation effects on proteins in high-flux synchrotron radiation circular dichroism spectroscopy. J Synchrotron Radiat 12(4): 517–523

Yeates TO, Fam BC (1999) Protein crystals and their evil twins. Structure 7(2):R25–R29

Zhu Y, Zhu L-N, Guo R, Cui H-J, Ye S, Fang Q (2014) Nanoliter-scale protein crystallization and screening with a microfluidic droplet robot. Sci Rep 4:5046

Exploiting Microbeams for Membrane Protein Structure Determination

8

Anna J. Warren, Danny Axford, Neil G. Paterson, and Robin L. Owen

Abstract

A reproducible, and sample independent means of predictably obtaining large, well-ordered crystals has proven elusive in macromolecular crystallography. In the structure determination pipeline, crystallisation often proves to be a rate-limiting step, and the process of obtaining even small or badly ordered crystals can prove time-consuming and laborious. This is particularly true in the field of membrane protein crystallography and this is reflected in the limited number of unique membrane protein structures deposited in the protein data bank (less than 650 by June 2016 – http://blanco.biomol.uci.edu/mpstruc). Over recent years the requirement for, and time and cost associated with obtaining, large crystals has been partially alleviated through the development of beamline instrumentation allowing data collection, and structure solution, from ever-smaller crystals. Advances in several areas have led to a step change in what might be considered achievable during a synchrotron trip over the last decade. This chapter will briefly review the current status of the field, the tools available to ease data collection and processing, and give some examples of exploitation of these for membrane protein microfocus macromolecular crystallography.

Keywords

Microfocus macromolecular crystallography • Membrane proteins • *In situ* data collection • Instrumentation • X-ray microtomography

The original version of this chapter has been revised. An erratum to this chapter can be found at DOI 10.1007/978-3-319-35072-1_13

A.J. Warren • D. Axford • N.G. Paterson
R.L. Owen (✉)
Diamond Light Source, Harwell Science and innovation Campus, Didcot OX11 0DE, UK
e-mail: anna.warren@diamond.ac.uk; danny.axford@diamond.ac.uk; neil.paterson@diamond.ac.uk; robin.owen@diamond.ac.uk

I. Moraes (ed.), *The Next Generation in Membrane Protein Structure Determination*,
Advances in Experimental Medicine and Biology 922, DOI 10.1007/978-3-319-35072-1_8

8.1 Recent Developments in Instrumentation and Data Collection

Beamlines at third generation synchrotrons provide small and highly intense X-ray beams. This has been made possible due to developments in both machine and beamline instrumentation. On the machine side, a decrease in the emittance of the electron beam to typically less than 5 nm rad in the horizontal and less than 0.1 nm rad in the vertical, represents a reduction of more than an order of magnitude when compared to second generation sources. Continuous top-up, pioneered at the Swiss Light Source (Ludeke and Munoz 2002), provides a steady X-ray flux over the duration of an experimental visit and provides a constant power load on beamline components and aids thermal stability.

The provision of low divergence, high brilliance X-ray beams means a number of approaches can be used to achieve a microbeam. A high flux density source means that apertures can be used to reduce the beamsize at the sample to 10 microns without the remaining flux dropping to zero. Alternatively, focusing elements can be used to achieve a microbeam containing all, or nearly all, of the full flux provided by the source. The different approaches for achieving microbeams are summarised further in (Evans et al. 2011b; Smith et al. 2012), but rapid developments in areas such as mirror fabrication mean that even beamlines nominally not dedicated to microfocus macromolecular crystallography can provide a beamsize that would have been considered small only a few years ago.

Reductions in beamsize place greater demands and constraints on beamline instrumentation and infrastructure. As the beamsize decreases, goniometry must become more precise and accurate, the resolving power of sample visualization must improve, the beam position must be tracked, with feedback implemented if necessary, and thermal stability becomes increasingly important. Commercial and in-house developments in all of these areas have allowed microfocus beamlines to flourish over the last decade.

Advances in other areas have also benefited microcrystallography. Automated sample exchange means that large numbers of crystals can be mounted in a single visit to the synchrotron. The benefits of being able to target a large number of samples are twofold. Firstly the screening and evaluation of large numbers of crystals is not time-prohibitive and secondly the requirement for a large number of crystals to form a single dataset is no longer a barrier. The limitation on the amount of data that can be collected from a single crystal is primarily imposed by radiation damage (Holton and Frankel 2010). Based on the expected rate of decay of crystals, and the desired outcome of the experiment (*i.e.* molecular replacement or experimental phasing), Holton and Frankel provide a means of estimating the minimum crystal size required to obtain a complete dataset or alternatively, given a particular crystal size, the number of crystals required for complete data.[1]

The extremely limited lifetime of microcrystals in an intense microbeam means that care must be taken to optimise data collection. The use of a strategy program to exploit the symmetry of the reciprocal lattice minimizes the total rotation range required to obtain a complete dataset (Dauter 1999). The use of a sacrificial crystal to determine the sample lifetime at the beamline being used can be extremely valuable (Krojer and von Delft 2011), and provide a good idea of whether it will be possible to collect even the minimal rotation range from a single crystal. For extremely small crystals it is often possible to collect only a few degrees of data and the optimal approach is to collect from a large number of crystals without pre-orientation as screening images may use a large fraction of the total lifetime in the beam.

The finite lifetime of crystals in the X-ray beam can be addressed through the use of many crystals or, in the case when crystals are larger than the X-ray beam, by introduction of new material from the crystal into the X-ray beam. A simple means of achieving this is through the intentional offset of the center of rotation so that

[1]Online calculator available at http://bl831.als.lbl.gov/xtalsize.html.

new material is continually introduced as the sample is rotated (Moukhametzianov et al. 2008). Another approach available at many beamlines is the line or helical scan. In this case a start and end point are defined and the crystal traces a helical path between these during data collection.

The advent of fast readout large area detectors has also facilitated throughput. The move from collecting a succession of single shuttered images to shutterless data collection represents a paradigm shift in macromolecular crystallographic data collection, reducing timing demands on instrumentation and reducing the duration of a dataset to a few tens of seconds (at most). While a strategy of collecting data with an oscillation range per image of half the crystal mosaicity was shown to be optimal a number of years ago (Pflugrath 1999), time and computing constraints meant that oscillation angles of 1° or more were typically used for data collection both in-house and at the synchrotron. The absence of a time penalty associated with collecting more frames per degree with fast readout detectors means that fine-phi slicing is now considered routine at synchrotron beamlines equipped with a pixel array detector (PAD). In addition to fast-readout, the characteristics of PADs such as small point spread function, lack of readout noise and dead-time during exposure mean data collection should be optimized to reflect these (Mueller et al. 2011).

A central tenet of the experimental setup of a diffraction experiment is the reduction of background scatter or noise. It has been shown that signal to noise in diffraction images can be improved by matching the beam size to the crystal size, this is particularly important for small micron-sized crystals. The gains made by matching the beamsize to the crystal can be dramatic as illustrated by data collection with two beamsizes from $5 \times 5 \times 5$ μm^3 polyhedra crystals (Evans et al. 2011a). By reducing the beamsize from a mismatched 8×8 μm^2 to 4.5×5 μm^2 data could be collected to higher resolution while at the same time reducing the absorbed dose.

Holton and Frankel (2010) describe how the presence of background scatter is an underlying reason for the large gap between the theoretical minimum crystal size (or minimum number of crystals required for a dataset) and the size of microcrystals from which structures have been determined. The approach described above of matching the beamsize to the crystal is one of the most obvious means of reducing background scatter but both the sample environment and sample preparation should also be optimized.

The benefits of minimizing the volume of solvent surrounding the crystal are twofold. Firstly, reducing the volume of non-diffracting material the X-ray beam passes through reduces diffuse scatter and secondly crystal alignment is simplified. Refractive effects can mean the apparent position of the crystal changes as the sample is rotated. For microcrystals this offset can be more than the size of the crystal or beamsize, making approaches such as diffraction based centering necessary. These approaches are described in more detail below but the need for them, or the time required to execute them, can be minimized by reducing the volume of solvent surrounding the crystal.

8.2 *In situ* Data Collection

The collection of data *in situ*, i.e. directly from the crystallisation platform, can provide a valuable tool in membrane protein crystallography. Given the increased numbers of reagents involved in the formation of crystals from membrane proteins and the typically extended optimisation process, the opportunity to interrogate with a synchrotron X-ray beam without recourse to sample manipulation and cooling is very appealing. This data collection method can thus dramatically speed up the feedback loop between initial crystal hits and ultimate structure determination. *In situ* data collection from SBS (Society for Biomolecular Sciences) standard footprint plates has been developed at a number of Macromolecular Crystallography (MX) beamlines at different synchrotron sources including the Swiss Light Source (Bingel-Erlenmeyer et al. 2011), European Synchrotron Radiation Facility (le Maire et al. 2011), Diamond Light Source (Axford et al. 2012) and Spring-8 (Kunio et al. 2013).

It has become clear that the technique is best used in conjunction with a microfocused X ray beam, since early crystal hits are typically small or poorly formed (Axford et al. 2012). Additionally, for the best information to be obtained, the crystallization platform needs to be designed with *in situ* data collection in mind in order to avoid excessive background scatter and to aid sample location. Increased uptake of the method has encouraged commercial suppliers to develop plates specifically designed for *in situ* crystallography, including the Greiner CrystalQuickX and the Mitegen *in situ*-1. These plates are designed for vapour diffusion experiments. A primary aspect of these designs is the use of thin, flat, low-X-ray-scatter materials to maximise the chance of measuring weak and short-lived diffraction at room temperature and to minimize the chance of optical aberrations that can lead to sample-beam misalignment. Crystallisation platforms designed around counter diffusion protocols can also be X ray compatible, with examples including CrystalSlide™ (Ng et al. 2008) and CrystalHarp™, the latter using quartz capillaries.

Although the vast majority of crystallisation plates are injection moulded from varieties of plastic, the most commonly used material for LCP crystallisation platforms is glass, in a simple sandwich of two sheets. Glass is popular due to its stiffness, smoothness and low cost. However, 1 mm of glass transmits only 16 % of 12.4 keV X-rays, rendering it completely unsuitable for *in situ* experiments. The use of plastic films is becoming more common for the cover sheet since it does not shatter when cut for crystal fishing. If UV permissive plastic is also used as the support material UV-fluorescence detection of crystals is possible in addition to *in situ* data collection. Hybrid designs (Swissci AG) are also available in the form of multi-layer plates. In this case a lower glass support, that provides stiffness during handling, can be removed prior to *in-situ* data collection on the remaining thin sandwich of plastic containing the sample. Complementing the growth of dedicated facilities for *in situ* data collection from standard plates the development of smaller crystallisation platforms allows *in situ* data collection on goniometry designed for conventional loops and pins. Examples of this include the X-CHIP (Kisselman et al. 2011) and modular plate designs, such as the CrystalSlide™ and the CrystalHarp™, where sections can be removed from a SBS format plate before mounting at the beamline, see Fig. 8.1.

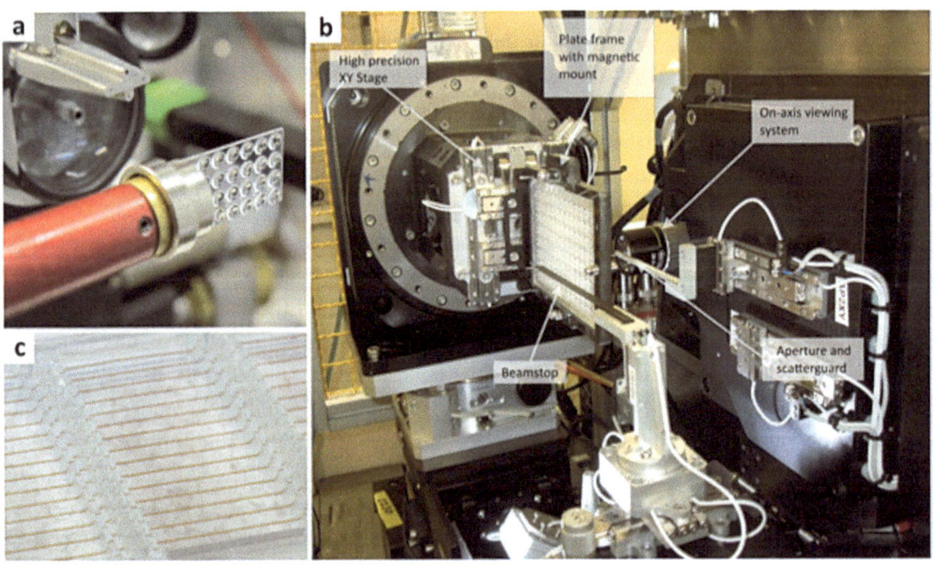

Fig. 8.1 Example of platforms for *in situ* data collection. (**a**) X-CHIP (X-ray Crystallization High-throughput Integrated Platform) mounted on the GM/CA beamline at the Advanced Photon Source USA. (**b**). A Goniometer designed for SBS standard plates, with associated end-station components labeled, on beamline I24 at Diamond Light Source UK. (**c**) A close up of a CrystalHarp™ plate displaying the array of detachable capillaries in which gradient diffusion crystallization can occur

In addition to sample screening, *in situ* data collection has the potential to be used for structure determination. Most recently, this has been demonstrated with the TecA integral membrane protein for which a structure was available from a conventional single crystal data set at 100 K. *In situ* crystallography enabled the determination of an equivalent room temperature structure via the merging of small wedges of data collected from 57 separate crystals but without the need for any sample manipulation (Axford et al. 2015).

8.3 Crystal Alignment and Identification

The accurate alignment of crystals in the X-ray beam is a critical step prior to data collection. This can be particularly challenging on microfocus beamlines where the crystal and beamsize are much less than the size of a standard loop and these are both comparable to the visual resolution limit of the viewing system. Lensing effects and refraction at the surface and poor optical properties of mother liquor surrounding crystals can also combine to make crystal identification and alignment difficult. Alignment errors of only a few microns are sufficient to ensure crystals move out of the X-ray beam while being rotated and thus a number of approaches have been developed to ease crystal alignment. Image recognition can be used to speed up alignment, or automatically identify crystals (Karain et al. 2002; Lavault et al. 2006; Pothineni et al. 2006); however these rely on edge detection or contrast to identify the position of crystals (Bern et al. 2004), which can be hindered for example by the presence of precipitate (Cumbaa et al. 2003). It is particularly difficult to apply these routines to microcrystals however, when crystals may only be few pixels in size on on-line viewing systems, lipidic cubic phase (LCP) provides an additional challenge. LCP provides a membrane-like environment for proteins (Landau and Rosenbusch 1996), and has proved to be a crucial tool in the crystallisation of membrane proteins. A drawback of LCP is that upon cryo-cooling to 100 K it becomes opaque making the identification and centering of crystals in loops difficult (Cherezov et al. 2009). The popularity of LCP and limitations of visual alignment have led to the development of a number of tools for crystal identification and alignment at microfocus beamlines. Current methods for the alignment of microcrystals, and some future directions are outlined below.

8.3.1 Current Methods for Crystal Alignment

8.3.1.1 Grid Scan

One of the most widely used methods of centring small or optically invisible samples is to use the diffraction grid scan (Cherezov et al. 2009; Song et al. 2007; Aishima et al. 2010; Bowler et al. 2010; Hilgart et al. 2011). This can be used to align small micro-crystals, crystals hidden in opaque LCP or to identify the best diffracting regions of larger crystals. The principle of this technique is to use a low dose X-ray beam to raster over the loop area. Diffraction images tagged with a position are then scored to indicate the best diffracting region(s). Diffraction images can be processed on the fly, and scored by, for example, the number of Bragg candidates present on the image using the spot finding software DISTL (Zhang et al. 2006). The scoring criteria look only at low resolution spots to avoid any ice rings present, and to help particularly with weakly diffracting samples. There are also options available to use a fluorescence raster scan, exploiting the presence of fluorescing elements present in the crystals (i.e. selenomethionine), with orders of magnitude lower intensity when compared to the diffraction grid scan. There are optimal strategies for exploiting the grid scan whilst minimizing the number of images collected and time required. The loop can first be scanned face on to determine the location of the crystal, with the crystal being centered to the best location. The loop can then be rotated through 90° with another scan being collected to determine the crystal location edge on. After being centered on both these positions the crystal is then centered and ready for data collection. It is vital that the lowest dose possible is used for these scans to ensure as little radiation damage as possible prior to the data collection. This approach has proven very successful for centring membrane protein crystals on a number of different beamlines. Other strategies for grid scan alignment of crystals are explained further in the examples below.

8.3.1.2 UV Imaging

Ultraviolet (UV) fluorescence has routinely been used as a non-invasive method for distinguishing protein crystals in well plates. In the UV range from 260 to 320 nm, tryptophan, tyrosine and phenylalanine absorb the light, and native fluorescence is detected from approximately 300–450 nm (Lakowicz 2013). It is possible to distinguish between salt and protein crystals, with the protein crystals fluorescing brightly to give high contrast against the background (Judge et al. 2005; Calero et al. 2014). Advances in the technique have allowed the technique to be exploited at beamlines at the ESRF (France) (Vernede et al. 2006) and the Photon Factory (Japan) (Chavas et al. 2011) In the latter case they used pulses of UV light to ensure minimal damage to the crystals.

8.3.2 Future Directions

8.3.2.1 X-ray Imaging

A technique that is being developed that works particularly well for membrane proteins is that of X-ray imaging. It is based on a well-studied technique where the absorption contrast of a material can be investigated using X-rays, but which isn't currently routinely being used for macromolecular crystallography. Brockhauser *et al.* showed that it was possible to collect X-ray microtomography data on soluble protein crystals on both high energy and MX beamlines at the ESRF (France). From this data they were able to reconstruct a three dimensional model of their crystal, surrounding liquor and loop to help with the addition of analytical absorption corrections (Brockhauser et al. 2008).

Warren et al. progressed this work by carrying out similar experiments on MX beamlines at the Diamond Light Source (UK) with membrane proteins in LCP to show how it was possible to image these loops and clearly distinguish where the crystals were mounted in the loop (Fig. 8.2). The paper described how this method was a lower dose method to the currently used grid scan and gives more information about the crystal size and shape to help determine how the beamsize should be altered for the subsequent data collection (Warren et al. 2013). Information on the three dimensional shape of crystals could also be used in the data scaling step, and this will prove to be particularly valuable for long wavelength data collection.

8.3.2.2 Non-linear Optical Imaging

Second-order nonlinear optical imaging of chiral crystals (SONICC) has also recently been explored for imaging of membrane proteins in cubic

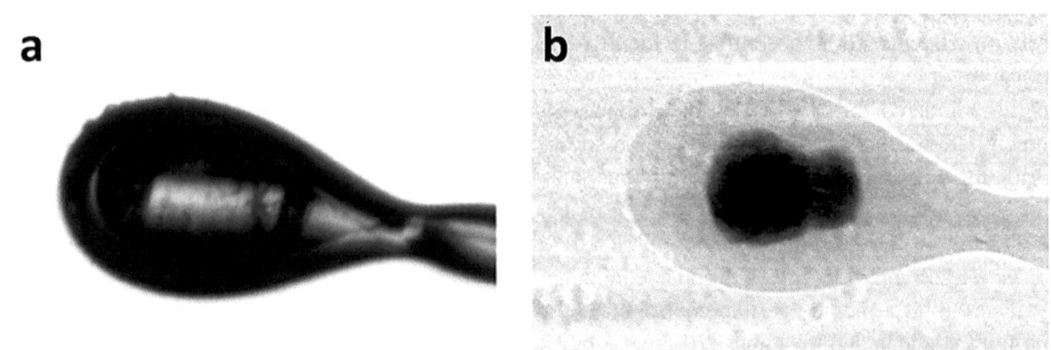

Fig. 8.2 View of membrane protein crystals, the human A_{2A} adenosine G-protein coupled receptor in lipidic cubic phase mounted on a nylon loop, (**a**) using a visible microscope and (**b**) the same orientation of sample viewed as a radiograph. It is unclear from image (**a**) where the crystals are located, while after X-ray imaging (**b**) the presence and location of two crystals held within the loop becomes evident (Warren et al. 2013)

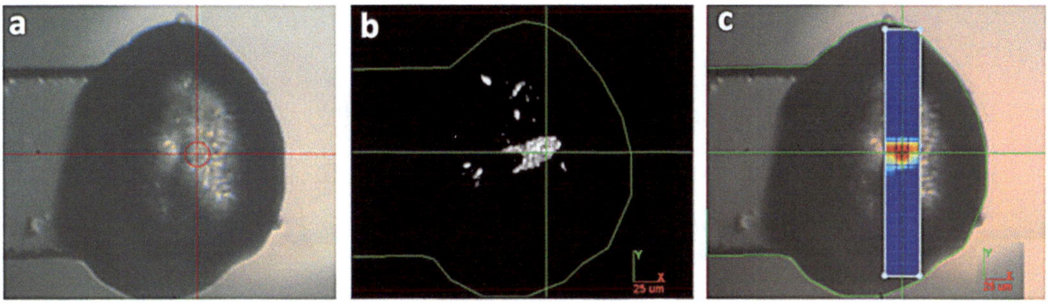

Fig. 8.3 View of GPCR membrane protein crystal in lipidic cubic phase. (**a**) Using a visible microscope. (**b**) The same orientation of sample viewed using SONICC and (**c**) the same sample located using grid scanning (Images provided courtesy of Garth Simpson, Purdue University)

phase. It works on the principles of second harmonic generation (SHG) in which the frequency of the light used is doubled. Constructive interference is observed from ordered chiral molecules, such as protein crystals, whereas crystals such as salt, or amorphous LCP will not produce a signal from the SHG (Kissick et al. 2010; DeWalt et al. 2013; Closser et al. 2013). SONICC has recently been installed on a beamline at the APS (USA) to image the crystals when mounted within loops. Here the method was compared to bright field images, and the current grid scan method to confirm it was possible to locate numerous membrane protein crystals within LCP (Madden et al. 2013), a visual comparison of the methods is shown in Fig. 8.3. The effects of radiation damage were also considered due to the laser exposure on the crystals, however no significant effect was observed (Kissick et al. 2013). For more details regarding this method please see Chap. 7 of this book.

8.4 Examples of Exploitation of Microbeams for Membrane Protein Structure Determination

The sections above briefly summarise some recent advances in synchrotron instrumentation and experimental approaches for making the most of microbeams. Below we describe some recent membrane structure determinations exploiting different aspects of these developments and highlighting best practice. Due to our affiliation these focus on work carried out at I24, Diamond Light Source as we can provide additional information on the experiments not necessarily found in the materials and methods sections. The approaches used and described are however broadly applicable to membrane protein structure determination at microfocus beamlines worldwide.

8.4.1 The Class B GPCR Corticotropin – Releasing Factor Receptor 1

The structure of the class B human GPCR corticotropin – releasing factor receptor 1 (CRF_1R) was determined at I24, Diamond Light Source. The structure and function of this GPCR is described by (Hollenstein et al. 2013); we here highlight some of the experimental approaches exploited to allow data to be collected to a resolution that allowed identification of an unexpected binding pocket.

Crystals of CRF_1R were grown in LCP with a maximum dimension of approximately 15 μm. Upon cryocooling the LCP became opaque and in order to align crystals to the X-ray beam the diffraction grid scan was used (Fig. 8.4). A single grid scan was typically sufficient to align larger crystals (approx. 10 microns in size or larger), but for small crystals (5 μm) an iterative approach was required. Following the initial gridscan, the beamsize was reduced to $5 \times 5\ \mu m^2$ to better

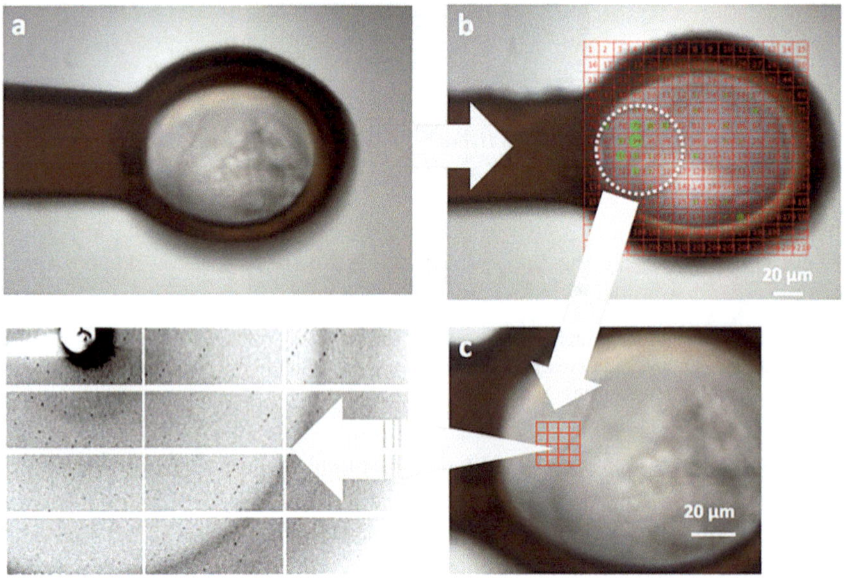

Fig. 8.4 Strategy for locating crystals in opaque. After the crystal is mounted (**a**), a grid with a pitch of 10 µm is drawn that covers the entire loop (**b**). Following on-the-fly scoring using DISTL promising sub-areas are identified. In the case of small crystals, a second grid can be drawn (**c**) and, with a concomitant decrease in the beamsize, the position of the crystal is more accurately determined. In the example shown here the pitch of the second grid is 5 µm. The resolution at the edge of the diffraction image shown is 3 Å

match the crystal size and a second grid scan carried out around any hits found in the first gridscan. For all gridscans the X-ray beam was attenuated to minimise radiation damage to crystals. Once centered in the X-ray beam CRF$_1$R crystals diffracted to 3 Å, but only with long exposure times (7.5 s per degree of oscillation). This corresponds to an absorbed dose of 6 MGy per second. Radiation damage was therefore a limiting factor and it was only possible to collect a small wedge (2–3°) of data from each crystal. A complete dataset was built up using the microdiffraction assembly method (Hanson et al. 2012). In this approach data from each crystal are split into 1° bins. These bins are then iteratively scaled and merged to a reference low resolution (4.3 Å) dataset using XSCALE. Using this approach a complete dataset to 2.97 Å was built up containing data from 30 crystals. This compares with a 3.15 Å dataset formed using a conventional assembly method.

8.4.2 Structures of ABCB10, a Human ATP-Binding Cassette Transporter in Apo- and Nucleotide-Bound States

The structure of ABCB10, one of three mitochondrial ATP-binding cassette (ABC) transporters has been determined using microfocus crystallography. These are located in the inner membrane of mitochondria and were found to adopt an unexpected conformation when complexed with nucleotide analogues, compared to previously reported structures. Further details about the differences in conformation, as well as a more detailed description of the structures are described by (Shintre et al. 2013). All structural data were collected on I24 at Diamond Light Source. Initial structure solution was achieved using plate-like crystals which were phased by isomorphous replacement with a single mercury derivative (SIRAS). These crystals were highly

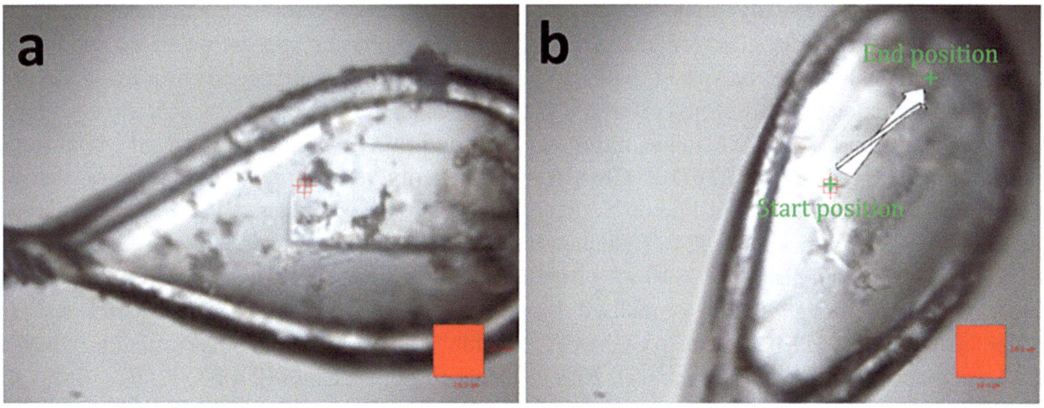

Fig. 8.5 Plate-like crystal of ABCB10 before (**a**) and after (**b**) the loop was bent for data collection. (**b**) Shows the centered start and end positions for the line scan data collection

anisotropic. Data isomorphism and diffraction quality was significantly improved by flash cooling crystals at 6 °C instead of their growth temperature of 20 °C, which led to reproducible diffraction to around 3.4 Å in the best direction and 4.4 Å in the worst. Over 500 crystals were screened to optimize the crystallisation and freezing conditions, followed by a further 300 crystals to find a suitable heavy atom derivative. Due to the anisotropy and disorder of the nucleotide binding domain, the model building could not be completed. A different rod-shaped crystal form was obtained with protein purified in the presence of lipid which diffracted isotropically to 2.9 Å. It was possible to phase these crystals using molecular replacement, using the plate form structure as a model and complete the structure.

For the plate-like crystals, all data collections utilised the helical line scan on I24. During this form of data collection the crystal is translated whilst it is rotating so that the entire length of the crystal is exploited and new material is continually introduced into the beam. This technique minimises crystal degradation and it may even be possible to outrun radiation damage. For sample alignment both ends of the crystal need to be well centred defining a path along the crystal for the data collection.

Due to a relatively long unit cell *c* axis, which coincided with the thinnest dimension of the plate, the mounting loop needed to be bent manually with a pipette tip to align the *c* axis closer to the rotation axis to minimise overlaps (Fig. 8.5). The start and end points of the helical scan were determined by two perpendicular line scans at each end of the crystal. Visual alignment was not possible due to the loop reorientation and limited thickness of the plates. In the majority of cases, the beam was shaped to a vertical letter box (30–50 μm high × 10 μm wide) to illuminate a larger portion of the crystal.

For the rod-shaped crystal form, broad grid scanning was used to identify the best diffracting regions of each crystal. In every case, the best diffraction was obtained from either end of the rods (dimensions $20 \times 20 \times 150$ μm^3) with much weaker diffraction from the central portion. Data to 2.85 Å were collected using a 10×10 μm^2 beam and rotating about a single portion of crystal.

8.4.3 Structure of the Integral Membrane Protein Diacylglycerol Kinase (DgkA)

DgkA is functionally unique enzyme catalysing the conversion of diacylglycerol and ATP to phosphatidic acid and ADP and enabling membrane phosolipid turnover (Li et al. 2013). After initial trials crystals were produced in lipidic cubic phase at 4 °C for wild type, a thermally stable mutant and a seleno-methionine (SeMet) thermally stable mutant. Maximum crystal dimensions were $50 \times 50 \times 50$ μm^3 bipyramids for the wildtype, $50 \times 75 \times 100$ μm^3

bypyramids for the thermally stable mutant and $10 \times 50 \times 150$ μm^3 rectangles for the SeMet derivative. Crystals were harvested with 30–100 μm micromounts (Mitegen) in a 4 °C cold room and flash-cooled in liquid nitrogen. Data were collected at the 23-ID-B GM/CA-CAT beamline at the Advanced Photon Source, Argonne, and also at beamline I24 at Diamond Light Source. On both beamlines a 10×10 μm^2 beam was used, though this was realised by different means at each beamline. At GM/CA-CAT a collimator was used (Fischetti et al. 2009), while at I24 a two-stage demagnification was used. In both cases, raster scanning with a tenfold attenuated beam was used to locate and identify the most ordered regions of a crystal. The highest resolution dataset was obtained by combining a complete low-resolution dataset (3 Å) from a single crystal with 18 wedges of 10° of high-resolution and 10° data from multiple crystals.

Xia2 was used to integrate diffraction data and the best wedges were identified by R-value and isomorphic cell parameters. Anomalous data were obtained in a similar way by merging 19 separate data collections from SeMet crystals to build up a 200 fold redundant dataset to 3.0 Å. Seventeen putative Se sites were identified using SHELX C/D/E (Sheldrick 2008) and phases extended using PHENIX (Adams et al. 2002) to 2.05 Å, representing the highest resolution native data.

A previously published solution phase NMR structure failed as a model for molecular replacement and was seen to be significantly different to the final crystal structure. Subsequent solid-state NMR spectroscopy studies within lipid bilayers have supported this crystallographic structure, highlighting the importance of working in effective membrane mimetic environments such as LCP (Murray et al. 2014).

8.4.4 Structural Basis for Outer Membrane Lipopolysaccharide Insertion

Beta barrel membrane proteins often prove difficult targets for molecular replacement due to the wide variation in strand number and propensity for motion in the barrel. This necessitates an experimental phasing approach with the initial map quality of key importance in ensuring successful main chain tracing.

The LptD/E complex performs the final stage in insertion of the mature lipopolysaccharide (LPS) into the outer leaflet of the outer membrane in Gram-negative bacteria (Dong et al. 2014). LptE sits inside the 26 β-strand LptD barrel acting as both a plug for the barrel and as guide for correct LPS orientation. SeMet labelled crystals ($110 \times 110 \times 170$ μm^3) of the LptD/E complex (87.2 kDa, 20 selenium atoms) were obtained in space group $I2$ with unit cell $a = 173.4$ Å, $b = 76.1$ Å, $c = 213.6$ Å and $\beta = 111.5$ °. The solvent content was approximately 67 % with two complexes per asu. The X-ray diffraction from these crystals was anisotropic with $CC_{1/2} > 0.5$ cutoffs of h,l = 2.80 Å and k = 3.9 Å.

The strategy used for structure solution was to collect multi-wavelength anomalous diffraction (MAD) data to provide a high quality initial map but the low symmetry of the $I2$ space group and weak diffraction meant that high anomalous multiplicity would be difficult to achieve without significant radiation damage on a single crystal. The approach chosen was to use the microfocus capabilities of beamline I24 at Diamond Light Source to divide the crystal into four regions with a 20×20 μm^2 beam and collect 360° of fine-sliced data (0.1° oscillation) at one of the four MAD wavelengths (peak, inflection, high remote and low remote) in each region (Fig. 8.6a). To achieve the best possible experimental map, this approach was repeated on three isomorphous crystals resulting in the collection of around 45,000 frames in total. All data were processed using XDS (Kabsch 2010) to 2.8 Å resolution and the individual wavelengths combined to give 16-fold anomalous multiplicity. All 40 selenium sites were identified using SHELXC/D through autoSHARP (Vonrhein and Schulz 2007). Subsequent density modification with DM gave a starting map with CC = 0.68 (Fig. 8.6b) into which all of LptE and most of the pore strand main chain from LptD could be automatically built with Buccaneer (Cowton 2006).

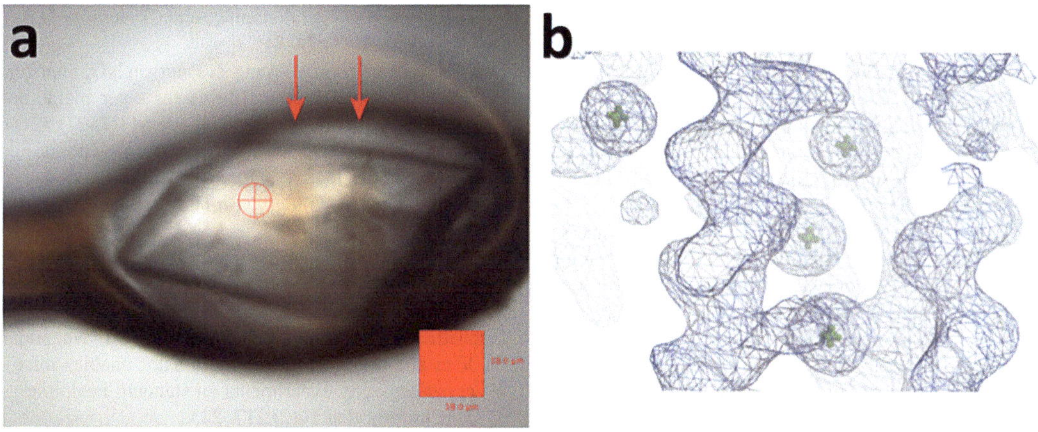

Fig. 8.6 (**a**) LptD/E crystal after two data collection slices. The exposed areas are indicated with *red arrows* and the 20×20 μm^2 beam by the *red circle* and *crosshairs*. (**b**) Section of the LptD/E density modified map contoured at 0.09 e $Å^{-3}$ (1.5 rmsd) showing part of the electron density for the LptE helix and strands. Selenium sites are indicated by a *green cross*

8.5 Concluding Remarks

The challenges of membrane protein crystallography often continue beyond the well-documented difficulties of crystallisation, with weak and often anisotropic X-ray diffraction and low overall resolution hindering structure solution. Difficulties with crystallisation can be partly addressed through the use of *in situ* data collection which removes confounding factors such as mechanical stress during loop mounting and damage during the cryocooling process when assessing the promise of initial crystal hits. A microfocussed beam allows this characterization to be done at an early stage of the crystal optimisation process.

The challenges associated with collecting high quality diffraction data from small or badly ordered crystals can be partially addressed through the use of a microbeam. Microbeams themselves give rise to challenges not least because of beam induced damage and the difficulties associated with sample alignment. These challenges can be addressed in part through the use of many crystals to form a single complete dataset and exploitation of on-line tools such as the grid scan or tomography for alignment.

The use of beamsizes of less than 10 microns is now the norm for many membrane protein crystallographers. Further developments in synchrotron sources and beamline instrumentation will see both a more widespread availability of microfocus beams worldwide and also the provision of even smaller, more intense microbeams. (Many) multi-crystal strategies for structure solution will become standard practice and the ever-decreasing lifetime of crystals in the beam will mean new modes of sample delivery such as liquid jets or fixed aperture arrays are ported from sources such as the free electron laser for use at synchrotrons. Taken as a whole, advances in instrumentation and methods will help ease the pathway to structure solution for ever more challenging targets in membrane protein crystallography at synchrotron sources.

Acknowledgments The authors would like to thank all users who helped in the development of some of the techniques and tools described in this review by providing useful feedback and suggestions. We would particularly like to thank the four groups whose work exploiting selection of these tools is described in more detail in this chapter.

References

Adams PD, Grosse-Kunstleve RW, Hung L-W, Ioerger TR, McCoy AJ, Moriarty NW et al (2002) PHENIX: building new software for automated crystallographic structure determination. Acta Crystallogr D Biol Crystallogr 58(11):1948–1954

Aishima J, Owen RL, Axford D, Shepherd E, Winter G, Levik K et al (2010) High-speed crystal detection and characterization using a fast-readout detector. Acta Crystallogr D Biol Crystallogr 66(9):1032–1035

Axford D, Owen RL, Aishima J, Foadi J, Morgan AW, Robinson JI et al (2012) *In situ* macromolecular crystallography using microbeams. Acta Crystallogr D Biol Crystallogr 68(5):592–600

Axford D, Foadi F, Hu N-J, Choudhury HG, Iwata S et al (2015) Structure determination of an integral membrane protein at room temperature from crystals in situ. Acta Crystallogr D Biol Crystallogr 71(6):1228–1237

Bern M, Goldberg D, Stevens RC, Kuhn P (2004) Automatic classification of protein crystallization images using a curve-tracking algorithm. J Appl Crystallogr 37(2):279–287

Bingel-Erlenmeyer R, Olieric V, Grimshaw JPA, Gabadinho J, Wang X et al (2011) SLS crystallization platform at beamline X06DA—a fully automated pipeline enabling *in situ* X-ray diffraction screening. Crys Growth Des 11(4):916–923

Bowler MW, Guijarro M, Petitdemange S, Baker I, Svensson O, Burghammer M et al (2010) Diffraction cartography: applying microbeams to macromolecular crystallography sample evaluation and data collection. Acta Crystallogr D Biol Crystallogr 66(8):855–864

Brockhauser S, Di Michiel M, McGeehan JE, McCarthy AA, Ravelli RBG (2008) X-ray tomographic reconstruction of macromolecular samples. J Appl Crystallogr 41(6):1057–1066

Calero G, Cohen AE, Luft JR, Newman J, Snell EH (2014) Identifying, studying and making good use of macromolecular crystals. Acta Crystallogr F Struct Biol Commun 70(8):993–1008

Chavas LMG, Yamada Y, Hiraki M, Igarashi N, Matsugaki N, Wakatsuki S (2011) UV LED lighting for automated crystal centring. J Synchrotron Rad 18(1):11–15

Cherezov V, Hanson MA, Griffith MT, Hilgart MC, Sanishvili R et al (2009) Rastering strategy for screening and centring of microcrystal samples of human membrane proteins with a sub-10 μm size X-ray synchrotron beam. J R Soci Interface 6(5):S587–S597

Closser RG, Gualtieri EJ, Newman JA, Simpson GJ (2013) Characterization of salt interferences in second-harmonic generation detection of protein crystals. J Appl Crystallogr 46(6):1903–1906

Cowton KD (2006) The Buccaneer software for automated model building. 1. Tracing protein chains. Acta Crystallogr D Biol Crystallogr 62(9):1002–1011

Cumbaa CA, Lauricella A, Fehrman N, Veatch C, Collins R et al (2003) Automatic classification of sub-microlitre protein-crystallization trials in 1536-well plates. Acta Crystallogr D Biol Crystallogr 59(9):1619–1627

Dauter Z (1999) Data-collection strategies. Acta Crystallogr D Biol Crystallogr 55:1703–1717

DeWalt EL, Begue VJ, Ronau JA, Sullivan SZ, Das C, Simpson GJ (2013) Polarization-resolved second-harmonic generation microscopy as a method to visualize protein-crystal domains. Acta Crystallogr D Biol Crystallogr 69(1):74–81

Dong H, Xiang Q, Gu Y, Wang Z, Paterson NG, Stansfeld PJ et al (2014) Structural basis for outer membrane lipopolysaccharide insertion. Nature 511(7507):52–56

Evans G, Axford D, Owen RL (2011a) The design of macromolecular crystallography diffraction experiments. Acta Crystallogr D Biol Crystallogr 67(4):261–270

Evans G, Axford D, Waterman D, Owen R (2011b) Macromolecular microcrystallography. Crystallogr Rev 17(2):105–142

Fischetti RF, Xu S, Yoder DW, Becker M, Nagarajan V et al (2009) Mini-beam collimator enables microcrystallography experiments on standard beamlines. J Synchrotron Rad 16(2):217–225

Hanson MA, Roth CB, Jo E, Griffith MT, Scott FL, Reinhart R et al (2012) Crystal structure of a lipid G protein-coupled receptor. Science 335(6070): 851–855

Hilgart MC, Sanishvili R, Ogata CM, Becker M, Venugopalan N, Stepanov S et al (2011) Automated sample-scanning methods for radiation damage mitigation and diffraction-based centering of macromolecular crystals. J Synchrotron Rad 18(5):717–722

Hollenstein K, Kean J, Bortolato A, Cheng RK, Dore AS et al (2013) Structure of class B GPCR corticotropin-releasing factor receptor 1. Nature 499:438–443

Holton JM, Frankel KA (2010) The minimum crystal size needed for a complete diffraction data set. Acta Crystallogr D Biol Crystallogr 66(4):393–408

Judge RA, Swift K, Gonzalez C (2005) An ultraviolet fluorescence-based method for identifying and distinguishing protein crystals. Acta Crystallogr D Biol Crystallogr 61(1):60–66

Kabsch W (2010) XDS. Acta Crystallogr D Biol Crystallogr D66(2):125–132

Karain WI, Bourenkov GP, Blume H, Bartunik HD (2002) Automated mounting, centering and screening of crystals for high-throughput protein crystallography. Acta Crystallogr D Biol Crystallogr 58(1):1519–1522

Kisselman G, Qiu W, Romanov V, Thompson CM, Lam R et al (2011) X-CHIP: an integrated platform for high-throughput protein crystallization and on-the-chip X-ray diffraction data collection. Acta Crystallogr D Biol Crystallogr 67(6):533–539

Kissick DJ, Gualtieri EJ, Simpson GJ, Cherezov V (2010) Nonlinear optical imaging of integral membrane protein crystals in lipidic mesophases. Anal Chem 82(2):491–497

Kissick DJ, Dettmar CM, Becker M, Mulichak AM, Cherezov V et al (2013) Towards protein-crystal centering using second-harmonic generation (SHG) microscopy. Acta Crystallogr D Biol Crystallogr 69(5):843–851

Krojer T, von Delft F (2011) Assessment of radiation damage behaviour in a large collection of empirically optimized datasets highlights the importance of unmeasured complicating effects. J Synchrotron Rad 18(3):387–397

Kunio H, Yoshiaki K, Go U, Koichi H, Hironori M, Kazuya H et al (2013) Achievement of protein micro-crystallography at SPring-8 beamline BL32XU. J Phys Conf Ser 425(1):12002

Lakowicz JR (ed) (2013) Principles of fluorescence spectroscopy. Springer Science & Business Media, New York

Landau EM, Rosenbusch JP (1996) Lipidic cubic phases: a novel concept for the crystallization of membrane proteins. Proc Natl Acad Sci U S A 93(25):14532–14535

Lavault B, Ravelli RBG, Cipriani F (2006) C3D: a program for the automated centring of cryocooled crystals. Acta Crystallogr D Biol Crystallogr 62(11):1348–1357

le Maire A, Gelin M, Pochet S, Hoh F, Pirocchi M, Guichou J-F et al (2011) In-plate protein crystallization, *in situ* ligand soaking and X-ray diffraction. Acta Crystallogr D Biol Crystallogr 67:747–755

Li D, Lyons JA, Pye VE, Vogeley L, Aragao D, Kenyon CP, Shah STA et al (2013) Crystal structure of the integral membrane diacylglycerol kinase. Nature 497:521–524

Ludeke A, Munoz M (2002) Top-up experience at the Swiss Light Source. Proc EPAC 17(2):721–723

Madden JT, Toth SJ, Dettmar CM, Newman JA, Oglesbee RA, Hedderich HG, Everly RM, Becker M, Ronau JA, Buchanan SK, Cherezov V, Morrow ME, Xu S, Ferguson D, Makarov O, Das C, Fischetti R, Simpson GJ (2013) Integrated nonlinear optical imaging microscope for on-axis crystal detection and centering at a synchrotron beamline. J Synchrotron Radiat 20(4):531–540

Moukhametzianov R, Burghammer M, Edwards PC, Petitdemange S, Popov D, Fransen M, McMullan G, Schertler GFX, Riekel C (2008) Protein crystallography with a micrometre-sized synchrotron-radiation beam. Acta Crystallogr D Biol Crystallogr 64(2):158–166

Mueller M, Wang M, Schulze-Briese C (2011) Optimal fine phi-slicing for single-photon-counting pixel detectors. Acta Crystallogr D Biol Crystallogr 68(1): 42–56

Murray DT, Li C, Gao FP, Qin H, Cross TA (2014) Membrane protein structural validation by oriented sample solid-state NMR: diacylglycerol kinase. Biophys J 106(8):1559–1569

Ng JD, Clark PJ, Stevens RC, Kuhn P (2008) In situ X-ray analysis of protein crystals in low-birefringent and X-ray transmissive plastic microchannels. Acta Crystallogr D Biol Crystallogr 64(2):189–197

Pflugrath J (1999) The finer things in X-ray diffraction data collection. Acta Crystallogr D Biol Crystallogr 55:1718–1725

Pothineni SB, Strutz T, Lamzin VS (2006) Automated detection and centring of cryocooled protein crystals. Acta Crystallogr D Biol Crystallogr 62(11):1358–1368

Sheldrick GM (2008) A short history of SHELX. Acta Crystallogr D Biol Crystallogr 64(1):112–122

Shintre CA, Pike ACW, Li Q, Kim J-I, Barr AJ, Goubin S et al (2013) Structures of ABCB10, a human ATP-binding cassette transporter in apo- and nucleotide-bound states. Proc Natl Acad Sci U S A 110(24):9710–9715

Smith JL, Fischetti RF, Yamamoto M (2012) Micro-crystallography comes of age. Curr Opin Struct Biol 22(5):602–612

Song J, Mathew D, Jacob SA, Corbett L, Moorhead P, Soltis SM (2007) Diffraction-based automated crystal centering. J Synchrotron Rad 14(2):191–195

Vernede X, Lavault B, Ohana J, Nurizzo D, Joly J et al (2006) UV laser-excited fluorescence as a tool for the visualization of protein crystals mounted in loops. Acta Crystallogr D Biol Crystallogr 62(3):253–261

Vonrhein C, Schulz GE (2007) Automated structure solution with autoSHARP. Methods Mol Biol 364:215–230

Warren AJ, Armour W, Axford D, Basham M, Connolley T et al (2013) Visualization of membrane protein crystals in lipid cubic phase using X-ray imaging. Acta Crystallogr D Biol Crystallogr 69(7):1252–1259

Zhang Z, Sauter NK, van den Bedem H, Snell G, Deacon AM (2006) Automated diffraction image analysis and spot searching for high-throughput crystal screening. J Appl Crystallogr 39(1):112–119

Applications of the *BLEND* Software to Crystallographic Data from Membrane Proteins

9

Pierre Aller, Tian Geng, Gwyndaf Evans, and James Foadi

Abstract

X-ray diffraction from crystals of membrane proteins very often yields incomplete datasets due to, among other things, severe radiation damage. Multiple crystals are thus required to form complete datasets, provided the crystals themselves are isomorphous. Selection and combination of data from multiple crystals is a difficult and tedious task that can be facilitated by purpose-built software. *BLEND*, in the CCP4 suite of programs for macromolecular crystallography (MX), has been created exactly for this reason. In this chapter the program is described and its workings illustrated by means of data from two membrane proteins.

Keywords

Blend • Multiple crystals • Membrane proteins • Data scaling • Data merging • Isomorphism • Cluster analysis • Radiation damage

9.1 Introduction

The correct interpretation of X-ray diffraction patterns for macromolecules is nowadays routine. A few computer programs have reached the status of mature and robust applications capable of suggesting the most likely symmetry groups and cell parameters for the protein under study as well as carry out data integration (Otwinowski and Minor 1997; Leslie and Powell 2007; Kabsch 2010; DIALS 2015). The subsequent scaling of integrated intensities and the extraction of structure factors, to be later used in the solution process, does not represent a problem if the diffraction data collected are complete and if unique intensities have sufficient multiplicity (French and Wilson 1978; Kabsch 2010; Evans and Murshudov 2013). Straightforward application of these packages, though, is typically limited to soluble proteins for which good-size crystals can be easily obtained and for which radiation damage does not restrict the collection rotation

P. Aller • G. Evans • J. Foadi (✉)
Diamond Light Source Ltd, Harwell Science and Innovation Campus, OX11 0DE Didcot, UK
e-mail: pierre.aller@diamond.ac.uk; gwyndaf.evans@diamond.ac.uk; james.foadi@diamond.ac.uk

T. Geng
Membrane Protein Laboratory, Diamond Light Source Ltd, Harwell Science and Innovation Campus, OX11 0DE Didcot, UK
e-mail: t.geng@imperial.ac.uk

I. Moraes (ed.), *The Next Generation in Membrane Protein Structure Determination*, Advances in Experimental Medicine and Biology 922, DOI 10.1007/978-3-319-35072-1_9

range. However, X-ray data from other classes of macromolecules, like membrane proteins or macromolecular complexes, are less easy to integrate and scale. Complicating factors such high-resolution diffraction spot weakness, due to the increased solvent content and lattice disorder, severe anisotropy, due to limitations in crystal packing along all spatial directions, and incomplete reciprocal space covering, due to crystal decay during X-ray exposure are the most common. The traditional way to deal with such difficulties is to spend a considerable amount of time optimising crystallisation conditions aiming at the formation of relatively large and homogeneous crystals, capable of scattering at high resolution and to withstand radiation damage for longer collection times. Progress at third generation synchrotrons has recently made it feasible to collect data from small crystals, even at room temperature, hence opening a new mode of collecting data from difficult crystals (Axford et al. 2012, 2015; Aller et al. 2015). In the case of membrane proteins, one tries to enhance spots signal over the scattering background by increasing X-ray beam flux and/or exposure time. This is especially made possible by the adoption of new high-frame-rate pixel-array detectors (Broennimann et al. 2006). The images produced will be easier to index and integrate, but the collection range is severely limited, as the crystal's diffracting power will rapidly decrease because of radiation damage. As a result, small wedges of reciprocal space are sampled, leading to inaccurate estimates of cell parameters and integrated intensities. If several isomorphous crystals are available, data accuracy can be improved by averaging contributions from different data collections on individual crystals. Close isomorphism does not, in general, occur and crystals will be formed with a varying degree of non-isomorphism. If cell parameters between two crystals do not change considerably, such crystals can be effectively considered as isomorphous. Therefore, data merging from the relative intensities can potentially provide a dataset with increased completeness. An important aspect in the new approach to collect data from membrane proteins crystals is, therefore, the availability of software to analyse data from multiple crystals and single out isomorphous groups in order to generate complete datasets.

In this chapter the program *BLEND* (Foadi et al. 2013), part of the CCP4 suite of programs (Winn et al. 2011), will be introduced and explained. The chapter is specifically centred on applications of the program to assemble complete data for membrane proteins.

9.2 Main Ideas and Structure of the *BLEND* Program

BLEND has been developed to help give guidance to users on merging data from the most isomorphous crystals using cluster analysis. The statistical descriptors for cluster analysis used in BLEND are the unit cell parameters (a, b, c, α, β, γ). Before running *BLEND* all single datasets from the different crystals must be integrated with either *XDS* (Kabsch 2010), *MOSFLM* (Leslie and Powell 2007) or DIALS (DIALS 2015). *BLEND* can read HKL files from *XDS* and unmerged MTZ files from *MOSFLM* or *DIALS*. *BLEND* can, at present, be executed in five different modes: *dendrogram-only*, *analysis*, *synthesis*, *combination*, and *graphics* modes.

Typically, the *analysis* mode is run first. The program reads integrated reflection files and runs cluster analysis on cell parameters by default. The analysis mode will generate a tree or a dendrogram to display which datasets are isomorphous. The user must be aware that dendrograms will be generated even in case of close isomorphism. Hence, caution is required when scrutinizing these dendrograms. The Linear Cell Variation (LCV), or its absolute equivalent (aLCV) values, calculated for all clusters (see *BLEND* documentation at http://www.ccp4.ac.uk/html/blend.html) might help in the interpretation of the tree. Indeed the LVC indicator shows the largest difference between two datasets; some non-isomorphism between datasets is generally to be expected with LCV values greater than 2 %. It often happens that the *analysis* mode shows strong outliers (high LCV) making the interpretation of the dendrogram difficult. Removing these outliers and re-running *BLEND* in *analysis* mode generally provides a clearer tree to interpret.

The second step is the execution in *synthesis* mode. During this step the user will decide which datasets to merge together based on the dendrogram's results. *POINTLESS* (Evans 2006) and *AIMLESS* (Evans and Murshudov 2013) will be run automatically on each selected cluster, and the statistics after scaling (R_{meas}, R_{pim}, and Completeness) are also displayed. Users can choose which datasets/clusters are the best fit for further work, based on the statistics provided. For example a cluster with reasonable completeness and lowest R_{meas}/R_{pim} can often be the best option. Some other times, weak merging statistics can be accommodated in favour of higher resolutions. All files generated by the *synthesis* mode are stored in a directory named "merged_files". These are reflection files in MTZ format before and after scaling, and the corresponding logs from *POINTLESS* and *AIMLESS* for each cluster can be found. When fine-tuned scaling options are needed, users are advised to execute *AIMLESS* jobs starting from the unscaled reflection files obtained from previous executions of *BLEND*.

The *combination* mode can be run any time after the *analysis* mode or the *synthesis* mode. This mode adds flexibility to the program. It allows the user to merge any datasets/clusters of choice. For instance, here the user might choose to merge together datasets shown to belong to distant clusters during the previous run *in synthesis* mode. *BLEND* merges these groups of datasets running *POINTLESS* and *AIMLESS*, even though the specific group was not contemplated during *synthesis* mode because no clusters in the dendrogram contained those specific datasets.

More details and explanations of the various modes and options available in *BLEND* can be found in the following sections.

9.3 Membrane Proteins Used in This Chapter

Data from crystals of two different membrane proteins have been included in this chapter to help demonstrate BLEND workings. The first is Hemophilus Influenza TehA (Chen et al. 2010; Axford et al. 2015) and the second is the Human Histamine Receptor H1R (Shimamura et al. 2011) in complex with a second-generation anti-histamine drug (Heifetz et al. 2015). These proteins have been chosen because data were readily available and because *BLEND* is applied differently to the two types of datasets, facilitating a broad display of its functionality. The data for TehA consist of 67 small reciprocal space wedges collected from 56 different crystals *in situ* at room temperature and the structure was solved at 2.3 Å resolution (Axford et al. 2015). The data for the H1R consist of 18 partially complete or complete datasets from 18 different crystals that were collected at 100 K. This structure was solved at 3 Å resolution. Data for both structures were collected at the microfocus beamline I24 of the Diamond Light Source synchrotron. All data were included in a directory referred here for convenience as $BTEST.

9.4 Dendrogram-Only Mode (–aDO): Quick Clustering and Data Bookkeeping

Cluster analysis on groups of cell parameters helps to form first impressions on collected data and to organise them for later processing. The starting input for *BLEND* is integrated data that have not yet been scaled and merged. As mentioned earlier, these can either be MTZ unmerged binary files (produced by the integration programs *MOSFLM and DIALS*) or HKL ASCII files produced by *XDS*. To further clarify this point, *XDS* yields two types of HKL files. The first is normally called "INTEGRATE.HKL" and it includes integrated and unscaled intensities for all non-unique reflections. The correct space group is normally assigned in "XDS_ASCII.HKL", but not within "INTEGRATE.HKL" where, in general, intensities are assumed to describe a structure without symmetry (P1). Scaling and estimation of space group is done by the *XDS* CORRECT module. Scaled (but unmerged) intensities with final space group information are included in "XDS_ASCII.HKL". Although *XDS* files of the "XDS_ASCII.HKL" type can be fed into *BLEND* without any apparent warning, this is not recommended because the program eventually tries to scale all data with *AIMLESS*. With the "XDS_ASCII.HKL" files this would mean to scale once more intensities that had already been scaled. Thus the advice is to use intensities in the "INTEGRATE.HKL" file types. All input data can be included in the same directory, or spread across a number of different directories. The command line to execute *BLEND* will vary in the two cases.

9.4.1 All Input Files in a Same Directory

9.4.1.1 TehA

Assuming that the TehA data in MTZ format are all included in a single directory called "$BTEST/TehA", a quick clustering is obtained by running *BLEND* in *dendrogram-only* mode (*-aDO*) using the following command line:

```
blend -aDO $BTEST/TehA
```

The program starts and halts immediately after, waiting for input keywords. Pressing the `Enter` key is equivalent to accepting default values for all keyworded procedures and parameters. After a few seconds the program ends successfully. Among all files produced by *BLEND* the most important to look at is the dendrogram produced. For the present protein case, this is shown in Fig. 9.1.

The dendrogram represents proximity between various crystals in the whole group of data under investigation. The red numbers are annotations for the top five clusters of the linear cell variation (LCV) and the absolute linear cell variation (aLCV, within brackets). They provide prompt information on crystals isomorphism. High LCV and aLCV values indicate non-isomorphism among crystals in the specific cluster. In Fig. 9.1 most datasets show LCV equal to 1.18 %; this means that the largest difference in size among all crystals in that cluster amounts to 1.18 % of crystals size. A few crystals, though, make the LCV increase to 38.06 % and 57.38 %. This, in general, indicates that the crystals are entirely foreign to the structure under investigation or, as the case presented here, the crystal datasets have been indexed incorrectly by the integration program. At this point one should analyse carefully the integration stage for these "dendrogram outliers" and either index them correctly, or discard them if indexing remains difficult to achieve. It turns out that datasets 64, 65, 66 and 67 had all been collected with the X-ray beam placed at the interface between contiguous crystals, giving rise to overlapping reciprocal lattices, not neatly interpretable as belonging to a specific space group. The choice in this specific case was to discard the four datasets with the remaining 63 datasets appearing to be substantially isomorphous (LCV = 1.18 %). These were divided into two main clusters with cell parameters showing smaller differences (LCV = 0.68 % for the left branch and LCV = 0.80 % for the right branch).

9.4.1.2 H1R

For this crystal structure, 18 datasets, many of them fairly complete, have been collected from 18 different cryo-cooled crystals in multiple-collection episodes. As none of the individual datasets was considered to be satisfactory in terms of data quality or resolution, it was, decided to carry out multiple-crystals analysis with the hope of improving resolution and obtaining an interpretable electron density.

A first quick analysis was performed with the following command line:

```
blend -aDO $BTEST/H1R
```

The dendrogram produced is shown in Fig. 9.2. The various groups in the tree are not shown at their usual cluster height, but, rather, at the merging level, where nodes at a same level correspond to clusters with equal number of datasets. For example, clusters 9, 10, 1, 4, 3, 7, 6 are plotted at the lowest level because they correspond to the first merging, when two individual datasets are joined into one cluster; clusters 2, 5, 12 are plotted at the second level because they correspond to the second merging (like cluster 6 and dataset 17 joining into cluster 12), with all three clusters being formed by three datasets. This type of annotated dendrogram is very useful as it gives an overview

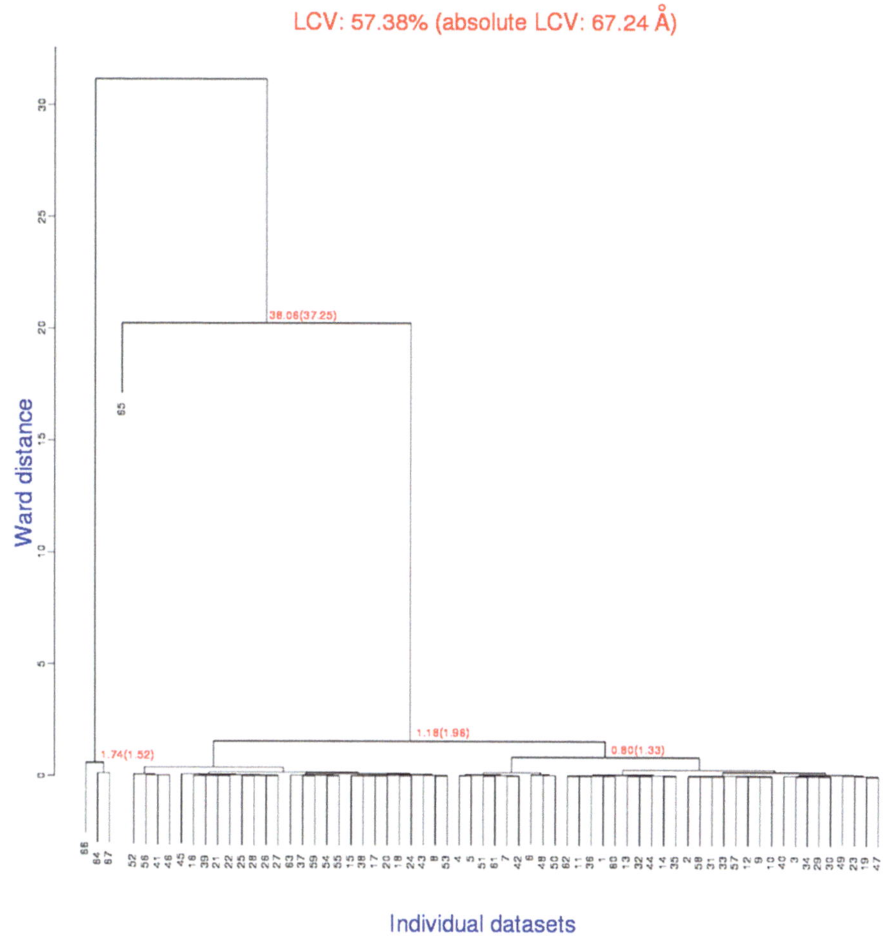

Fig. 9.1 Dendrogram produced by *BLEND* for the TehA test data. The *red numbers* are LCV values with aLCV values *within brackets*. Four of the 67 datasets analysed have unusually large LCV values. This, typically, indicates incorrect indexing, normally caused by crystal quality or data collection specific issues

of crystals isomorphism at the unit cell level. In this case, for example, it is clear that the three top clusters 13, 14 and 15 have an acceptable level of isomorphism at 3 Å resolution, but their union into larger clusters is bound to introduce some considerable structural differences, resulting in map artefacts or distortions.

9.4.2 Input Files in Different Directories

In the H1R case, data have been collected in multiple instances at different times. Such data had to be copied into a same directory in order to run *BLEND* as described in the previous section. Alternatively, data can be left in the original location and an ASCII file can be written to list their relative computer locations (paths). This file becomes the new input for *BLEND*.

9.4.2.1 TehA

As previously shown, datasets 64, 65, 66 and 67 are outliers. A new ASCII file can be prepared in which these datasets are excluded. This is easily done if files are available from a previous run. More specifically, the `mtz_names.dat` file that was produced by the previous run in *dendrogram-only* mode and contains paths to all

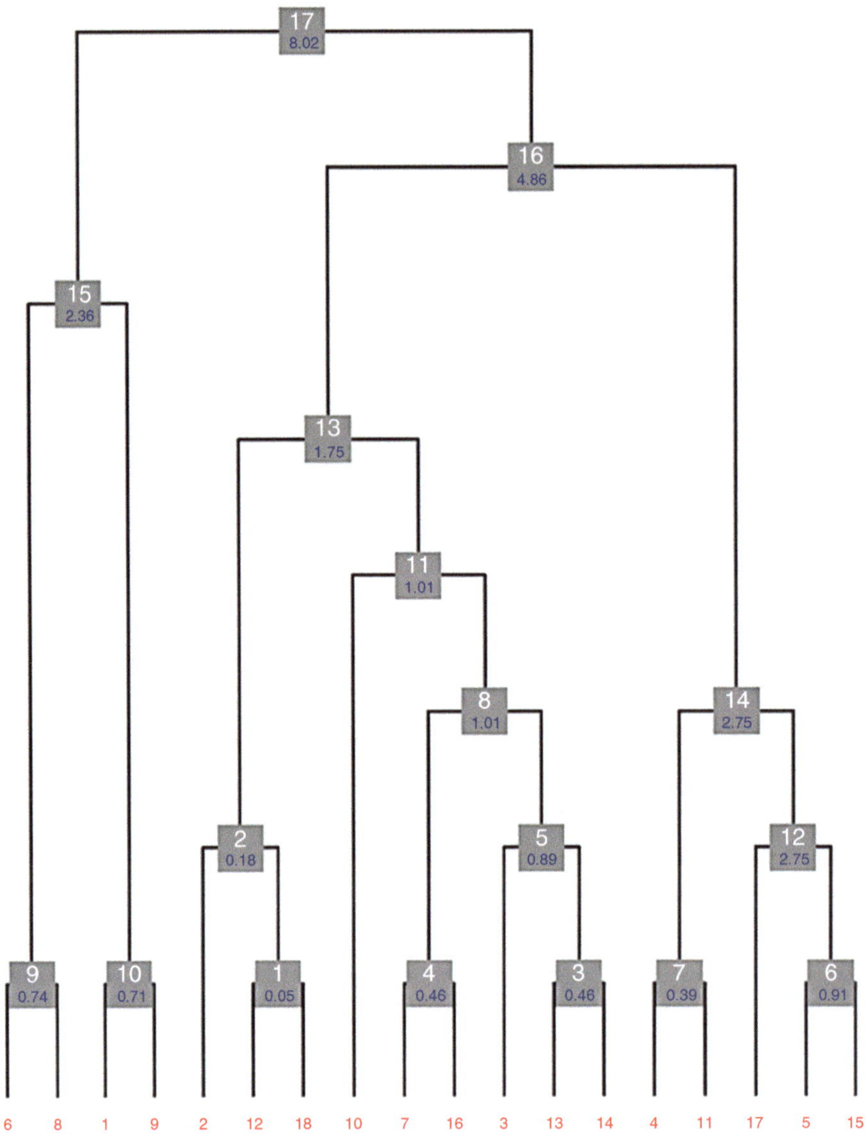

Fig. 9.2 Dendrogram with annotated aLCV values for H1R datasets. In this dendrogram the various nodes, here represented as *grey boxes*, are not displayed at their usual cluster heights but, rather, at their merging level. *Boxes* at a same level describe clusters with equal numbers of datasets

67 datasets can be copied, given a new name, say "`original_TehA.dat`" and edited in order to remove the unwanted datasets. *BLEND* can be subsequently executed with the following command line

```
blend -aDO original_TehA.dat
```

pressing the Enter key as required (see Sect. 9.4.1.1)

9.4.2.2 H1R

The `mtz_names.dat` file created by the previous run in *dendrogram-only* mode (see Sect. 9.4.1.2) can be copied to a new file, let us name it "`original_H1R.dat`", in which some of the least homogeneous datasets are excluded. The user then can execute *BLEND* in *dendrogram-only* mode in the following way:

```
blend -aDO original_H1R.dat
```

The result will be a tree slightly changed. The procedure can be repeated again with the exclusion of the least isomorphous datasets, until convergence to low values of aLCV are reached.

Later in this chapter we will follow a different approach by calculating merged datasets out of all those clusters having aLCV values smaller than a predefined value (see Sect. 9.6).

9.5 Analysis Mode (-a): Clustering, Radiation Damage and Resolution Estimate

Complete analysis of input data is achieved by running *BLEND* in *analysis* mode (*−a*). This is also a necessary step for later executions in *synthesis* and *combination* modes. The starting point for this section will be either the `original_TehA.dat` or `original_H1R.dat` files that are modified copies of the `mtz_names.dat` files obtained during previous runs in *dendrogram-only* mode.

9.5.1 TehA

Only 63 datasets are listed in the `original_TehA.dat` file and the command line for its execution in *analysis* mode is

```
blend -a original_TehA.dat
```

The dendrogram derived from this run is shown in Fig. 9.3. Essentially, the dendrogram produced here is a subsection of the dendrogram shown in Fig. 9.1. Differently from the execution in *dendrogram-only* mode (Sect. 9.4), two additional and important tasks were carried out this time; (1) radiation damage analysis and (2) estimate of resolution. These are crude procedures with the purpose of filtering out potentially noisy data.

In the *radiation damage procedure*, the average intensity in each resolution shell is analysed with respect to image number. For each resolution shell, the intensity is expected to decrease with increasing image number in particular for crystals affected by radiation damage. The observed intensity "decay" is more rapid at high resolutions. Dependencies on image number and resolution are statistically modelled as a linear exponential whose parameters are found using least squares. If the decay parameter is found to be increasing significantly with resolution, then the dataset in study is flagged as being affected by radiation damage. Knowledge of the modelled parameters can be subsequently used to estimate from which image the intensity severely drops, on average, to a predefined fraction (keyword RADFRAC). All the dataset images after that particular image will be automatically discarded from all the following analyses, unless the user decides otherwise. More details on this procedure can be found in Axford et al. (2015) where is also explained the rigorous validity of the method especially for cases where crystals are small and totally bathed in the X-ray beam (so that its exposed volume does not change during rotation), and the rotation range is small to avoid sizable fluctuations in the primary X-ray beam. All these conditions are applicable to the case study in this section (TehA) where crystals were small and matching the beam size. This procedure is also valid for longer rotation sweeps, provided that the beam is stable, the crystal is not too large, and the exposed volume does not change considerably during rotation.

Averaging intensities in resolution shells (this time including all images) are computed to estimate resolution cutoff. Averages are also computed for intensity errors. Ratios of the calculated averages form the starting points for a 10° polynomial interpolation. The resolution at which the polynomial falls below 1.5 is, by default, taken as the suggested highest resolution for the dataset under investigation. The numerical choice of 1.5 is based on tests on several data and is in general a conservative choice. It is important to remember that all the averages, at this stage, are calculated with unscaled data, and scaling in *BLEND* happens later, when multiple datasets are joined together using *AIMLESS*. The "Mean((I)/sd(I))" quantity used in *AIMLESS* to judge data quality at various resolution ranges was mostly found to

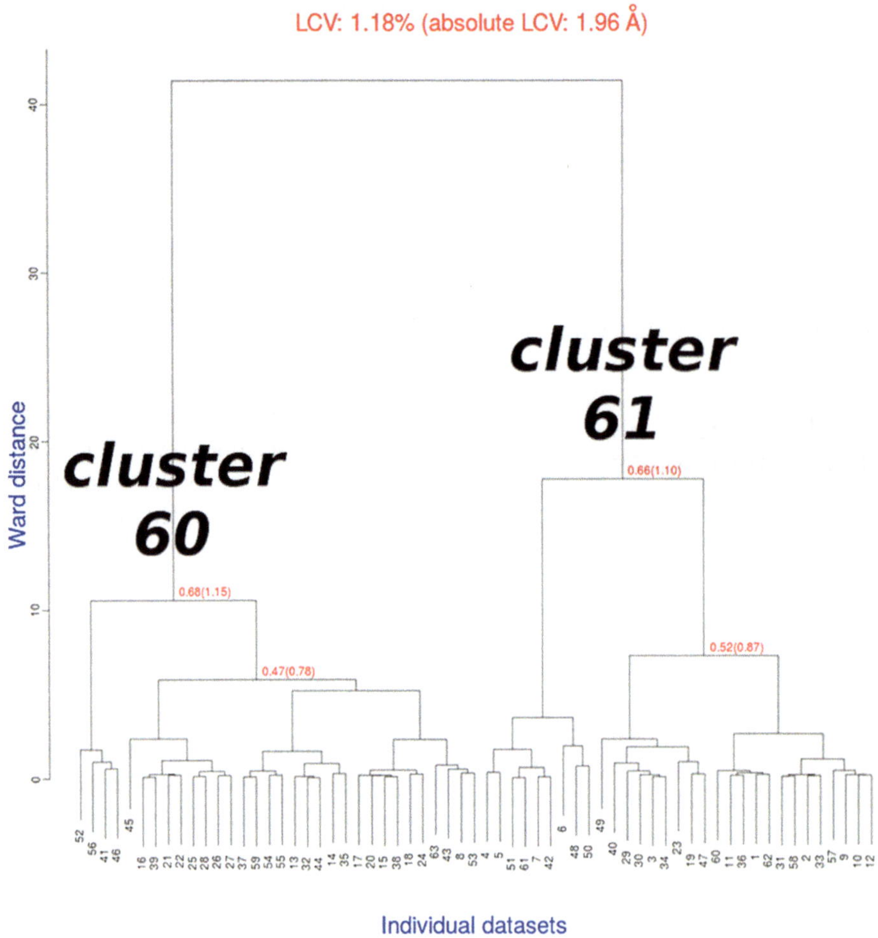

Fig. 9.3 Dendrogram produced by *BLEND* (*analysis* mode) for TehA test data in which datasets 64, 65, 66 and 67 have been removed

be better than 2 for the greatest majority of tests in *BLEND*. The 1.5 value can, in any case, be changed through keyword ISIGI by the user.

The results from the two procedures described above produced by *BLEND* is found in a file named

`FINAL_list_of_files.dat`

The ASCII file is divided into six columns and for each dataset includes the following information:

1. Path to MTZ integrated data
2. Dataset serial number (same as the one used in the dendrogram)
3. Number of last accepted image, as suggested by the radiation damage procedure
4. Number of first image
5. Number of last image
6. Applied resolution cutoff

The first 12 lines of the `FINAL_list_of_files.dat` file for the TehA data are shown in Table 9.1.

Both suggested cutoffs, as estimated by the *radiation damage* and by the *resolution* procedures can be modified by the user, either acting on keywords RADFRAC and ISIGI or during the *synthesis and combination* modes, using specific keywords for *POINTLESS* and *AIMLESS*. As

Table 9.1 The first 12 lines of file `FINAL_list_of_files.dat` produced by BLEND ran in *analysis* mode for the TehA datasets

File name	Dataset serial number	Cutoff image	First image	Last image	Suggested resolution
integrate05.mtz	1	17	1	30	2.614
integrate06.mtz	2	19	1	30	2.146
integrate07.mtz	3	17	1	30	2.105
integrate08.mtz	4	30	1	30	3.230
integrate09.mtz	5	30	1	30	3.450
integrate10.mtz	6	30	1	30	3.855
integrate11.mtz	7	30	1	30	3.290
integrate12.mtz	8	19	1	30	2.089
integrate12_2.mtz	9	21	1	40	2.132
integrate13.mtz	10	19	1	30	2.597
integrate14.mtz	11	18	1	30	2.098
integrate15.mtz	12	18	1	30	2.552

Cutoff images and suggested resolutions are calculated using the procedures described in the main text of Sect. 9.5

observed in Table 9.1, images cutoff have not been suggested for some of the datasets (datasets 4, 5, 6, 7). It is most probable that radiation damage has also occurred for these crystals, but either it was not too severe, or the conditions for the algorithm to work were not valid, resulting in the procedure not being applied to the specific dataset.

9.5.2 H1R

Here, *BLEND* was executed in *analysis* mode using the file prepared in Sect. 9.4 through the command line below

```
blend -a original_H1R.dat
```

During data collection, most of the rotation angle ranges were relatively wide and the crystals not small. Therefore, the procedure to determine radiation damaged parts of data collection was expected not to work well. For this run the values of the FINAL_list_of_files.dat file are listed in Table 9.2.

As observed in Table 9.2, most of the datasets did not present suggested cutoffs. However, a chance to prune damaged parts of the collection for each dataset is given after scaling, using the "pruning" variant of the *BLEND* combination mode (see Sect. 9.7).

9.6 Synthesis Mode (−s): Obtaining Complete Data

9.6.1 TehA

The dendrogram in Fig. 9.3 resulted from the *analysis* mode execution has two branches. The aLCV is nearly 2 Å for the two main branches while for each of the secondary branches is just above 1 Å. This occurrence points to a slight non-isomorphism between the two related groups of datasets. Validation of isomorphism in *BLEND* is achieved with its execution in *synthesis* mode. Details of the various syntax expressions for this mode are included in the official documentation. The aim of the *synthesis* mode execution is to obtain scaled and merged datasets for each cluster of the dendrogram. The top cluster, obtained by merging all datasets together, has height 41.395 (as seen in the file `CLUSTERS.txt`, produced by *BLEND* earlier in *analysis* mode); thus any number higher than this value will cause *BLEND* to scale and merge datasets for all 62 nodes of the tree. The command line is as follows (the input value "42" has been chosen arbitrarily as a numeric value higher than height 41.395)

```
blend -s 42 < bkeys.dat
```

where the ASCII file `bkeys.dat` includes all keywords needed. As this is the first job in

Table 9.2 Values extracted from the `FINAL_list_of_files.dat` file for the H1R datasets

File name	Dataset serial number	Cutoff image	First image	Last image	Suggested resolution
dataset_003.mtz	1	300	2	300	3.087
dataset_007.mtz	2	107	3	107	2.835
dataset_008.mtz	3	26	1	40	3.849
dataset_009.mtz	4	40	1	40	2.877
dataset_016.mtz	5	100	1	100	3.014
dataset_017.mtz	6	100	1	100	2.894
dataset_019.mtz	7	79	1	79	3.047
dataset_020.mtz	8	100	1	100	3.101
dataset_021.mtz	9	100	1	100	3.135
dataset_023.mtz	10	25	1	40	4.238
dataset_024.mtz	11	35	1	35	2.861
dataset_026.mtz	12	450	1	450	2.588
dataset_027.mtz	13	450	1	450	2.593
dataset_028.mtz	14	422	1	422	2.621
dataset_029.mtz	15	449	1	449	2.546
dataset_030.mtz	16	450	1	450	2.478
dataset_033.mtz	17	224	1	224	2.600
dataset_035.mtz	18	277	1	277	2.565

synthesis mode, data resolution is still not known (this is normally estimated after scaling). It is, therefore, worth imposing a somewhat ambitious value, selecting the highest resolution among all datasets, as suggested by the run in *analysis* mode, and analyse data after scaling to assess more realistic values. In this specific case study, *BLEND* suggests a resolution of 2.076 for dataset 42 (the highest among all datasets), thus the keywords file `bkeys.dat` contains the following line:

```
RESOLUTION HIGH 2.076
```

All files produced after execution in *synthesis* mode are grouped in the directory "merged_files". The first lines of the *summary of overall statistics* listed in file `MERGING_STATISTICS.info`, regarding to the TehA *synthesis* mode run are shown in Table 9.3.

The two main branches of the dendrogram, cluster 60 and cluster 61, do not seem to provide good-quality data. This can be due to the resolution of 2.076 Å being too high and low data completeness (88.1 % for cluster 60 and 69.8 % for cluster 61). Furthermore, some of the individual datasets composing the clusters could have an inherent bad quality and, thus, needed to be filtered out. Filtering will be dealt with in the next section on combination and graphics modes. Here we will look at ways to improve statistics and completeness with respect to resolution. Plots of $CC_{1/2}$ versus resolution (Evans 2006) for clusters 60 and 61 are displayed in Fig. 9.4.

Resolution seems better for cluster 61. This, though, can change after specific datasets have been filtered out. The overall $CC_{1/2}$ seems to suggest that scaling at a resolution around 2.5 Å should work sensibly. A new execution of *BLEND* in *synthesis* mode, this time at resolution 2.5 Å, is reflected in the statistics summary in Table 9.4.

9.6.2 H1R

The low isomorphism of many of the clusters is measured by the relatively high values of the aLCV parameter. For the H1R structure, we were trying to achieve good resolution, let us say at around 3 Å, hence it is appropriate to merge datasets with an aLCV value smaller than 3–4 Å. Resolution during scaling can also be fixed at 3 Å so that comparison among the various merged

Table 9.3 Statistics after execution of BLEND in synthesis mode for the most complete merged datasets of TehA at 2.08 Å resolution

Cluster number	R_{meas}	R_{pim}	Completeness	Multi-plicity	Resolution $CC_{1/2}$	Resolution Mn(I/sd)	Resolution Max
62	1.884	0.739	90.60	8.90	2.53	2.15	2.08
60	6.518	3.283	88.10	4.60	3.45	2.16	2.08
58	0.205	0.089	82.90	4.20	2.38	2.15	2.08
57	0.106	0.053	80.10	3.00	2.08	2.16	2.08
61	0.550	0.221	69.80	5.70	2.57	2.28	2.08
52	0.106	0.063	68.00	1.90	2.08	2.17	2.08
59	0.352	0.169	63.80	4.00	2.63	2.25	2.08
56	1.415	0.816	59.20	2.40	3.13	2.37	2.08
47	0.098	0.060	58.10	1.90	2.08	2.19	2.08
55	0.126	0.067	55.10	2.60	2.18	2.25	2.08
53	0.694	0.416	53.60	2.10	2.96	2.38	2.08
54	0.644	0.399	52.70	2.10	3.01	2.24	2.08
48	1.893	1.201	50.80	1.90	3.29	2.47	2.08
50	0.217	0.132	50.70	1.90	2.47	2.40	2.08
49	665.899	461.930	48.70	1.30	6.40	2.41	2.08

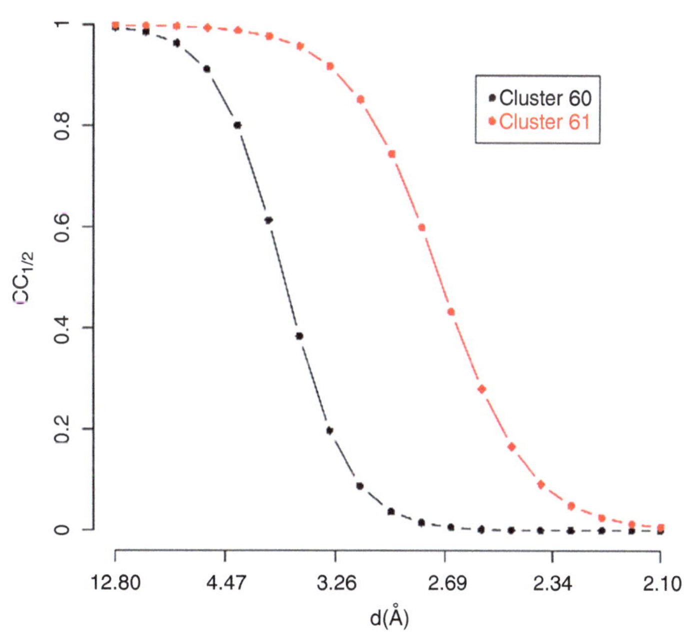

Fig. 9.4 $CC_{1/2}$ curves for the two datasets assembled with data from each of the two branches of the dendrogram for TehA

datasets becomes easier. In addition, resolution can still be extended in later runs of the program. The "bkeys.dat" keywords file used for this *BLEND* job includes the following lines:

```
RESOLUTION HIGH 3.0
TOLERANCE 100
```

The second keyword ("TOLERANCE") has been added to stop *POINTLESS* from halting execution. When cell parameters are very different *POINTLESS* halts execution because it suspects crystals might come from different structures. Its tolerance has a default value of 2 and larger numbers increase this tolerance. "TOLERANCE

Table 9.4 Statistics after execution of BLEND in *synthesis* mode for the most complete merged datasets of TehA at 2.50 Å resolution

Cluster number	R_{meas}	R_{pim}	Completeness	Multi-plicity	Resolution $CC_{1/2}$	Resolution Mn(I/sd)	Resolution Max
62	0.586	0.157	93.60	10.90	2.56	2.50	2.50
60	1.535	0.598	93.20	5.50	3.21	2.50	2.50
58	0.154	0.061	88.20	4.90	2.50	2.50	2.50
57	0.086	0.042	87.60	3.40	2.50	2.50	2.50
52	0.087	0.051	78.90	2.00	2.50	2.50	2.50
61	0.514	0.185	70.90	7.10	2.75	2.50	2.50
47	0.083	0.050	66.00	2.10	2.50	2.50	2.50
59	0.228	0.098	65.90	4.90	2.68	2.50	2.50
56	0.977	0.531	64.80	2.70	3.15	2.50	2.50
55	0.103	0.053	60.90	3.00	2.50	2.50	2.50
49	31.153	21.159	59.70	1.30	6.30	2.50	2.50
54	0.325	0.184	59.30	2.30	3.01	2.50	2.50
53	0.394	0.220	59.20	2.40	2.94	2.50	2.50
40	0.089	0.060	58.60	1.30	2.50	2.50	2.50
48	1.194	0.726	57.30	2.10	3.30	2.50	2.50

100" basically tells *POINTLESS* not to halt, even though crystals are very different. The *BLEND* run for this specific group of datasets in *synthesis* mode is started with the following command line:

```
blend -saLCV 3 < bkeys.dat
```

All files and statistics connected to this BLEND run are found in the directory "merged_files". In this study, the three clusters with largest acceptable values of aLCV (less than 3) are clusters 13, 14 and 15. Although complete, these clusters have alarming merging statistics values and more work is needed to improve results. This is achieved by using the *combination* and *graphics* modes in BLEND as described in see Sect. 9.7.

9.7 Combination and Graphics Modes (−c,−g): Improving Results from Synthesis Using Combination, Filtering and Graphics

Results obtained from running *BLEND* in *synthesis* mode, as said before, can be quickly surveyed by looking at the `MERGING_STATISTICS.info` file. Logs from all *POINTLESS* and *AIMLESS* jobs associated with each scaling and merging are also saved in the "merged_files" directory and therefore accessible if specific details are needed. If some of the new datasets show sufficient completeness, resolution and satisfactory data quality as described by R_{meas} and R_{pim}, then the associated scaled and merged MTZ files in the same directory can be used for phasing and model building. However, in those cases where the above criteria are not met, it will be necessary to create new datasets that are not represented by the dendrogram nodes using *BLEND* in *combination* mode. At this stage it is also of great help to visualise dendrogram's connections and associated merging statistics through *BLEND graphics* mode. It is practical to have completeness, R_{meas} and $CC_{1/2}$ resolution visually associated with each node of the dendrogram in order to easily be able to judge the goodness of specific dataset combinations. Yet, the resulted annotated dendrogram would though appear cluttered with numbers, most of which likely not to be readable. For this reason, it has been found convenient in *BLEND* to introduce a *graphics* mode producing only parts of the annotated dendrogram focusing on specific clusters. Graphics files in PNG and PS format are produced and stored under a directory called "graphics" every time a run in *graphics* mode is executed. In the command line, the only required fields are the specific cluster number and the number of levels from the specified cluster that the user would like to visualise. The higher the number of clusters the more packed the annotated dendrogram will appear.

The other mode described in this section, the *combination* mode, is needed for all groupings not present in the dendrogram. Although the grouping suggested by *BLEND*, using cell parameters, tends to provide optimal datasets in terms of isomorphism and merging statistics, still there are many factors (quality of individual datasets, insufficient coverage of the reciprocal space, etc.) that preclude such grouping to be the best possible (Foadi et al. 2013). Therefore, it is ideal for the user to be allowed to try different dataset combinations, alternative to the ones represented by the clusters. This is the main reason why the *combination* mode was created in *BLEND*. All files and statistics produced by *BLEND* in *combination* mode are sequentially stored in a directory called "combined_files".

9.7.1 TehA

In the TehA data, cluster 60 is the only one with reasonable completeness, however its quality, at least in terms of resolution and R_{meas}, is not great. One could ask if there are rogue individual datasets, part of cluster 60, responsible for the bad statistics. This can be easily investigated by running *BLEND* in *graphics* mode, focusing on cluster 60, and demanding sufficiently high number of cluster levels. The syntax command do this is

```
blend -g D 60 5
```

where ("−g") means execution in *graphics* mode of the annotated dendrogram type focusing on cluster 60 to 5 levels of merging. The letter "D" in the syntax command at present is a default letter (it is envisaged that other types of plots will be added to future versions of*BLEND*. For these other types the "−g" of the graphics mode will be followed by letters different from "D"). The resulting annotated dendrogram is displayed in Fig. 9.5.

It is reasonably clear from Fig. 9.5 that cluster 49 (the small branch on the left) is the main responsible for the deterioration of data quality in cluster 60. Cluster 49 was composed of datasets 41, 46, 52 and 56.

To check if any of these datasets causes data quality to deteriorate, *BLEND* was executed in *combination* mode, starting from cluster 49 and subtracting one dataset at a time, using a special syntax developed for this purpose (see *BLEND* documentation at http://www.ccp4.ac.uk/html/blend.html)

```
blend -c [49] [[41]] < bkeys.dat
blend -c [49] [[46]] < bkeys.dat
blend -c [49] [[52]] < bkeys.dat
blend -c [49] [[56]] < bkeys.dat
```

Results from these 4 runs are shown in the first 4 lines of Table 9.5.

From the data shown in Table 9.5, it is clear that dataset 46 is a rogue dataset. Furthermore, it is also clear from the annotated dendrogram in Fig. 9.5 that dataset 45 also deteriorates statistics quality. Although cluster 45 has a reasonable R_{meas} value of 0.092 it jumps to a high value of 0.325 when the addition of dataset 45 turning cluster 45 into cluster 54. Thus, statistics were recalculated for cluster 60 without datasets 45 and 46 by the following command

```
blend -c [60] [[45,46]]
  < bkeys.dat
```

Results from this run are shown in the fifth row of Table 9.5. The improvement observed is indisputable. It also was observed that without datasets 45 and 46 the resolution could be extended to 2.2 Å (row 6 in Table 9.5). Electron density maps obtained using the data just described, limited to 2.3 Å for comparison with data from a single cryo-cooled crystal, are shown in Axford et al. (2015).

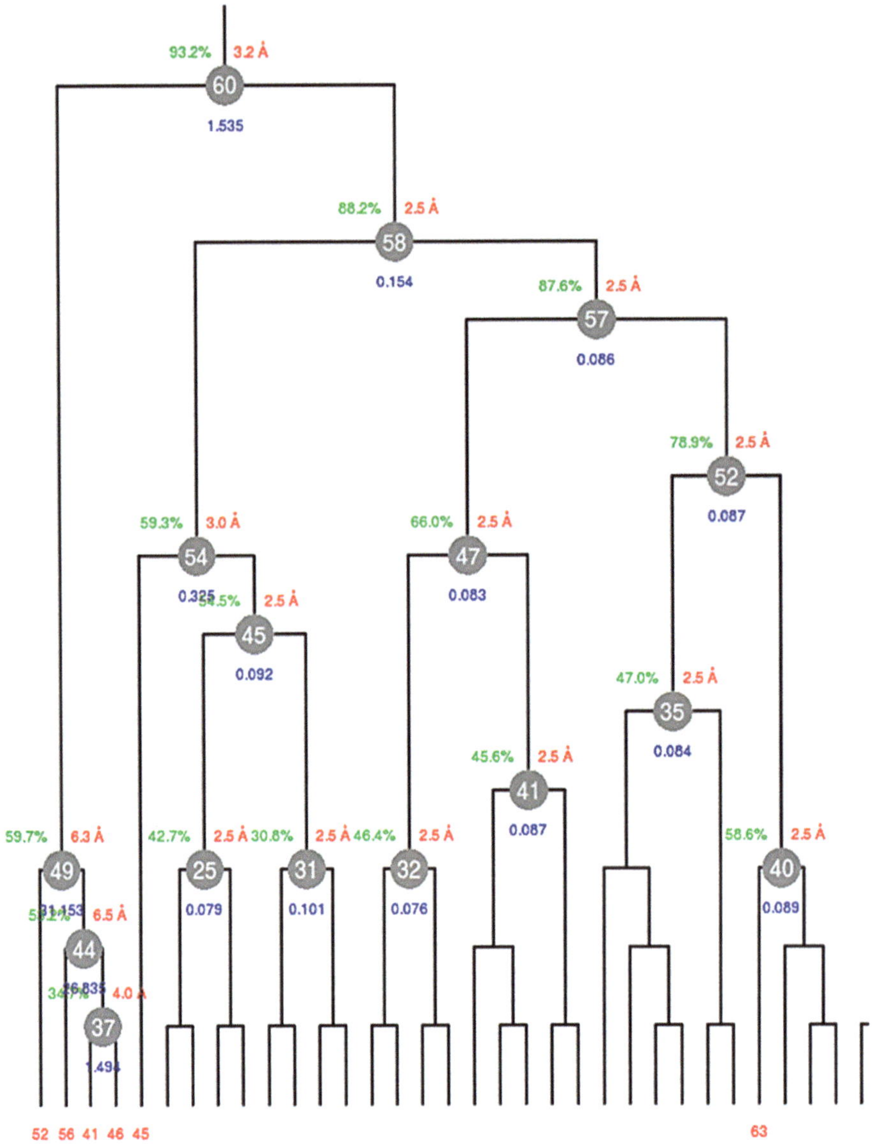

Fig. 9.5 Annotated dendrogram created by BLEND in graphics mode around cluster 60, with 5 levels of annotation ("blend −g D 60 5")

9.7.2 H1R

Three useful and visual summaries from statistics produced by *BLEND* in *synthesis* mode (see Sect. 9.6) were obtained in *graphics* mode with the following three command lines,

```
blend -g D 13 10
blend -g D 14 10
blend -g D 15 10
```

where the three clusters selected (13, 14, and 15) were the ones with acceptable largest values of the aLCV parameter, as explained in Sect. 9.6. The three plots produced are shown in Fig. 9.6 (see *BLEND* documentation for details of executions in graphics mode – http://www.ccp4.ac.uk/html/blend.html).

None of the statistics were satisfactory, but some of the estimated resolution values along

Table 9.5 Overall statistics obtained running BLEND in combination mode for the TehA case

Datasets	R_{meas}	R_{pim}	Completeness	Multi-plicity	Resolution $CC_{1/2}$	Resolution Mn(I/sd)	Resolution Max
46,52,56	107.055	72.605	47.40	1.30	11.59	2.71	2.50
41,52,56	0.112	0.077	49.10	1.20	2.50	2.50	2.50
41,46,56	26.835	18.425	53.20	1.20	6.47	2.50	2.50
41,46,52	12.127	8.429	42.40	1.20	8.57	2.50	2.50
A	0.093	0.037	93.20	5.10	2.50	2.50	2.50
B	0.113	0.047	91.70	4.60	2.20	2.20	2.20

A cluster 60 without datasets 45 and 46 with resolution limited to 2.50 Å, *B* same as A with resolution extended to 2.20 Å

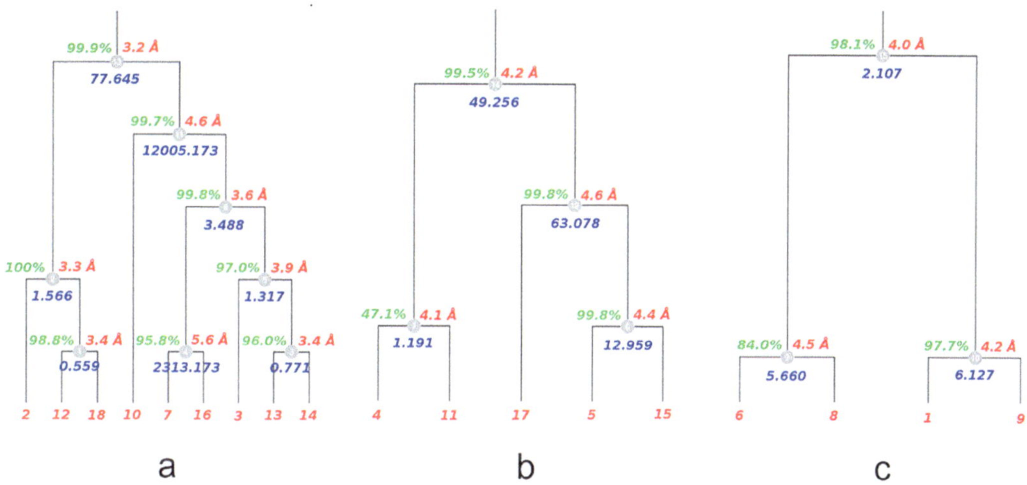

Fig. 9.6 Annotated trees with statistics for clusters 13 (**a**), 14 (**b**) and 15 (**c**) obtained running BLEND*in synthesis* mode for H1R data

with the $CC_{1/2}$ parameter were interestingly high and certainly could be improved. The low values of aLCV and the relatively good resolution for many of the groups contrasts with the bad values of the merging statistics (R_{meas}, R_{pim}). This is normally due to situations where full convergence in the scaling process has not been reached. In the H1R study, this was mainly caused by the poor diffracting quality of the images or radiation damage present in the datasets. With help from the annotated trees displayed in Fig. 9.6, and running *BLEND* in *combination* mode, datasets filtering could be tried in order to improve statistics, similarly to what was done for the TehA case study case. However, here a different approach was tried because of the bad quality data, mostly due to the effects of radiation damage. Therefore, for each dataset forming a given cluster, images collected towards the end of the rotation sweep were eliminated so that R_{meas} and in particular R_{pim} values decreased while, completeness was not allowed to go below a given threshold. In *BLEND* this can be done automatically using a variant of the *combination* mode, " -cP", where "P" stands for "pruning". In other words, during each cycle, specific fractions of images are eliminated from the dataset with the highest mean value of R_{merge}, until target completeness (default is 95 %) or maximum number of cycles is reached. Pruning cycles are also halted if a whole dataset eventually turns out to be fully eliminated; this option can, in fact, be carried out differently with the simple combination mode, filtering out the specified dataset. Once cycling is completed, *BLEND* selects the final results for the cycle with lowest R_{pim}. When the pruning variant of the *combination* mode is applied to clusters 13, 14 and 15, statistics generally seems to improve.

Table 9.6 Statistics for clusters 13, 14, 15 in the H1R case before and after the elimination of the images with the pruning variant of the *combination* mode in *BLEND* ("−cP")

		R_{meas}		R_{pim}		Resolution (CC1/2 > 0.3)		Completeness (%)		Multiplicity	
Clusters	Images pruned	Before	After	Before	After	Before	After	Before	After	Before	After
13	404	77.645	0.729	13.153	0.146	3.2	3.0	99.9	99.9	31.6	22.4
14	186	49.256	1.819	16.933	0.598	4.2	3.9	99.5	98.3	11.8	9.0
15	0	2.107	2.107	0.675	0.675	4.0	4.0	98.1	98.1	8.7	8.7

The procedure automatically selects the cycle with the best R_{pim} Improvements were evident for clusters 13 and 14. However, for cluster 15, any pruning cycle has returned data with worst statistics therefore results are unchanged in the table for this cluster

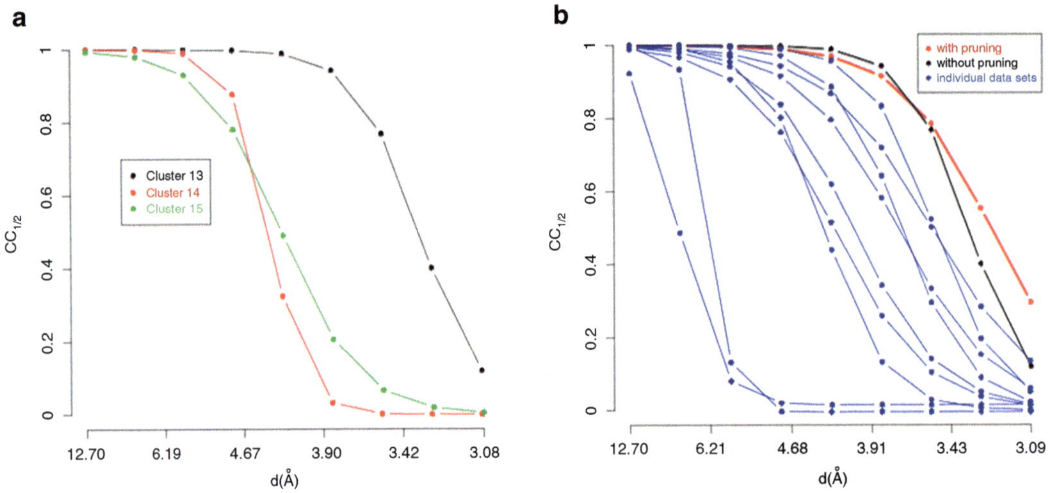

Fig. 9.7 Beneficial effects of multiple-crystals datasets on resolution, as quantified by $CC_{1/2}$ curves. (**a**) Resolution for cluster 13 is far better than resolution for clusters 14 and 15. (**b**) Merging data from multiple isomorphous crystals helps to extend resolution. Each *blue curve* describes the $CC_{1/2}$ value for all individual datasets composing cluster 13 (*black curve*). Resolution is further extended when parts of data affected by radiation damage are eliminated using the pruning variant of BLEND*combination* mode (*red curve*)

Details are displayed in Table 9.6 where a comparison between the results from the *synthesis* mode and *combination* mode are shown.

Perhaps the most interesting result concerning the processing of these data is an increase of the final resolution, as illustrated by the improved $CC_{1/2}$ curves (see Fig. 9.7 for details).

9.8 Conclusions

In this chapter it has been shown how *BLEND* can be used effectively to analyse and manage X-ray diffraction data from multiple crystals of membrane proteins. Its three main modes (*analysis, synthesis and combination*), together with specific variants, enable users to explore data isomorphism and data quality, and provide a flexible and easy-to-use platform to create complete datasets to be used in the follow up stages of phasing, model building and refinement.

BLEND cluster analysis, based on unit cell differences, is particularly useful for membrane protein data analysis, because of the high-content solvent and detergent-aided packing, making this class of proteins particularly susceptible to large cell size variations. A quantitative and practical measure of such variability is provided by LCV and aLCV. Groups of data with small values of these quantities can normally be merged into more complete and redundant data, using *synthesis* and *combination* modes. In particular, unwanted datasets can be filtered out of specific groups using the *combination* mode.

One last but important comment, concerns *BLEND*'s flexibility as represented by its propensity to adapt to both easy and difficult data. While the software can produce complete datasets with minimal user's intervention when data have inherently good quality, a more interactive use of the software can lead to complete datasets in those cases where traditional data merging produce unacceptable or unusable data.

Acknowledgments The authors wish to acknowledge support from the Wellcome Trust (grant 099165/Z/12/Z) and Diamond Light Source Ltd.

References

Aller P, Sanchez-Weatherby J, Foadi J, Winter G, Lobley CM et al (2015) Application of in situ diffraction in high-throughput structure determination platforms. Methods Mol Biol 1261:233–253

Axford D, Foadi J, Hu NJ, Choudhury HG, Iwata S et al (2015) Structure determination of an integral membrane protein at room temperature from crystals in situ. Acta Crystallogr D Biol Crystallogr 71(6):1228–1237

Axford D, Owen RL, Aishima J, Foadi J, Morgan AW et al (2012) In situ macromolecular crystallography using microbeams. Acta Crystallogr D Biol Crystallogr 68(5):592–600

Broennimann C, Eikenberry EF, Henrich B, Horisberger R et al (2006) The PILATUS 1M detector. J Synchrotron Radiat 13(2):120–130

Chen YH, Hu L, Punta M, Bruni R, Hillerich B (2010) Homologue structure of the SLAC1 anion channel for closing stomata in leaves. Nature 467(7319): 1074–1080

DIALS (2015) Diffraction integration for advanced light sources. From http://dials.diamond.ac.uk/

Evans PR (2006) Scaling and assessment of data quality. Acta Crystallogr D Biol Crystallogr 62: 72–82

Evans PR, Murshudov GN (2013) How good are my data and what is the resolution? Acta Crystallogr D Biol Crystallogr 69(7):1204–1214

Foadi J, Aller P, Alguel Y, Cameron A, Axford D (2013) Clustering procedures for the optimal selection of data sets from multiple crystals in macromolecular crystallography. Acta Crystallogr D Biol Crystallogr 69(8):1617–1632

French S, Wilson K (1978) On the treatment of negative intensity observations. Acta Crystallogr A Found Crystallogr 34:517–525

Heifetz A, Schertler GFX, Seifert R, Tate CG, Sexton PM et al. (2015) *GPCR* structure, function, drug discovery and crystallography: report from academia-industry international conference (UK Royal Society) Chicheley Hall, 1–2 September 2014. Naunyn-Schmiedeberg's Archives of Pharmacology 388(8)

Kabsch W (2010) XDS. Acta Crystallogr D Biol Crystallogr 66(2):125–132

Leslie AGW, Powell HR (2007) Processing diffraction data with Mosflm. In: Evolving methods for macromolecular crystallography. Springer, Dordrecht, pp 41–51

Otwinowski Z, Minor W (1997) Processing of X-ray diffraction data collected in oscillation mode. In: Carter CW, Sweet RM (eds) Methods in enzymology, Vol. 276: Macromolecular crystallography, part A. Academic, New York, pp 307–326

Shimamura T, Shiroishi M, Weyand S, Tsujimoto H, Winter G et al (2011) Structure of the human histamine H1 receptor complex with doxepin. Nature 475(7354): 65–70

Winn MD, Ballard CC, Cowtan KD, Dodson EJ, Emsley P et al (2011) Overview of the CCP4 suite and current developments. Acta Crystallogr D Biol Crystallogr 67(4):235–242

Serial Millisecond Crystallography of Membrane Proteins

10

Kathrin Jaeger, Florian Dworkowski, Przemyslaw Nogly, Christopher Milne, Meitian Wang, and Joerg Standfuss

Abstract

Serial femtosecond crystallography (SFX) at X-ray free-electron lasers (XFELs) is a powerful method to determine high-resolution structures of pharmaceutically relevant membrane proteins. Recently, the technology has been adapted to carry out serial millisecond crystallography (SMX) at synchrotron sources, where beamtime is more abundant. In an injector-based approach, crystals grown in lipidic cubic phase (LCP) or embedded in viscous medium are delivered directly into the unattenuated beam of a microfocus beamline. Pilot experiments show the application of microjet-based SMX for solving the structure of a membrane protein and compatibility of the method with *de novo* phasing. Planned synchrotron upgrades, faster detectors and software developments will go hand-in-hand with developments at free-electron lasers to provide a powerful methodology for solving structures from microcrystals at room temperature, ligand screening or crystal optimization for time-resolved studies with minimal or no radiation damage.

Keywords

Serial crystallography • High viscosity injector • LCP injector • Synchrotron • Membrane proteins

K. Jaeger • P. Nogly • J. Standfuss (✉)
Laboratory of Biomolecular Research, Paul Scherrer Institute, 5232 Villigen PSI, Switzerland
e-mail: kathrin.jaeger@psi.ch; przemyslaw.nogly@psi.ch; joerg.standfuss@psi.ch

F. Dworkowski • C. Milne • M. Wang
Swiss Light Source, Paul Scherrer Institute, 5232 Villigen PSI, Switzerland
e-mail: florian.dworkowski@psi.ch; chris.milne@psi.ch; meitian.wang@psi.ch

10.1 Introduction

10.1.1 Difficulties Associated with Membrane Protein Crystallography

Membrane proteins are essential for various cellular functions and regulate vital physiological tasks. They make up about ~30 % of the

I. Moraes (ed.), *The Next Generation in Membrane Protein Structure Determination*, Advances in Experimental Medicine and Biology 922, DOI 10.1007/978-3-319-35072-1_10

proteome in most organisms and are potent drug targets. Over 60 % of current human drug targets are membrane proteins (Yildirim et al. 2007).

Although they are highly abundant and have been studied for many years, the structural information on membrane proteins lags well behind that of soluble proteins (White 2004). This is mainly due to the fact that membrane proteins are challenging to crystallise for a number of reasons, including their relatively hydrophobic surface, lack of stability outside of their native membranes, and high structural flexibility. These properties lead to difficulties at all levels of study, including expression, solubilisation, purification, crystallisation, X-ray diffraction data collection and structure solution (Carpenter et al. 2008).

From a bio-medical perspective, the high-resolution structures of pharmaceutically relevant eukaryotic proteins are essential for rational drug design and our understanding of core functional mechanisms. Therefore, much effort has been invested into developing innovative approaches and better methods for membrane protein crystallization over the last years. This chapter will be focused on the principle behind microjet-based serial millisecond crystallography using synchrotron radiation and illustrate its recent application to membrane proteins.

10.1.2 Methods for Improving Membrane Protein Crystallography

Lipidic cubic phase (LCP) crystallisation was invented about 18 years ago (Landau and Rosenbusch 1996), and improvements in crystallisation tools, robotics and imaging have led to wide application of the technology for the crystallisation of membrane proteins (Cherezov 2011). Its implementation has been critical for determining high-resolution structures of several membrane protein families, such as G protein-coupled receptors, microbial rhodopsins, ion channels, transporters and enzymes (Caffrey 2015). LCP crystallisation is an *in meso* method, meaning that the protein is reconstituted into a lipid mesophase bilayer and is thereby removed from the necessity for a purification detergent micelle environment. Thus, crystallogenesis takes place in a more native-like lipidic bilayer (Caffrey and Cherezov 2009). Typically, LCP crystallisation produces highly ordered crystals, but they are often limited in size – sometimes in the range of only a few micrometres. The optimisation of very small crystals in order to obtain sufficiently large crystals, which are better at withstanding radiation damage, can be difficult and time consuming (Cherezov et al. 2009). Notably, prominent human drug targets such as some G protein-coupled receptors are known to develop only microcrystals (Liu et al. 2013; Caffrey and Cherezov 2009).

The use of high-intensity microfocused X-ray beams ($>10^9$ photons s^{-1} μm^{-2}) and long exposure times are necessary to achieve enough signal from weakly diffracting microcrystals to obtain high-resolution data. Since small crystals are very sensitive towards radiation damage and quickly lose diffraction power, a complete data set can often only be obtained by merging data from several crystals (Cherezov et al. 2009). In the general approach, crystals are measured under cryogenic conditions since the life-dose of protein crystals is one to two magnitudes higher at cryogenic temperature (Garman and Schneider 1997). However, the handling of tiny crystals during cryo-cooling poses challenges because the LCP turns opaque upon freezing and scanning techniques are usually needed to align the X-ray beam on a particular crystal. Moreover, there is always the question whether crystallographic structures obtained at temperatures far from the physiological may display artifacts.

10.2 Serial Crystallography

Serial crystallography mitigates radiation damage and avoids the need for cryo-cooling by using many small crystals for data collection. The unexposed crystals are continuously replenished, therefore the total radiation dose is spread over the entire sample enabling data collection at room temperature and eliminating the need for suitable cryo-protectants. As a result, a complete data set can be obtained by merging the diffraction patterns from many single crystals.

10.2.1 Serial Femtosecond Crystallography at X-Ray Free-Electron Lasers (SFX)

Serial crystallography has been proven successful for structure determination at synchrotrons (Stellato et al. 2014; Heymann et al. 2014; Nogly et al. 2015; Botha et al. 2015) and X-ray free-electron lasers (XFELs) (Chapman et al. 2011; Boutet et al. 2012; Johansson et al. 2012; Redecke et al. 2013; Kern et al. 2013; Demirci et al. 2013; Johansson et al. 2013; Barends et al. 2014; Kern et al. 2014; Hirata et al. 2014). Originally, it had however been developed for the use at X-ray free-electron lasers where it relies on short X-ray pulses in the femtosecond time scale and has therefore been named serial femtosecond crystallography (SFX). If pulses are short enough (<20 fs) that they terminate before radiation damage processes have manifested themselves (Neutze et al. 2000; Lomb et al. 2011; Nass et al. 2015), the collection of single diffraction patterns from individual crystals is possible with high radiation doses sufficient to turn crystals into plasma (Chapman et al. 2011). Since the X-ray pulses are very bright, even submicron-sized crystals can yield high-quality diffraction data (Boutet et al. 2012; Redecke et al. 2013) with each dataset consisting of thousands of merged individual snapshots.

The first injector system used in such XFEL experiments was a gas dynamic virtual nozzle (GDVN) (DePonte et al. 2008; Weierstall et al. 2012). In this system, crystals are delivered in their crystallisation mother liquor at a speed of about 10 m/s and flow rate of about 10 μl/min. Due to the high flow rate and the currently available XFEL pulse repetition rates (120 Hz at LCLS) (Park et al. 2013) only about 1 out of 10000 crystals (0.01 %) is hit by an X-ray pulse and large amounts of sample are needed (Park et al. 2013), especially for a *de novo* structure determination (Barends et al. 2014). For example, in the first pioneering SFX experiments the collection of a complete data set required about 100 mg of protein, which cannot feasibly be produced for most membrane proteins (Chapman et al. 2011). High viscosity injectors using LCP or other viscous delivery media solved this problem and dramatically reduced the amount of sample required (Weierstall et al. 2014; Botha et al. 2015; Sugahara et al. 2015) as they can effectively operate at much lower flow rates. The LCP injector for instance, can extrude LCP at flow rates of 1–300 nl/min while still providing a continuous 10–50 μm diameter stream of LCP containing crystals, thus reducing sample consumption about 50–100 fold, compared to the GDVN (Weierstall et al. 2014). Assuming convergence of data from 10,000 indexed patterns into a complete dataset usable for molecular replacement, structures can be solved from as little as 50 μg of protein crystallised in LCP. Therefore LCP grown microcrystals are especially well suited for SFX measurements.

Among the first examples for membrane protein structures solved from sub-10 μm-sized crystals were two challenging GPCRs, the human serotonin (5-HT_{2B}) receptor (Liu et al. 2013) and the human smoothened (SMO) receptor in complex with cyclopamine (Weierstall et al. 2014), which were solved from crystals produced with less than 0.5 mg of protein. Another example is the Diacylglycerol kinase (DgkA), a membrane kinase crystallised in LCP, which was solved using only 0.2 mg of protein (Caffrey et al. 2014). Recently, another GPCR structure of the human angiotensin receptor (AT_1R) (Zhang et al. 2015) was solved by LCP-SFX to a resolution of 2.9 Å using LCP grown micro-crystals ($10 \times 2 \times 2$ μm^3). For this receptor, the usage of LCP-SFX was necessary since larger AT_1R crystals ($40 \times 4 \times 4$ μm^3) could not be further size-optimised and diffracted only to 4 Å resolution at a synchrotron source under cryogenic conditions. But the use of high viscosity injectors is not limited to crystals grown in LCP. Recently, it was shown that *in surfo* grown crystals can be transferred into viscous medium (such as mineral greases or petroleum jelly) and can be used for SFX data collection (Sugahara et al. 2015). Nevertheless, a limiting factor regarding the use of SFX still remains the availability of XFEL sources, which stimulated the interest in transferring SFX principles to synchrotron beamlines.

10.2.2 Serial Crystallography at Synchrotron Sources

10.2.2.1 Serial Millisecond Crystallography (SMX)

The limited available beamtime at XFELs and planned upgrades to existing synchrotron sources support the great interest in performing serial room-temperature measurements at synchrotron microfocus beamlines (Fig. 10.1). Adapting novel crystal supply methods, like the LCP jet (Weierstall et al. 2014) or the similar High-Viscosity Extrusion (HVE) injector (Botha et al. 2015) operating at low sample delivery speed, enables similar sample usage compared to XFEL setups. Due to lower photon flux at synchrotron sources and the quasi-continuous beam, the optimal exposure times per image are at present in the millisecond range. As a result the primary radiation damage cannot be neglected as in SFX experiments, even though radiation damage due to radical formation and diffusion can still be greatly reduced (Owen et al. 2012). Nevertheless, due to the lower X-ray doses compared to XFELs and continuous crystal supply, the total radiation dose in a dataset is spread over thousands of crystals, reducing the dose per crystal to 0.2-1 MGy, which is sufficiently low for data collection at room temperature (Garman 2010; Owen et al. 2012).

It has been shown recently, that SFX approaches can be adapted and successfully transferred to synchrotron sources. In a first room-temperature serial crystallography experiment at a synchrotron, the structure of lysozyme was solved to 2.1 Å (Stellato et al. 2014). For this experiment a continuous flow of microcrystals passing through the X-ray beam in their mother liquor was established using a thin-walled capillary. The beam was not shuttered between exposures, therefore the time in the beam and dose a crystal receives is determined by the time it takes the crystal to cross the X-ray focus. In this way, millions of detector frames were collected at a constant rate from randomly injected microcrystals at room temperature.

A different approach utilizes raster scanning of solid support systems (Coquelle et al. 2015) or in situ scanning of crystals in crystallization plates (Axford et al. 2015). The position of single

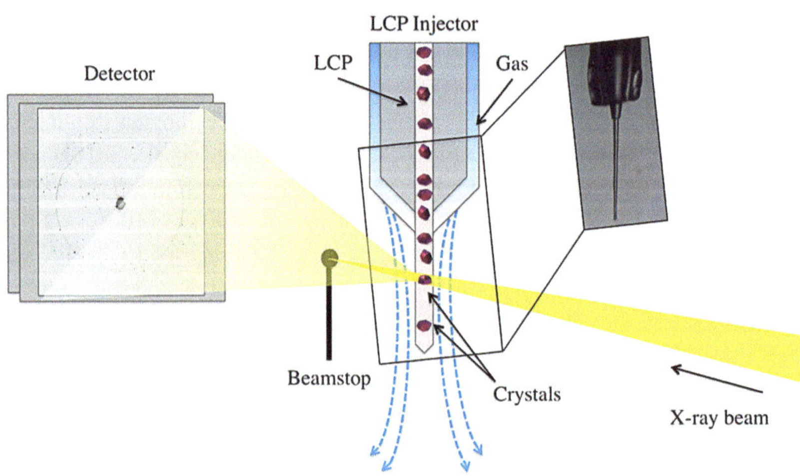

Fig. 10.1 The experimental setup for LCP injector-based SMX data collection at a synchrotron microfocus beamline. The LCP containing protein crystals is delivered as a continuous stream (20–50 μm in diameter) to the crosspoint with the microfocused X-ray beam. A co-axial gas flow (*light blue dashed curved lines*) stabilizes the LCP stream. Diffraction patterns are collected using a standard beamline detector

crystals is bookmarked in a grid and the plate translated to expose each marked crystal. The IMISX (*In Meso In Situ* Crystallography) method combines such a scanning approach with a three-layer sandwich plate, in which the crystals are grown directly in LCP and then the whole plate is transferred into the X-ray beam. With this method the structure of two model membrane proteins were solved successfully (Huang et al. 2015).

Nearly simultaneously, two publications demonstrated LCP microjet-based serial millisecond crystallography, including an example of a membrane protein. One of them describes how LCP crystals of the light-driven proton-pump bacteriorhodopsin (bR) were used to determine the room temperature structure at 2.4 Å using an LCP-injector at a synchrotron source (Nogly et al. 2015). The other one illustrates that lysozyme crystals produced with only 0.3 mg of protein and transferred into LCP can be used for room-temperature structure determination using a modified LCP injector called High-Viscosity Extrusion (HVE) injector at a synchrotron beamline (Botha et al. 2015). Instead of LCP, alternative viscous carrier media such as vaseline and MeBiol gel (Botha et al. 2015) or mineral grease (Sugahara et al. 2015) can be used. The development of further viscous carrier media with high biocompatibility and low X-ray background are an intense focus of current research and more options will likely become available in the near future.

10.2.2.2 Sample Preparation for LCP Injector-Based SMX

Important for a successful SMX experiment is a high density of crystals within the carrier medium to achieve high hit rates for efficient data collection. The crystals should be further homogenous in size with the lower end of the size range dependent on crystal order and beam intensity. Ideally their smallest dimension should be slightly larger than the focal spot of the X-ray beam. The largest dimension of the crystals is determined by the diameter of the nozzle used to extrude the carrier medium. Typically the nozzle of the LCP jet produces a carrier medium column (e.g. LCP) of 20–50 μm with smaller diameters generally being blocked by precipitant, larger crystals or dust particles within the sample and the larger diameters producing higher background scattering or X-ray absorption while providing stable operation over several hours. In cases where crystals are bigger than 50 μm, extensive mixing through a LCP syringe coupler can break them into smaller pieces that are able to pass through the 50 μm nozzle.

In the work of Nogly et al. (Nogly et al. 2015), bacteriorhodopsin crystals for SMX were grown in tube-shaped LCP formed in a gas-tight syringe filled with crystallisation precipitant (Fig. 10.2a). When crystals reach the desired size, the precipitant is removed and the LCP containing crystals is mixed with additional monoolein to transform the residual precipitant into homogeneous LCP (Fig. 10.2b). Extensive mixing through a LCP syringe coupler was used to break crystals bigger than 50 μm into smaller pieces (<50 μm).

Crystals were delivered into the beam with the LCP injector at a flow rate of 20–60 nl min^{-1} generating a continuous LCP column with a diameter of about 50 μm containing crystals in random orientations. Data collection was performed at room temperature at the ID13 microfocus beamline (ESRF) with a beamsize of 2×3 μm, and a mechanical shutter interrupting the X-ray beam to allow collection of diffraction images with exposure times of 10–50 ms at a flux of up to 9.1×10^{11} photons s^{-1}. The dead time between exposures (55 ms) plus data-transfer rate led to data acquisition at 10–17 Hz. In that way, 1,343,092 images were collected of which 12982 were classified as hits using *Cheetah* (Barty et al. 2014) resulting in a hit rate of 1 %. The software *CrystFEL* (White et al. 2012) was used to successfully index and integrate 5691 images out of the 12982 hits. To minimize model bias the structure was solved using sensory rhodopsin (Royant et al. 2001) as a molecular replacement search model in *Phaser* (McCoy et al. 2007) followed by automatic model building with *Phenix.autobuild* (Adams et al. 2002). The auto build model was completed by manual building in *Coot* (Emsley and Cowtan 2004) and further refinement using *PDBREDO* (Joosten et al. 2012). In a final model building and refinement round,

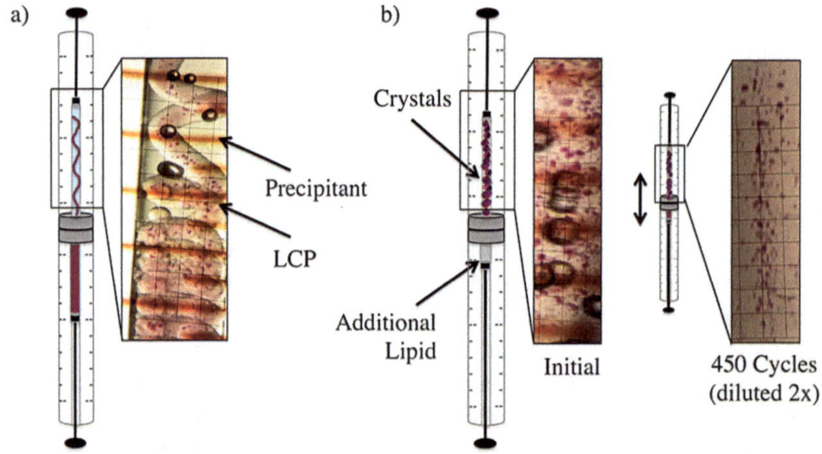

Fig. 10.2 Sample Preparation for SMX using an LCP Injector. (**a**) First, the protein is reconstituted in LCP and up to 20 μl of the LCP mixture is transferred into a second 100 μl gas-tight syringe containing precipitant solution. The first syringe is disconnected and the second syringe sealed to prevent drying out. Crystals are growing inside the LCP tube after incubation at a desired temperature for hours up to several days. (**b**) Before the samples can be loaded into the LCP injector, the remaining precipitant solution has to be removed and additional fresh lipid has to be added until a clear cubic phase is obtained. To homogenise the samples, crystals are passed through the coupler until they reach a homogenous size distribution. In this way, large crystals are broken down into smaller pieces that do not block the LCP injector nozzle. In order to adjust the crystal density, the crystal containing LCP can be diluted with LCP prepared from mixing protein buffer and lipid. A detailed description of a similar crystal preparation procedure for serial crystallography can be found in Liu et al. 2014 (Liu et al. 2014)

Refmac5 (Murshudov et al. 2011) was used to complete the model with water molecules, lipid fragments and all-trans retinal. The refined model contained 225 out of 249 bR residues. Overall, only 0.8 mg of sample in approximately 200 μl LCP was necessary to obtain the data necessary to solve the membrane protein structure using the SMX approach. Importantly, the resulting room temperature structure was very similar to a bR cryostructure (with a RMSD of Cα atoms of 0.54 Å) determined from the same batch of crystals. In this particular case the differences due to cryo-cooling were thus mostly limited to rotamer changes in several amino acid side chains even though the functional properties of proteins can vary significantly at non-physiological temperatures (Weik and Colletier 2010).

In the second microjet-based serial millisecond work (Botha et al. 2015), it was demonstrated that it is possible to apply *de novo* phasing techniques like multiple isomorphous replacement with anomalous scattering (MIRAS) in the SMX approach. A similar high-viscosity extrusion (HVE) injector was used to extrude native and heavy-atom-derivatized lysozyme crystals embedded in various viscous media, including LCP and petroleum jelly, at the Swiss Light Source beamline X10SA. They could show that diffraction quality did not differ for lysozyme crystals grown in LCP directly or being embedded in LCP after growth. Since the control of crystal size during growth is much easier *ex meso*, the following experiments were always performed in that way. A successful MIRAS experiment was performed, using similar lysozyme crystals for a native dataset collected at 9.4 keV, and gold- and iodid-derivatives, soaked in $KAuCl_4$ and potassium iodide and collected at 12.4 keV and 6.5 keV, respectively.

While crystals of the average dimension of 10 × 10 × 30 μm were used for the native data collection and heavy-atom soaks, due to the increased absorption at 6.0 keV crystals of 15 × 15 × 60 μm were used for the sulfur-SAD experiment. The inner diameter of the capillary of the HVE was chosen between 25–100 μm to accommodate crystal size and avoid damage to the crystals, and crystals embedded in LCP were extruded at an average pressure of 6500 psi, resulting in a jet velocity between 100 and 300 $\mu m\ s^{-1}$, depending on capillary diameter and crystal concentration. To match the crystal size and concentration, the X-ray beamsize was kept at $30 \times 10\ \mu m$ (h × v) and data was collected with 10 Hz continuously, reducing the probability to expose more than one crystal per frame, resulting in an average hitrate of 27 %. All data was processed with both dedicated SFX/SMX software packages, *CrystFEL* (White et al. 2012) and *nXDS* (Kabsch 2014), to allow for a direct comparison of their performance. MIRAS phasing was done using *autoSHARP* (Vonrhein et al. 2007). Data processed with *nXDS* showed poorer figures of merit compared to the *CrystFEL* data; however, the *nXDS* data allowed for an almost complete automatically build model using *ARP/wARP* (Langer et al. 2008), while the *CrystFEL* model was only about 50 % complete. They conclude, that the inclusion of more and weaker data in the *CrystFEL* software is preferable for the phasing, while the more conventional intensity contribution of the *nXDS* data is more suitable for model-building.

10.2.2.3 SMX Compared to Standard Crystallography and SFX

These recent examples demonstrate that it is feasible to combine serial crystallography at a synchrotron source with an injector-provided crystal supply. The advantages obtained by performing SMX at synchrotrons in comparison to standard crystallography using cryo-cooled crystals are manifold. First is the greatly simplified crystal handling since there is no difficult freezing procedure required and mounting of crystals in a loop is not necessary. Furthermore, long optimization that may be necessary to grow crystals of sufficient size to allow conventional crystallography is avoided. Importantly, only a single or at most very few snapshots per crystal are taken and many thousands of crystals are merged for one data set. Thus, the radiation dose per crystal stays well below the levels where radiation damage becomes apparent. The continuous crystal flow ensures that thousands of crystals can be screened within a short time and time-consuming crystal alignment in LCP on the goniometer is not necessary. The crystals used for collection of a whole data set may be produced with less than a milligram of protein (Nogly et al. 2015; Botha et al. 2015). What is more, data collection at room temperature and, at least for membrane proteins, native-like lipidic environment will more accurately catch the conformational state of the protein under native conditions and may prevent structural artifacts or altered dynamic behavior. The latter is particularly important for time-resolved pump-probe measurements where proteins within the crystals are required to remain functional properties close to the ones in solution or a lipidic membrane.

In comparison to SMX, SFX experiments profit from a short X-ray pulse that terminates before most radiation damage processes have manifested themself and therefore circumvents radiation damage almost completely. Each crystal is destroyed after data collection, but the high intensity allows collection of high-resolution data even from sub-micron crystals. The intensity of a synchrotron beam is limited by the available photon flux and longer exposure times are necessary to obtain high-resolution data. Therefore, radiation damage still plays a role within SMX experiments and the crystals have to be larger compared to SFX experiments. Another difference with SMX measurements is that the crystals do not stand still during the exposure on millisecond timescale, therefore tumbling of crystals might be taken into account during data processing by introducing a small oscillation range.

In addition, SMX may yield better quality of data (Botha et al. 2015) as the X-ray beam produced by a synchrotron is very stable in comparison to the several fold variation in the intensity of XFEL pulses and so fewer random errors need to be accounted for. However, the main sources of error in serial crystallography are not well understood. Recent studies suggest that variations in XFEL pulses and crystal-to-crystal variation play only a minor role (Barends et al. 2015). Indeed already the first example for a membrane protein structure that could not be solved using conventional crystallography has been recently

Table 10.1 Characteristics of SFX, SMX and conventional oscillation-based cryo-crystallography

	SFX	SMX	Conventional
Sample			
Protein per data set	<0.5 mg	<0.5 mg	<0.1 mg
Crystal size	~1–40 μm	~5–40 μm	Typically >30 μm
Number of crystals	Thousands	Thousands	Single to few
Delivery medium	Liquid & viscous	Viscous	None or cryo-protectant
X-ray beam			
Focus size	1–5 μm	1–10 μm	5–50 μm
Intensity	Very high	High	Medium to high
Availability	Low	Medium	High
Radiation damage			
Dose per crystal	High	Low	Medium to high
Global	None	Low	Low to high
Site specific	None to low	Low	Low to high
Data collection			
Oscillation range	None	Negligible	Variable
Temperature	RT	RT	Cryo
Exposure time	fs	ms	ms to s
Time needed	Hours	Hours	Minutes
Time-resolved pump probe compatibility	High (fs – ms)	Medium (ms)	Limited to radiation hard crystals with low mosaicity
Structure determination			
Molecular replacement (number of patterns)	5000–10000	5000–10000	Hundreds
De novo structure determination	Derivative, native-SAD	Derivative, native-SAD	Derivative, native-SAD
Grade of implementation			
Experimental stations	Dedicated stations	Pilot experiments	Well established and optimized
Data integration	Monte Carlo	Monte Carlo	Optimized
Structure refinement	Advanced	Advanced	Highly optimized

described (Zhang et al. 2015). Development of more robust data integration routines may thus very well lead to data quality comparable to that collected on highly optimized beamlines for conventional crystallography. The different crystallographic methods – SFX, SMX and conventional oscillation-based cryo-crystallography are summarized and compared in Table 10.1

10.2.2.4 SMX Technical Details and Future Developments

Development of a novel approach like SF/MX does not come without its own set of challenges and many problems will have to be solved until it becomes a standard crystallography method. Especially, the specialized software to evaluate and process serial crystallographic data is in rapid development. As each crystal is used for only one short exposure and not rotated in controlled fashion as in conventional crystallography, each diffraction pattern contains only partially recorded intensities. Furthermore, there is no spatial relationship between the individually collected patterns and even the same Bragg intensities will vary between different crystals due to variations in crystal size, shape, and diffraction quality. The Monte Carlo method of integration (Kirian et al. 2010) has been developed to overcome this problem but relies on very high redundancy to integrate and merge the collected patterns into a cohesive dataset. Sorting into empty images and images with diffraction patterns using software such as *Cheetah* (Barty et al. 2014) allows to reduce data volume to levels that can be handled using semi-automated processing software such as *CrystFEL* (White et al. 2012), *cctb.xfel* (Sauter et al. 2013) or *nXDS* (Kabsch 2014). An essential improvement were computational methods to solve the indexing ambiguity arising from merging single shot diffraction patterns from crystals with different indexing possibilities (Brehm and Diederichs 2014; Liu and Spence 2014), something which would have made serial crystallography unsuitable for 27 out of the possible 65 crystallographic space groups. Another important milestone was the demonstration that *de novo* phasing is feasible using SFX (Barends et al. 2014) and SMX data (Botha et al. 2015).

The number of diffraction patterns merged into a typical serial crystallographic dataset depends to a large extent on what method is used for structure determination, i.e. molecular replacement needs fewer diffraction patterns compared to *de novo* phasing with the latter being also dependent on the phasing power of the used derivative. Merging of more additional images improves the resolution to which a dataset is usable. Recently a series of post-refinement routines (Uervirojnangkoorn et al. 2015; Ginn et al. 2015; White 2014) has been published that introduce improved methods to model serial crystallographic data and thus lead to convergence of a dataset with fewer images. Implementation of these techniques will lead to data collection times much closer to the ones common for conventional crystallography and likely faster than scanning microcrystallography techniques used to solve structures from crystals embedded in LCP (Cherezov et al. 2009). Future development of algorithms to select the subset of images collected from the best diffracting crystals will allow to make best use of inherently high-throughput injector based serial crystallography to achieve higher resolution.

Many of the advantages of serial crystallography depend on the ability to collect data from many crystals in a fast and efficient way. In the pilot SMX experiments (Botha et al. 2015; Nogly et al. 2015) the available detectors limited the rate of data collection to 10–20 Hz. With faster detectors, as for example the Eiger (Johnson et al. 2012, 2014) (Dectris Ltd.), which are currently being installed at many macromolecular crystallography beamlines, data can in principle be collected at rates around 100–750 Hz. This will decrease the exposure time for a single diffraction image to maximal 10–1.3 ms and thus the intensity of the X-ray beamlines and the diffraction power of individual crystals will become the limiting factor that determines the effective data collection rate. However, a solution for this problem seems to be on the horizon. Pioneered by the MaxIV synchrotron currently under construction in Lund, Sweden, many of the 3rd generation synchrotrons, including ESRF, DLS, APS and SLS, plan to upgrade their lattice to a multi bend achromat (MBA) design, bringing them close to a diffraction limited storage ring (Hettel 2014). The diffraction limited storage ring will significantly reduce the emittance, giving a several orders of magnitude higher flux density at the sample position, while exhibiting reduced divergence at the same time (Fig. 10.3). Combined with an air-free beampath, either provided by transferring the beamline into vacuum, or purging the beampath with helium, the signal-to-noise ratio can be significantly improved. After such improvements have been implemented, microjet-based SMX experiments could offer many of the advantages of SFX at XFELs, but can be performed at microfocus beamlines at synchrotron sources where beamtime is more abundant.

10.3 Future Perspectives of SMX

Although there are still some technical difficulties to be overcome in order to use SMX as a standard crystallography approach, investment in this technique may well be worth it. SMX will reduce the time for conducting crystal optimisation, practicing crystal handling and freezing, especially for membrane proteins crystallised in LCP where manual handling is still a major

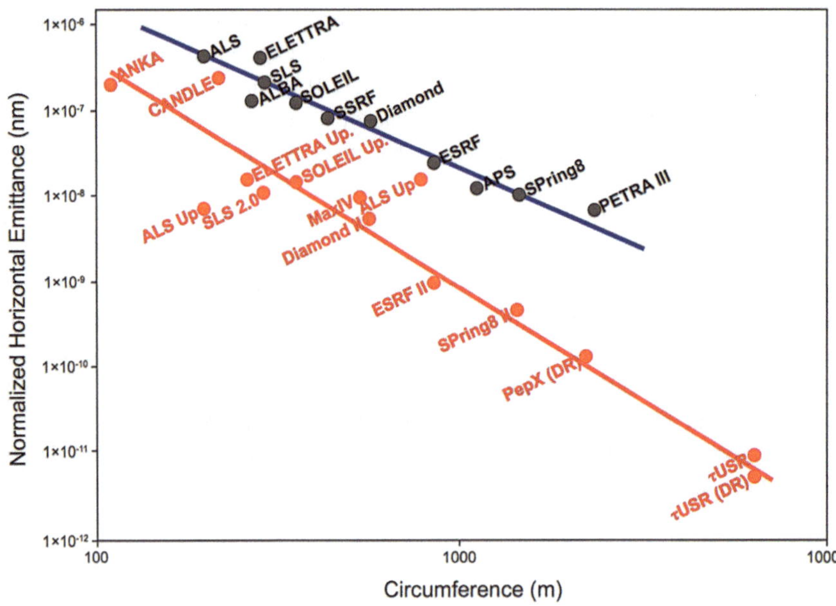

Fig. 10.3 Comparison of 3rd to 4th generation synchrotrons. The energy normalized horizontal emittance of existing and planned storage rings around the world in regards to their circumference. A clear trend of significant reduction in the horziontal emittance is visible, clearly distinguishing 3rd generation (*blue*) from 4th generation (*red*) machines (Adapted from R. Bartolini)

bottleneck. Structure based drug design of membrane proteins is often limited by the solubility of the small molecular compounds which need to be soaked into crystals. Solubilisation of these compounds is usually achieved by using DMSO or other solubilising agents, which commonly harms the fragile high-solvent crystal packing of membrane proteins. What is more, in prominent drug-targets like G protein-coupled receptors, the ligand-binding pocket is often only accessible from the membrane site and binding ligands are highly hydrophobic. In these cases, hydrophobic delivery agents used in injector based serial crystallography might be an efficient delivery agent. Small molecular drugs could be directly added to the gas-tight syringes used for crystal growth and then loaded into the injector. In this way, many compounds might be screened within a short time and tedious soaking experiments could be avoided. Therefore, SMX has the potential to develop into a higher throughput method for the determination of ligand-protein complexes.

Since the amount of protein used for SMX is low and thousands of crystals can be screened within a short time, also poorly expressing and hard to crystallise membrane proteins may yield a structure. Furthermore, the data collection at room temperature may more accurately present the conformational state of a protein under native conditions or reflects its dynamics. With the discussed synchrotron upgrades and implementation of appropriate pump-probe setups, time-resolved diffraction studies on the 100 ps to millisecond scale may also become more routine, complementing the faster time regimes, which will likely remain the domain of free-electron lasers (Schotte et al. 2003; Graber et al. 2011; Schotte et al. 2012).

A special synergy between XFELs and synchrotrons might be established on sites where both X-ray sources are available. SMX could be used as a more standard approach for solving structures from microcrystals at room temperature, high throughput ligand or pre-crystal screening. In contrast, the SFX beamtime could be more dedicated to time-resolved studies and weakly diffracting submicron sized crystals. SMX and SFX going hand-in hand may promise a brilliant future in (membrane protein) crystallography.

Acknowledgments The work was financially supported by A Co-fund PSI Fellowship (to P.N.) and the SNSF project grant 31003A_141235 and 31003A_159558 (to J.S.).

References

Adams PD, Grosse-Kunstleve RW, Hung L-W, Ioerger TR, McCoy AJ et al (2002) PHENIX: building new software for automated crystallographic structure determination. Acta Crystallogr D Biol Crystallogr 58(11):1948–1954

Axford D, Foadi J, Hu N-J, Choudhury HG, Iwata S et al (2015) Structure determination of an integral membrane protein at room temperature from crystals in situ. Acta Crystallogr D Biol Crystallogr 71(6):1228–1237

Barends TR, Foucar L, Botha S, Doak RB, Shoeman RL et al (2014) De novo protein crystal structure determination from X-ray free-electron laser data. Nature 505:244–247

Barends T, White TA, Barty A, Foucar L, Messerschmidt M et al (2015) Effects of self-seeding and crystal post-selection on the quality of Monte Carlo-integrated SFX data. J Synchrotron Radiat 22(3):644–652

Barty A, Kirian RA, Maia FR, Hantke M, Yoon CH et al (2014) Cheetah: software for high-throughput reduction and analysis of serial femtosecond X-ray diffraction data. J Appl Crystallogr 47(3):1118–1131

Botha S, Nass K, Barends TR, Kabsch W, Latz B et al (2015) Room-temperature serial crystallography at synchrotron X-ray sources using slowly flowing free-standing high-viscosity microstreams. Acta Crystallogr D Biol Crystallogr 71(2):387–397

Boutet S, Lomb L, Williams GJ, Barends TR, Aquila A et al (2012) High-resolution protein structure determination by serial femtosecond crystallography. Science 337:362–364

Brehm W, Diederichs K (2014) Breaking the indexing ambiguity in serial crystallography. Acta Crystallogr D Biol Crystallogr 70(1):101–109

Caffrey M (2015) A comprehensive review of the lipid cubic phase or in meso method for crystallizing membrane and soluble proteins and complexes. Acta Crystallogr Sect F Struct Biol Cryst Commun 71(1):3–18

Caffrey M, Cherezov V (2009) Crystallizing membrane proteins using lipidic mesophases. Nat Protoc 4(5):706–731

Caffrey M, Li D, Howe N, Shah ST (2014) 'Hit and run' serial femtosecond crystallography of a membrane kinase in the lipid cubic phase. Philos Trans R Soc Lond Ser B 369:20130621

Carpenter EP, Beis K, Cameron AD, Iwata S (2008) Overcoming the challenges of membrane protein crystallography. Curr Opin Struct Biol 18(5):581–586

Chapman HN, Fromme P, Barty A, White TA, Kirian RA et al (2011) Femtosecond X-ray protein nanocrystallography. Nature 470:73–77

Cherezov V (2011) Lipidic cubic phase technologies for membrane protein structural studies. Curr Opin Struct Biol 21(4):559–566

Cherezov V, Hanson MA, Griffith MT, Hilgart MC, Sanishvili R et al (2009) Rastering strategy for screening and centring of microcrystal samples of human membrane proteins with a sub-10 microm size X-ray synchrotron beam. J R Soc Interface 6: S587–S597

Coquelle N, Brewster AS, Kapp U, Shilova A, Weinhausen B et al (2015) Raster-scanning serial protein crystallography using micro- and nano-focused synchrotron beams. Acta Crystallogr D Biol Crystallogr 71(5):1184–1196

Demirci H, Sierra RG, Laksmono H, Shoeman RL, Botha S, Barends TRM, Nass K, Schlichting I, Doak RB, Gati C, Williams GJ, Boutet S, Messerschmidt M, Jogl G, Dahlberg AE, Gregory ST, Bogan MJ (2013) Serial femtosecond X-ray diffraction of 30S ribosomal subunit microcrystals in liquid suspension at ambient temperature using an X-ray free-electron laser. Acta Crystallogr Sect F Struct Biol Cryst Commun 69(9):1066–1069

DePonte DP, Weierstall U, Schmidt K, Warner J, Starodub D et al (2008) Gas dynamic virtual nozzle for generation of microscopic droplet streams. J Phys D Appl Phys 41(19):195505

Emsley P, Cowtan K (2004) Coot: model-building tools for molecular graphics. Acta Crystallogr D Biol Crystallogr 60(12):2126–2132

Garman EF (2010) Radiation damage in macromolecular crystallography: what is it and why should we care? Acta Crystallogr D Biol Crystallogr 66(4): 339–351

Garman EF, Schneider TR (1997) Macromolecular cryocrystallography. J Appl Crystallogr 30:211–237

Ginn HM, Brewster AS, Hattne J, Evans G, Wagner A et al (2015) A revised partiality model and post-refinement algorithm for X-ray free-electron laser data. Acta Crystallogr D Biol Crystallogr 71(6):1400–1410

Graber T, Anderson S, Brewer H, Chen YS, Cho HS et al (2011) BioCARS: a synchrotron resource for time-resolved X-ray science. J Synchrotron Radiat 18(4):658–670

Hettel R (2014) DLSR design and plans: an international overview. J Synchrotron Radiat 21(5):843–855

Heymann M, Opthalage A, Wierman JL, Akella S, Szebenyi DM et al (2014) Room-temperature serial crystallography using a kinetically optimized microfluidic device for protein crystallization and on-chip X-ray diffraction. IUCrJ 1(5):349–360

Hirata K, Shinzawa-Itoh K, Yano N, Takemura S, Kato K et al (2014) Determination of damage-free crystal structure of an X-ray-sensitive protein using an XFEL. Nat Methods 11(7):734–736

Huang C-Y, Olieric V, Ma P, Panepucci E, Diederichs K et al (2015) In meso in situ serial X-ray crystallography of soluble and membrane proteins. Acta Crystallogr D Biol Crystallogr 71(6):1399–0047

Johansson LC, Arnlund D, White TA, Katona G, Deponte DP et al (2012) Lipidic phase membrane protein serial femtosecond crystallography. Nat Methods 9(3): 263–265

Johansson LC, Arnlund D, Katona G, White TA, Barty A et al (2013) Structure of a photosynthetic reaction centre determined by serial femtosecond crystallography. Nat Commun 4:2911

Johnson I, Bergamaschi A, Buitenhuis J, Dinapoli R, Greiffenberg D et al (2012) Capturing dynamics with Eiger, a fast-framing X-ray detector. J Synchrotron Radiat 19(6):1001–1005

Johnson I, Bergamaschia A, Billich H, Cartier S, Dinapoli R, Greiffenberg D, Guizar-Sicairos M, Henrich B, Jungmann J, Mezza D, Mozzanica A, Schmitt B, Shi X, Tinti G (2014) Eiger: a single-photon counting x784 ray detector. J Instrum 9

Joosten RP, Joosten K, Murshudov GN, Perrakis A (2012) PDB_REDO: constructive validation, more than just looking for errors. Acta Crystallogr D Biol Crystallogr 68(4):484–496

Kabsch W (2014) Processing of X-ray snapshots from crystals in random orientations. Acta Crystallogr D Biol Crystallogr 70(8):2204–2216

Kern J, Alonso-Mori R, Tran R, Hattne J, Gildea RJ et al (2013) Simultaneous femtosecond X-ray spectroscopy and diffraction of photosystem II at room temperature. Science 340:491–495

Kern J, Tran R, Alonso-Mori R, Koroidov S, Echols N et al (2014) Taking snapshots of photosynthetic water oxidation using femtosecond X-ray diffraction and spectroscopy. Nat Commun 5:4371

Kirian RA, Wang X, Weierstall U, Schmidt KE, Spence JC et al (2010) Femtosecond protein nanocrystallography-data analysis methods. Opt Express 18:5713–5723

Landau EM, Rosenbusch JP (1996) Lipidic cubic phases: a novel concept for the crystallization of membrane proteins. Proc Natl Acad Sci U S A 93: 14532–14535

Langer G, Cohen SX, Lamzin VS, Perrakis A (2008) Automated macromolecular model building for X-ray crystallography using ARP/wARP version 7. Nat Protoc 3(7):1171–1179

Liu H, Spence JC (2014) The indexing ambiguity in serial femtosecond crystallography (SFX) resolved using an expectation maximization algorithm. IUCrJ 1(6): 393–401

Liu W, Wacker D, Gati C, Han GW, James D et al (2013) Serial femtosecond crystallography of G protein-coupled receptors. Science 342:1521–1524

Liu W, Ishchenko A, Cherezov V (2014) Preparation of microcrystals in lipidic cubic phase for serial femtosecond crystallography. Nat Protoc 9: 2123–2134

Lomb L, Barends TR, Kassemeyer S, Aquila A, Epp SW et al (2011) Radiation damage in protein serial femtosecond crystallography using an x-ray free-electron laser. Phys Rev B Condens Matter Mater Phys 84(21):214111

McCoy AJ, Grosse-Kunstleve RW, Adams PD, Winn MD, Storoni LC, Read RJ (2007) Phaser crystallographic software. J Appl Crystallogr 40(4): 658–674

Murshudov GN, Skubak P, Lebedev AA, Pannu NS, Steiner RA, Nicholls RA, Winn MD, Long F, Vagin AA (2011) REFMAC5 for the refinement of macromolecular crystal structures. Acta Crystallogr D Biol Crystallogr 67(4):355–367

Nass K, Foucar L, Barends TR, Hartmann E, Botha s et al (2015) Indications of radiation damage in ferredoxin microcrystals using high-intensity X-FEL beams. J Synchrotron Radiat 22(2):225–238

Neutze R, Wouts R, van der Spoel D, Weckert E, Hajdu J (2000) Potential for biomolecular imaging with femtosecond X-ray pulses. Nature 406:752–757

Nogly P, James D, Wang D, White T, Zatsepin N et al (2015) Lipidic cubic phase serial millisecond crystallography using synchrotron radiation. IUCrJ 2(2):168–176

Owen RL, Axford D, Nettleship JE, Owens RJ, Robinson JI et al (2012) Outrunning free radicals in room-temperature macromolecular crystallography. Acta Crystallogr D Biol Crystallogr 68(7): 810–818

Park J, Joti Y, Ishikawa T, Song C (2013) Monte Carlo study for optimal conditions in single-shot imaging with femtosecond x-ray laser pulses. Appl Phys Lett 103(26):264101

Redecke L, Nass K, DePonte DP, White TA, Rehders D et al (2013) Natively inhibited Trypanosoma brucei cathepsin B structure determined by using an X-ray laser. Science 339:227–230

Royant A, Nollert P, Edman K, Neutze R, Landau EM et al (2001) X-ray structure of sensory rhodopsin II at 2.1-A resolution. Proc Natl Acad Sci U S A 98: 10131–10136

Sauter NK, Hattne J, Grosse-Kunstleve RW, Echols N (2013) New Python-based methods for data processing. Acta Crystallogr D Biol Crystallogr 69(7):1274–1282

Schotte F, Lim M, Jackson TA, Smirnov AV, Soman J et al (2003) Watching a protein as it functions with 150-ps time-resolved x-ray crystallography. Science 300:1944–1947

Schotte F, Cho HS, Kaila VR, Kamikubo H, Dashdorj N et al (2012) Watching a signaling protein function in real time via 100-ps time-resolved Laue crystallography. Proc Natl Acad Sci U S A 109:19256–19261

Stellato F, Oberthur D, Liang M, Bean R, Gati C et al (2014) Room-temperature macromolecular serial crystallography using synchrotron radiation. IUCrJ 1(4):204–212

Sugahara M, Mizohata E, Nango E, Suzuki M, Tanaka T et al (2015) Grease matrix as a versatile carrier of proteins for serial crystallography. Nat Methods 12:61–63

Uervirojnangkoorn M, Zeldin OB, Lyubimov AY, Hattne J, Brewster AS et al (2015) Enabling X-ray free electron laser crystallography for challenging biological systems from a limited number of crystals. Elife 4:e05421

Vonrhein C, Blanc E, Roversi P, Bricogne G (2007) Automated structure solution with autoSHARP. Methods Mol Biol 364:215–230

Weierstall U, Spence JC, Doak RB (2012) Injector for scattering measurements on fully solvated biospecies. Rev Sci Instrum 83(3):35108

Weierstall U, James D, Wang C, White TA, Wang D et al (2014) Lipidic cubic phase injector facilitates membrane protein serial femtosecond crystallography. Nat Commun 5:3309

Weik M, Colletier JP (2010) Temperature-dependent macromolecular X-ray crystallography. Acta Crystallogr D Biol Crystallogr 66(4):437–446

White SH (2004) The progress of membrane protein structure determination. Protein Sci 13(7):1948–1949

White TA (2014) Post-refinement method for snapshot serial crystallography. Philos Trans R Soc Lond Ser B 369:20130330

White TA, Kirian RA, Martin AV, Aquila A, Nass K et al (2012) CrystFEL: a software suite for snapshot serial crystallography. J Appl Crystallogr 45(2):335–341

Yildirim MA, Goh KI, Cusick ME, Barabasi AL, Vidal M (2007) Drug-target network. Nat Biotechnol 25: 1119–1126

Zhang H, Unal H, Gati C, Han GW, Liu W et al (2015) Structure of the Angiotensin receptor revealed by serial femtosecond crystallography. Cell 161: 833–844

Serial Femtosecond Crystallography of Membrane Proteins

11

Lan Zhu, Uwe Weierstall, Vadim Cherezov, and Wei Liu

Abstract

Membrane proteins, including G protein-coupled receptors (GPCRs), constitute the most important drug targets. The increasing number of targets requires new structural information, which has proven tremendously challenging due to the difficulties in growing diffraction-quality crystals. Recent developments of serial femtosecond crystallography at X-ray free electron lasers combined with the use of membrane-mimetic gel-like matrix of lipidic cubic phase (LCP-SFX) for crystal growth and delivery hold significant promise to accelerate structural studies of membrane proteins. This chapter describes the development and current status of the LCP-SFX technology and elaborates its future role in structural biology of membrane proteins.

Keywords

Serial femtosecond crystallography • X-ray free electron laser • Lipidic cubic phase • LCP-SFX • LCP injector • Membrane proteins • G protein-coupled receptors

11.1 Introduction

Membrane proteins, accounting for approximately 30 % of all proteins in eukaryotic cells (Wallin and von Heijne 1998), play crucial roles in multiple cellular processes, from transmitting extracellular signals, transporting cellular

L. Zhu • W. Liu (✉)
School of Molecular Sciences, Arizona State University, Tempe, AZ 85287, USA

Center for Applied Structural Discovery at the Biodesign Institute, Arizona State University, Tempe, AZ 85287-1604, USA
e-mail: lan.zhu@asu.edu; w.liu@asu.edu

U. Weierstall
Center for Applied Structural Discovery at the Biodesign Institute, Arizona State University, Tempe, AZ 85287-1604, USA

Department of Physics, Arizona State University, Tempe, AZ 85287, USA
e-mail: weier@asu.edu

V. Cherezov
Bridge Institute, University of Southern California, Los Angeles, CA 90089, USA

Department of Chemistry, University of Southern California, Los Angeles, CA 90089, USA
e-mail: cherezov@usc.edu

I. Moraes (ed.), *The Next Generation in Membrane Protein Structure Determination*,
Advances in Experimental Medicine and Biology 922, DOI 10.1007/978-3-319-35072-1_11

components across the membrane, and catalysing chemical reactions, to mediating cellular interactions. Currently, approximately 60 % of approved drugs target membrane proteins, not only due to their vital roles, but also because of the easiness in accessing them at the cell surface (Yildirim et al. 2007). Structure characterization of membrane proteins is fundamental for the comprehensive understanding of their mechanisms of action and biological functions. However, their residence within a lipid bilayer renders great difficulties for the isolation, purification, stabilisation and crystallisation of membrane proteins, and thus makes their structure determination challenging.

The first high-resolution three-dimensional structure of a membrane protein was solved by X-ray crystallographic analysis in 1985 (Deisenhofer et al. 1985). After decades of efforts to develop effective approaches for structure determination, X-ray crystallography stands out as the most efficient tool for high-resolution three-dimensional structure determination of membrane proteins, attributing to over 600 unique membrane protein structures deposited in the Protein Data Bank (http://www.pdb.org). However, compared with soluble proteins, the progress for structure determination of membrane proteins lags far behind. Among the notoriously difficult membrane protein families, G protein-coupled receptors (GPCRs) have yielded about 30 unique structures so far (Shonberg et al. 2014).

GPCRs contain more than 800 members in humans, with about 200 of them being validated drug targets. Due to the low expression and stability of these receptors as well as their highly dynamic nature, they have resisted structural determination for many years and the first high-resolution structure of a GPCR bound to a diffusible ligand was solved by X-ray crystallography only recently in 2007 (Cherezov et al. 2007). This structure, as well as most GPCR structures solved since then, were derived from small but well-ordered crystals grown in a lipidic cubic phase (LCP) (Landau and Rosenbusch 1996; Caffrey and Cherezov 2009). However, the high X-ray flux required to collect high-resolution data from small crystals results in radiation damage (Meents et al. 2010), even in cryo-cooled samples. Radiation damage at synchrotron sources, together with the low-expression yield of membrane proteins and difficulties with growing diffraction quality crystals, remain the major bottleneck in this field (Juers and Matthews 2004).

Radiation damage can however be outrun at X-ray free-electron lasers (XFELs). An XFEL provides ultrashort X-ray pulses with a peak brilliance that is over nine orders of magnitude higher than from third-generation synchrotron sources, enabling structure determination from micrometer- or sub-micrometer-size nanocrystals with minimal radiation damage, based on the principle of "diffraction-before-destruction" (Neutze et al. 2000). The method of serial femtosecond crystallography (SFX) is able to minimize the radiation damage by using those intense and ultrashort pulses to collect a series of snapshot diffraction patterns from microcrystals at room temperature, eliminating the steps of crystal harvesting and cryo-cooling (Chapman et al. 2011; Boutet et al. 2012; Redecke et al. 2013). The diffraction snapshots originate from different crystals in random orientation and differing in size. These diffraction patterns constitute "stills", i.e. slices through the Ewald sphere containing only partial Bragg reflections. Therefore tens of thousands snapshots from crystals in random orientation are typically needed to adequately sample reciprocal space.

In this chapter, we will review the recent evolution of serial femtosecond crystallography, the application of SFX on membrane proteins, and an innovative modified LCP-SFX approach combining SFX and LCP crystallisation.

11.2 Challenges in Structure Determination of Membrane Proteins

Membrane proteins are relatively unstable once isolated from their native lipid bilayer membrane environment. Their crystallization in detergent micelles often leads to large but poorly diffracting crystals. In contrast, crystals grown in LCP are typically smaller in size, more ordered and contain fewer defects that contribute to mosaicity (Cherezov 2011). The high X-ray dose at conventional synchrotron sources leads to radiation damage and associated structure modification. To address this issue, cryogenic cooling of microcrystalline samples is used (Meents et al. 2010), where the crystal is cooled to 100 K to reduce secondary radiation damage whereas primary ionization events cannot be prevented (Meents et al. 2010; Juers and Matthews 2004).

Overall, structural studies of membrane proteins are hampered by the initial crystal quality as well as the limited growth capacity to large well-ordered crystals withstanding radiation damage.

11.3 History of Serial Femtosecond Crystallography

To overcome X-ray-induced radiation damage, it was proposed to use extremely intense X-ray pulses with short enough pulse duration that diffraction data can be recorded before the structural destruction occurs (Solem 1986). The first hard X-ray FEL, the Linac Coherent Light Source (LCLS) at SLAC in Stanford allowed to test these ideas for the first time (Emma et al. 2010). Together with fast-readout detectors and sample delivery in a liquid stream, short-wavelength X-ray pulses with high intensity at LCLS enabled for the first time high-resolution structure determination of protein molecules at room temperature from nano- and microcrystals. Photosystem I (PSI) was used as the first crystallographic experimental model at LCLS (Chapman et al. 2011). In this experiment, randomly orientated hydrated microcrystals in crystallisation solution were delivered in a liquid jet (DePonte et al. 2008), and intersected with the X-ray beam in vacuum. Thousands of single femtosecond snapshot diffraction patterns from microcrystals were collected in a serial fashion. New data analysis methods had to be developed to merge the resulting "still" images into a 3D data set (Kirian et al. 2010). Although the X-ray wavelength was limited at the time to 6.9 Å, which in turn limited the resolution of the PSI structure to 8.4 Å, it was a first step towards atomic resolution SFX with an XFEL. Data collection to a resolution better than 2 Å became possible with the commissioning of the LCLS Coherent X-ray Imaging (CXI) instrument. The model system for this study was hen egg-white lysozyme and the structure could be solved from micron sized crystals to a resolution of 1.9 Å with an X-ray wavelength of 1.32 Å (Boutet et al. 2012). Since then, the LCLS has been used for SFX structure determination of soluble and membrane proteins (Aquila et al. 2012; Johansson et al. 2012; Kern et al. 2012; Liu et al. 2013; Kupitz et al. 2014; Weierstall et al. 2014).

11.4 The Emergence of Serial Femtosecond Crystallography as a Tool for Structural Studies of Membrane Proteins

In serial femtosecond crystallography, extremely short X-ray pulses with high energy intersect a stream of protein microcrystals in random orientation and variable size (Chapman et al. 2011). At the LCLS, depending on crystal concentration, up to 120 "still" diffraction patterns from different crystals are recorded per second. After recording and indexing thousands of patterns, a Monte Carlo integration yields the structure factors in a format suitable for standard crystallography software (Kirian et al. 2011).

Following the first success of SFX on PSI crystallized in detergent solution (Chapman et al. 2011). SFX was applied to the structural studies of other membranes proteins, such as the photosynthetic reaction center from *Blastochloris viridis* (RCvir) that was crystallised in a liquid-like lipidic sponge phase (Johansson et al. 2012). This study was still limited by the wavelength of 6.17 Å that was available at the time, resulting in an 8.2 Å resolution structure of RCvir. A follow up SFX study on the same protein, but at higher X-ray energy of 9.34 keV corresponding to a wavelength of 1.32 Å led to a 3.5 Å resolution structure (Johansson et al. 2013).

In order to minimise microcrystal sample waste and achieve high hit rate, a well-designed sample delivery system is essential, which streams microcrystals in their crystallisation environment into the intersection with the pulsed XFEL beam.

Currently, two liquid sample delivery systems have been developed and used for SFX data collection. The most used so far is the gas dynamic virtual nozzle (GDVN) injector, which generates a continuous liquid stream of 1–20 μm diameter (DePonte et al. 2008). In the GDVN, the liquid containing protein microcrystals is focused by a sheath gas to a small diameter ("virtual aperture"), thereby avoiding clogging problems associated with "real" apertures (DePonte et al. 2008). The flow rate necessary to sustain continuous jetting with a GDVN nozzle is about 7 $\mu l \cdot min^{-1}$, which results in high sample consumption and sample waste, since at the current LCLS repetition rate of 120 Hz most sample runs to waste in between X-ray pulses, requiring 10–100 mg of crystallised protein for a complete dataset collection. Nevertheless, this type of injector has been successfully used for several structural studies of membrane proteins (Chapman et al. 2011; Johansson et al. 2012, 2013; Kupitz et al. 2014).

A second injector uses the principle of electrospinning, where the liquid is subjected to a 3.4–5.4 $kV \cdot cm^{-1}$ electric field. The high field generates a liquid meniscus (Taylor cone) similar as in Electrospray, but in electrospinning droplet formation is delayed by adding glycerol and/or polyethylene glycol and an unbroken stream of charged liquid is produced. This system has been used to deliver microcrystals inside a vacuum chamber with a flow rate of 0.17–3.1 $\mu l\ min^{-1}$ (Kern et al. 2012; Sierra et al. 2012). A caveat of this method is, that it requires the addition of a cryoprotectant to prevent freezing of the liquid in vacuum.

11.5 Serial Femtosecond Crystallography of Membrane Proteins in Lipidic Cubic Phase

Lipidic cubic phase represents a membrane-mimetic material that supports crystal growth of membrane proteins. While crystals that grow in LCP are typically small, they are usually well-ordered. Optimisation of crystallisation condition in LCP to obtain sufficiently large crystals suitable for conventional synchrotron X-ray crystallography is a time-consuming process. Small crystals in LCP are therefore ideal for serial femtosecond crystallography at XFELs, which eliminates many crystal optimisation steps and crystal harvesting.

The high viscosity and gel-like consistency of LCP allows efficient microcrystal delivery without much sample waste in a stream of LCP produced by a special injector, which is described next in the chapter (Weierstall et al. 2014). Acquisition of a complete dataset by the LCP-SFX method requires several tens of microlitres of LCP with homogenously dispersed microcrystals at high density. An optimized protocol of microcrystal preparation for LCP-SFX was recently published (Liu et al. 2014). Initial crystal hits are acquired from the nanoliter volume high-throughput robotic screening of LCP crystallization conditions in 96-well glass sandwich plates (Cherezov et al. 2004). Successful conditions producing high-density of small crystals are then scaled up by approximately 1000-times inside gas-tight Hamilton syringes (Fig. 11.1).

The most common host lipid for LCP crystallisation is monoolein, or 9.9 MAG (a monoacylglycerol with 9 hydrocarbons before and 9 after a single double bond). It was observed that using monoolein-based LCP might lead to a problem when sample is injected into a vacuum chamber. Evaporative cooling and dehydration transforms LCP into a lamellar crystalline phase, which produces strong diffraction rings that can damage the detector. To overcome this, 7.9 MAG lipid that was designed for low temperature crystallisation (Misquitta et al. 2004) can be added to the sample after crystals have grown (Liu et al. 2014).

11.6 Microcrystal Sample Delivery System in LCP-SFX

In order to generate a micron sized stream of LCP which has a dynamic viscosity of ~48.3 Pas (Perry et al. 2009), it was necessary to develop a new device, since the GDVN nozzle cannot be used with such high viscosity materials. The

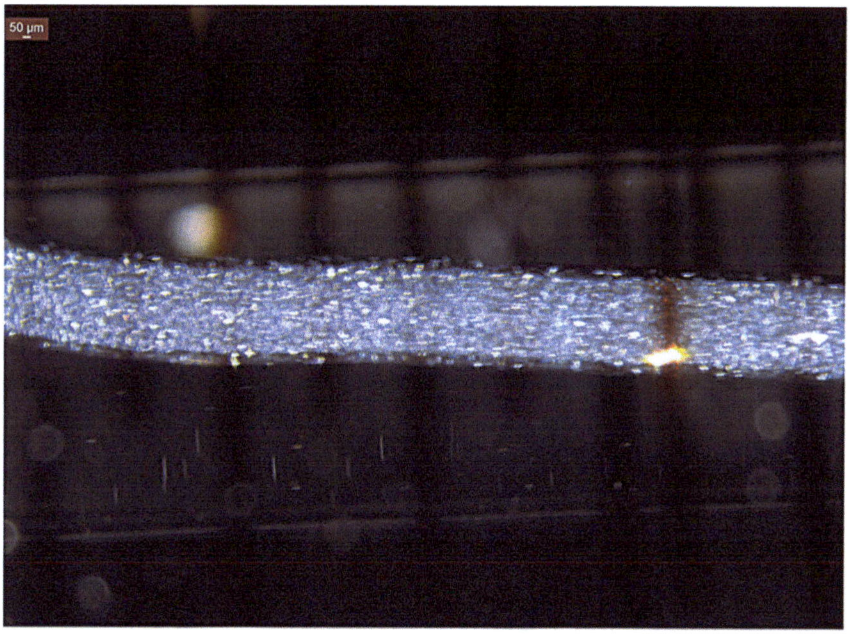

Fig. 11.1 Cross-polarized image of Angiotensin II receptor type 1 (AT1R) LCP microcrystals grown in gas-tight Hamilton syringes

goal was to design a system that can reliably extrude a viscous stream of 10–50 µm diameter at variable speed adjustable to the X-ray pulse repetition rate. The injection process has to be reliable at both vacuum and ambient pressures (e.g. in Helium atmosphere), since XFEL experiments are often conducted in vacuum to reduce background scattering. The stream speed has to be adjustable in the range of about 10 µm/s to 3 mm/s to allow the damaged sample to be replenished by fresh sample for X-ray pulse repetition rates of 10–120 Hz.

The final design of the LCP injector is shown in Fig. 11.2 and consists of a hydraulic stage, a sample reservoir and a nozzle. The hydraulic stage comprises a syringe body containing a sealed solid piston. The injector piston, which is driven forward with water pushed by an HPLC pump, compresses a pair of Teflon balls, which driven by a pressure-amplification stage, extrude the LCP sample through a 10–50 µm inner diameter (ID) capillary. The hydraulic piston has a large diameter on the inlet side and a much smaller diameter on the outlet side. The ratio of the respective bore areas gives a nominal pressure amplification factor of 34, delivering 10,200 psi to the LCP reservoir when hydraulic fluid (water) on the inlet side is pressurized to 300 psi. Two different size sample reservoirs are available, which can hold up to 25(45) µl of LCP. The reservoir bore is filled with LCP via a Hamilton syringe. On the piston side, the reservoir is sealed by a Teflon ball. On the nozzle side, the sample reservoir is connected to a fused silica capillary with 10–50 µm inner diameter. The capillary is kept as short as possible (6 cm) to reduce the pressure necessary for LCP extrusion.

The LCP is extruded out of this capillary into an evacuated sample chamber (or a helium filled chamber); sample extrusion requires a pressure of 2,000–10,000 psi depending on the nozzle diameter, flow speed, and sample viscosity. Shear force exerted by a co-flowing gas (helium or nitrogen at 300–500 psi supply pressure) keeps the LCP stream on axis. To generate the co-flowing gas stream, the tapered end of the sample capillary is inserted into a flame-polished square glass tube and protrudes out of the square exit aperture, so that gas can flow through the open corners at a rate adequate for LCP extrusion (Fig. 11.3).

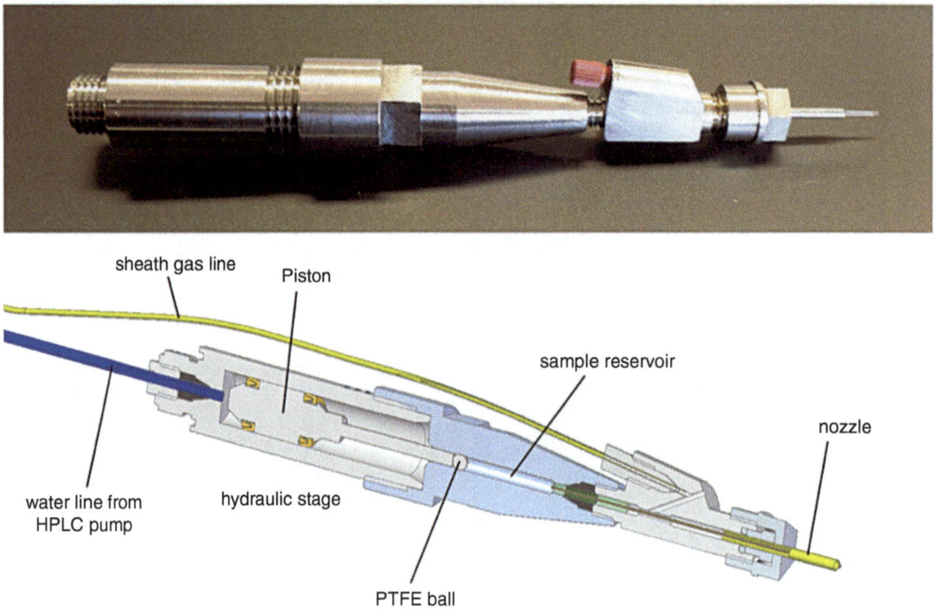

Fig. 11.2 *Top*: LCP injector. The device is about 12 cm long. *Bottom*: Cross-sectional view of the LCP injector

Fig. 11.3 LCPnozzle: (**a**) Square gas aperture before melting (**b**) after melting (**c**) with sample capillary of 50 μm ID protruding out of the melted gas aperture

The LCP flow rate can be optimized for the 10–120 Hz pulse repetition rate of the LCLS so that between X-ray pulses, the stream advances only the distance needed to expose fresh sample to the next pulse. The used repetition rate depends on the detector used, e.g. the CSPAD detector (Hart et al. 2012) currently in use at CXI is highly sensitive and has low dynamic range, but allows the use of the maximum 120 Hz repetition rate. The necessary distance the LCP stream has to advance between pulses (the 'damage diameter') depends on the X-ray beam diameter, LCP stream diameter and pulse energy. Thus, for example, at an X-ray energy of 9.5 keV, a pulse energy of 50 mJ at the sample and a beam diameter of 1.5 μm, this distance is 10–30 μm. Consequently, for a flow speed where the LCP stream travels 10–30 μm between X-ray pulses (1.2–3.6 mm/s), little, if any, sample is wasted and sample consumption is reduced dramatically compared with GDVN injection. Constant LCP flow rates of 1–300 nL/min can be achieved by adjusting the flow rate setting on the HPLC pump.

11.7 LCP-SFX Structures of Membrane Proteins

The LCP-SFX method was initially validated by comparing structures of the human serotonin 5-HT_{2B} receptor bound to an agonist ergotamine (ERG), solved by using traditional microcrystallography at a synchrotron source (5-HT_{2B}-SYN) (Wacker et al. 2013) and by SFX at the Coherent X-ray Imaging (CXI) instrument of LCLS (5-HT_{2B}-XFEL) (Liu et al. 2013). Both structures were determined at a similar resolution, 2.8 Å for the 5-HT_{2B}-XFEL structure and 2.7 Å for the 5-HT_{2B}-SYN, however, major differences were involved in the way how the data were collected and processed. The 5-HT_{2B}-XFEL data were collected at room temperature using microcrystals with an average size of $5 \times 5 \times 5$ μm streamed inside LCP medium, in which they have been grown, and intersecting in random orientations with the 1.5 μm in diameter XFEL pulses of 50 fs duration and 120 Hz repetition rate. Overall, 32,819 microcrystals contributed to the final dataset that was processed by the CrystFEL program (White et al. 2013). By comparison, the 5-HT_{2B}-SYN dataset was collected from cryo-cooled crystals of an average size of $80 \times 20 \times 10$ μm by taking oscillation patterns upon crystal exposure with a 10 μm in diameter synchrotron beam. Due to the crystal sensitivity to radiation damage, 17 crystals were required to obtain a complete dataset, and the data were processed by the HKL2000 program (Otwinowski and Minor 1997). Both structures were solved by molecular replacement (MR) and refined by Phenix (Adams et al. 2010) using the same parameters, and displayed high-quality electron density maps for most residues of 5-HT_{2B}, the agonist ergotamine, covalently attached palmitic acid, cholesterol and several lipid and water molecules.

Despite the very different methods of data acquisition and processing, both structures displayed almost identical backbone trace (Cα atoms RMSD = 0.46 Å excluding flexible N-terminus and parts of extracellular loop 2), validating the LCP-SFX protocols. While overall similar, several deviations between these two structures were noticed. First, the unit cell volume of 5-HT_{2B}-SYN was 2.1 % smaller than the one of 5-HT_{2B}-XFEL, consistent with the shrinkage of crystal lattice induced by cryo-cooling. Several side chains showed different rotamer conformations between these two structures, potentially due to the cryo-cooling-induced partial remodelling (Fraser et al. 2011). The average B-factor of 5-HT_{2B}-XFEL structure was 21 $Å^2$ larger than that of 5-HT_{2B}-SYN structure, consistent with the increased thermal motions at room temperature. Furthermore, the extracellular tip of helix II formed an α-helix instead of a water-stabilized kink in 5-HT_{2B}-SYN structure. Most of these discrepancies observed between the 5-HT_{2B}-XFEL and 5-HT_{2B}-SYN structures were attributed to the differences in data collection temperature (Juers and Matthews 2004). The LCP-SFX therefore enabled determination of the first room temperature GPCR structure, providing more accurate insights into receptor structure and function at close to native conditions.

After successful validation, the LCP-SFX method was applied to solve the structure of the human smoothened receptor (SMO) in complex with cyclopamine (Weierstall et al. 2014). SMO belongs to the Frizzled family GPCRs and participates in the embryonic development and growth of cancer cells. Blockade of the SMO receptor by small molecule antagonists is considered a promising strategy for the treatment of certain tumors. Cyclopamine is a small molecule, SMO antagonist (Chen et al. 2002), produced by corn lily, which was discovered during an epidemic of cyclopia in newborn lambs at an Idaho farm. In efforts to determine the structure of SMO in complex with cyclopamine, relatively large crystals were produced in LCP and tested at a synchrotron beamline. These crystals, however, produced poor diffraction with high mosaicity and anisotropy. On the contrary, much smaller crystals grown in similar conditions did not suffer from high mosaicity and were used to determine the SMO-cyclopamine structure at 3.2 Å resolution by LCP-SFX at LCLS. The electron density clearly revealed the location of cyclopamine in the long and narrow crevice inside SMO.

In early 2015 this method was used to determine the structure of the human δ-opioid receptor bound to a bifunctional peptide DIPP-NH_2 at 2.7 Å resolution (Fenalti et al. 2015). Bifunctional ligands acting as agonists towards μ-opioid receptor and antagonist towards δ-opioid receptor are promising alternatives to opiate painkillers, frequent administration of which do not lead to acquiring tolerance and dependency. Finally, most recently the LCP-SFX method enabled structure determination of the first novel GPCR, the human angiotensin II receptor type 1 (AT1R), a blood pressure regulator, in complex with an angiotensin receptor blocker (Zhang et al. 2015).

11.8 Summary and Outlook for Serial Femtosecond Crystallography in LCP

Within relatively short time since the emergence of the first hard XFEL source, it is becoming increasingly clear that serial femtosecond crystallography has the potential to accelerate structural studies of membrane proteins under more native conditions, which is highly critical for identifying structure-function relationships of membrane proteins. The innovative approach described in this chapter of using LCP crystallisation to obtain well-ordered high-quality microcrystals suitable for data collection by serial femtosecond crystallography with highly intense and ultrashort XFEL pulses has already yielded several room-temperature high-resolution structures of challenging membrane proteins (Liu et al. 2013; Weierstall et al. 2014; Fenalti et al. 2015; Zhang et al. 2015).

Compared with traditional X-ray crystallography, SFX practically eliminates radiation damage that usually occurs when working with small crystals. SFX approach accurately captures structural details at room temperature under physiological conditions, and LCP provides environmental conditions close to the native environment of membrane proteins. Thus, the combination of XFEL (Emma et al. 2010; Boutet and Williams 2010), LCP crystallisation (Liu et al. 2014), LCP injector (Weierstall et al. 2014) and fast-readout CSPAD detector (Hart et al. 2012) brings forth a unique approach enabling structural studies of difficult to crystallise membrane proteins.

Continued developments of LCP-SFX will further decrease sample consumption and required crystal size by possibly narrowing the diameter of the LCP stream to reduce the background scattering. This will require continued injector development and better control over the crystallisation process in LCP to obtain a more homogeneous crystal size distribution and thereby avoiding clogs from large crystals. New data processing software developments are already underway which should result in faster data collection using less diffraction patterns (Uervirojnangkoorn et al. 2015; Ginn et al. 2015). Improvements in detectors will allow to use higher X-ray fluence, since current detectors have limited dynamic range and require significant attenuation of the X-ray intensity to prevent saturation by strong Bragg peaks. Finally, it should be possible to look at conformational changes in protein crystals delivered in LCP (molecular movies) similar to the recent seminal studies with liquid injectors (Tenboer et al. 2014).

Acknowledgments This work was supported in parts by the NIH grants R01 GM108635 and U54 GM094618 (V.C.), NSF BioXFEL Science and Technology center grant 1231306 (U.W. and W.L.). Further supports were provided by the Arizona State University (ASU) Biodesign Seed Grant Program and ASU-Mayo Seed Grant Program (L.Z. and W.L.). Parts of this research were carried out at the LCLS, a National User Facility operated by Stanford University on behalf of the U.S. Department of Energy, Office of Basic Energy Sciences, and at the GM/CA CAT of the Argonne Photon Source, Argonne National Laboratory.

References

Adams PD, Afonine PV, Bunkoczi G, Chen VB, Davis IW, Echols N et al (2010) PHENIX: a comprehensive Python-based system for macromolecular structure solution. Acta Crystallogr D Biol Crystallogr 66(2):213–221

Aquila A, Hunter MS, Doak RB, Kirian RA, Fromme P et al (2012) Time-resolved protein nanocrystallography using an X-ray free-electron laser. Opt Express 20(3):2706–2716

Boutet S, Williams GJ (2010) The coherent X-ray imaging (CXI) instrument at the Linac coherent light source (LCLS). New J Phys 12:035024

Boutet S, Lomb L, Williams GJ, Barends TR, Aquila A et al (2012) High-resolution protein structure determination by serial femtosecond crystallography. Science 337:362–364

Caffrey M, Cherezov V (2009) Crystallizing membrane proteins using lipidic mesophases. Nat Protocols 4:706–731

Chapman HN, Fromme P, Barty A, White TA, Kirian RA, Aquila A et al (2011) Femtosecond X-ray protein nanocrystallography. Nature 470:773–777

Chen JK, Taipale J, Cooper MK, Beachy PA (2002) Inhibition of Hedgehog signaling by direct binding of cyclopamine to Smoothened. Genes Dev 16:2743–2748

Cherezov V (2011) Lipidic cubic phase technologies for membrane protein structural studies. Curr Opin Struct Biol 21(4):559–566

Cherezov V, Peddi A, Muthusubramaniam L, Zheng YF, Caffrey M (2004) A robotic system for crystallizing membrane and soluble proteins in lipidic mesophases. Acta Crystallogr D Biol Crystallogr 60(10):1795–1807

Cherezov V, Rosenbaum DM, Hanson MA, Rasmussen SG, Thian FS, Kobilka TS et al (2007) High-resolution crystal structure of an engineered human beta2-adrenergic G protein-coupled receptor. Science 318:1258–1265

Deisenhofer J, Epp O, Miki K, Huber R, Michel H (1985) Structure of the protein subunits in the photosynthetic reaction centre of Rhodopseudomonas viridis at 3A resolution. Nature 318:618–624

DePonte DP, Weierstall U, Schmidt K, Warner J, Starodub D, Spence JCH, Doak RB (2008) Gas dynamic virtual nozzle for generation of microscopic droplet streams. J Phys D Appl Phys 41:195505

Emma P, Akre R, Arthur J, Bionta R, Bostedt C, Bozek J et al (2010) First lasing and operation of an ångstrom-wavelength free-electron laser. Nat Photonics 4:641–647

Fenalti G, Zatsepin NA, Betti C, Giguere P, Han GW, Ishchenko A et al (2015) Structural basis for bi-functional peptide recognition at human delta-opioid receptor. Nat Struct Mol Biol 22:265–268

Fraser JS, van den Bedem H, Samelson AJ, Lang PT, Holton JM, Echols N, Alber T (2011) Accessing protein conformational ensembles using room-temperature X-ray crystallography. Proc Natl Acad Sci U S A 108:16247–16252

Ginn HM, Messerschmidt M, Ji X, Zhang H, Axford D, Gildea RJ et al (2015) Structure of CPV17 polyhedrin determined by the improved analysis of serial femtosecond crystallographic data. Nat Commun 6:6435

Hart P, Boutet S, Carml G, Dragone A, Duda B, Freytag D et al (2012) The Cornell-SLAC pixel array detector at LCLS. Ieee nuclear science symposium and medical imaging conference record. Symposium and Medical Imaging Conference (NSS/MIC), IEEE, pp 538–541

Johansson LC, Arnlund D, White TA, Katona G, DePonte DP, Weierstall U et al (2012) Lipidic phase membrane protein serial femtosecond crystallography. Nat Methods 9:263–265

Johansson LC, Arnlund D, Katona G, White TA, Barty A, DePonte DP et al (2013) Structure of a photosynthetic reaction centre determined by serial femtosecond crystallography. Nat Commun 4:2911

Juers DH, Matthews BW (2004) Cryo-cooling in macromolecular crystallography: advantages, disadvantages and optimization. Q Rev Biophys 37:105–119

Kern J, Alonso-Mori R, Hellmich J, Tran R, Hattne J, Laksmono H et al (2012) Room temperature femtosecond X-ray diffraction of photosystem II microcrystals. Proc Natl Acad Sci U S A 109:9721–9726

Kirian RA, Wang X, Weierstall U, Schmidt KE, Spence JC, Hunter M et al (2010) Femtosecond protein nanocrystallography-data analysis methods. Opt Express 18:5713–5723

Kirian RA, White TA, Holton JM, Chapman HN, Fromme P, Barty A et al (2011) Structure-factor analysis of femtosecond microdiffraction patterns from protein nanocrystals. Acta Crystallogr Sect A 67(2):131–140

Kupitz C, Basu S, Grotjohann I, Fromme R, Zatsepin NA, Rendek KN et al (2014) Serial time-resolved crystallography of photosystem II using a femtosecond X-ray laser. Nature 513:261–265

Landau EM, Rosenbusch JP (1996) Lipidic cubic phases: a novel concept for the crystallization of membrane proteins. Proc Natl Acad Sci U S A 93:14532–14535

Liu W, Chun E, Thompson AA, Chubukov P, Xu F, Katritch V et al (2012) Structural basis for allosteric regulation of GPCRs by sodium ions. Science 337:232–236

Liu W, Wacker D, Gati C, Han GW, James D, Wang D et al (2013) Serial femtosecond crystallography of G protein-coupled receptors. Science 342:1521–1524

Liu W, Ishchenko A, Cherezov V (2014) Preparation of microcrystals in lipidic cubic phase for serial femtosecond crystallography. Nat Protoc 9:2123–2134

Meents A, Gutmann S, Wagner A, Schulze-Briese C (2010) Origin and temperature dependence of radiation damage in biological samples at cryogenic temperatures. Proc Natl Acad Sci U S A 107:1094–1099

Misquitta Y, Cherezov V, Havas F, Patterson S, Mohan JM, Wells AJ et al (2004) Rational design of lipid for membrane protein crystallization. J Struct Biol 148:169 175

Neutze R, Wouts R, van der Spoel D, Weckert E, Hajdu J et al (2000) Potential for biomolecular imaging with femtosecond X-ray pulses. Nature 406:752–757

Otwinowski Z, Minor W (1997) Processing of X-ray diffraction data collected in oscillation mode. Methods Enzymol (Academic Press) 276:307–326

Perry SL, Roberts GW, Tice JD, Gennis RB, Kenis PJ (2009) Microfluidic generation of lipidic mesophases for membrane protein crystallization. Cryst Growth Des 9:2566–2569

Redecke L, Nass K, DePonte DP, White TA, Rehders D, Barty A et al (2013) Natively inhibited Trypanosoma brucei cathepsin B structure determined by using an X-ray laser. Science 339:227–230

Shonberg J, Kling RC, Gmeiner P, Löber S (2014) GPCR crystal structures: medicinal chemistry in the pocket. Bioorg Med Chem 23:3880–3906

Sierra RG, Laksmono H, Kern J, Tran R, Hattne J, Alonso-Mori R et al (2012) Nanoflow electrospinning serial femtosecond crystallography. Acta Crystallogr D Biol Crystallogr 68:1584–1587

Solem JC (1986) Imaging biological specimens with high-intensity soft x rays. J Opt Soc Am B 3:1551–1565

Tenboer J, Basu S, Zatsepin N, Pande K, Milathianaki D, Frank M et al (2014) Time-resolved serial crystallography captures high-resolution intermediates of photoactive yellow protein. Science 346: 1242–1246

Uervirojnangkoorn M, Zeldin OB, Lyubimov AY, Hattne J, Brewster A et al (2015) Enabling X-ray free electron laser crystallography for challenging biological systems from a limited number of crystals. Elife 4:05421

Wacker D, Wang C, Katritch V, Han GW, Huang XP, Vardy E et al (2013) Structural features for functional selectivity at serotonin receptors. Science 340: 615–619

Wallin E, von Heijne G (1998) Genome-wide analysis of integral membrane proteins from eubacterial, archaean, and eukaryotic organisms. Protein Sci 7:1029–1038

Weierstall U, James D, Wang C, White TA, Wang D, Liu W et al (2014) Lipidic cubic phase injector facilitates membrane protein serial femtosecond crystallography. Nat Commun 5:3309

White TA, Barty A, Stellato F, Holton JM, Kirian RA, Zatsepin NA, Chapman HN (2013) Crystallographic data processing for free-electron laser sources. Acta Crystallogr D Biol Crystallogr 69(7): 1231–1240

Yildirim MA, Goh KI, Cusick ME, Barabási AL, Vidal M (2007) Drug-target network. Nat Biotechnol 25:1119–1126

Zhang H, Unal H, Gati C, Han GW, Liu W, Zatsepin NA et al (2015) Structure of the Angiotensin receptor revealed by serial femtosecond crystallography. Cell 161:833–844

Beyond Membrane Protein Structure: Drug Discovery, Dynamics and Difficulties

12

Philip C. Biggin, Matteo Aldeghi, Michael J. Bodkin, and Alexander Heifetz

Abstract

Most of the previous content of this book has focused on obtaining the structures of membrane proteins. In this chapter we explore how those structures can be further used in two key ways. The first is their use in structure based drug design (SBDD) and the second is how they can be used to extend our understanding of their functional activity via the use of molecular dynamics. Both aspects now heavily rely on computations. This area is vast, and alas, too large to consider in depth in a single book chapter. Thus where appropriate we have referred the reader to recent reviews for deeper assessment of the field. We discuss progress via the use of examples from two main drug target areas; G-protein coupled receptors (GPCRs) and ion channels. We end with a discussion of some of the main challenges in the area.

Keywords

Structure based drug design • Molecular dynamics • Simulation • Ion channels • Glutamate receptor • G-protein coupled receptor • Docking • Virtual screening

12.1 Structures in Drug Design

One of the most active areas where recent crystal structures have begun to inform drug design in a significant way has been the GPCR renaissance (Jazayeri et al. 2015). In the past, structure based drug design (SBDD) has been the preserve of soluble proteins that readily crystallise. Membrane proteins have remained problematic because of the difficulties associated with generating high-resolution data. However, as described elsewhere in this book, various advances have

P.C. Biggin (✉) • M. Aldeghi
Department of Biochemistry, University of Oxford, South Parks Road, Oxford OX1 3QU, UK
e-mail: philip.biggin@bioch.ox.ac.uk; matteo.aldeghi@lincoln.ox.ac.uk

M.J. Bodkin • A. Heifetz
Evotec Ltd, 114 Innovation Drive, Milton Park, Abingdon, Oxfordshire OX14 4RZ, UK
e-mail: mike.bodkin@evotec.com; Alex.Heifetz@evotec.com

I. Moraes (ed.), *The Next Generation in Membrane Protein Structure Determination*, Advances in Experimental Medicine and Biology 922, DOI 10.1007/978-3-319-35072-1_12

begun to make SBDD a real possibility for many membrane protein targets including GPCRs (Tate 2012). Structures of ion channels, many of which are key therapeutic targets, have also started to appear.

One structure on its own can be tremendously informative, but the appearance of many related structures allows comparative analysis to be done, which can help to define conserved motifs and highlight functionally relevant positions. This is exactly what has been done for GPCRs and has shown conserved interaction networks and characteristic features of ligand binding (Venkatakrishnan et al. 2013). It can also lead to unifying theories of activation (Tehan et al. 2014). Moreover, such analysis is tremendously helpful in developing tools that can be used to assess the quality of models derived from simulation and modelling (Heifetz et al. 2013a). This kind of approach, in conjunction with the huge wealth of mutagenesis data (Isberg et al. 2014), is part of the reason for the renewed interest in GPCRs as drug targets. Today, crystal structures exist for almost every representative sub-branch of the GPCR genome.

A similar approach has been adopted for the analysis of the cys-loop family of ion channels (Spurny et al. 2015) although the variation in the structures, and the fact that many are bacterial in origin, makes cross-comparative work like this more difficult to interpret. The cys-loop family of receptors contains a large number of therapeutic targets, particularly for the treatment of central nervous system (CNS) related disorders. For example, the homomeric α7 nicotinic acetylcholine receptor has been considered a target for the treatment of many disease states associated with cognitive impairment (Uteshev 2014) and there have been several reports of allosteric modulators which may offer therapeutic potential in the future (Young and Geyer 2013).

Ionotropic glutamate receptors on the other hand have the distinct advantage that the ligand-binding domain can be expressed as a soluble construct (that maintains almost wild type binding affinity (Armstrong and Gouaux 2000)) and is readily crystallisable. The AMPA subtype has been recognised as a therapeutic target for neurological disorders including schizophrenia (Ward et al. 2010, 2015; Partin 2015), and structural information has been an important factor in drug design considerations (Harms et al. 2013). Indeed, crystallography has already been integrated with lead optimization methods for the development of new positive allosteric modulators of AMPA receptors (Ward et al. 2010, 2015). A number of patents in this area have been filed (Pirotte et al. 2013) and it also appears to be an area in which academic laboratories are also starting to contribute directly to (Partin 2015; Harms et al. 2013; Caldwell et al. 2015; Chen et al. 2013; Jamieson et al. 2011; Timm et al. 2011; Weeks et al. 2014). Another type of glutamate receptor, the NMDA receptor, can also be modulated in an allosteric fashion at different sites as recently reviewed by Strong et al. (2014). Indeed there is renewed interest in NMDA receptors as therapeutic targets and structural information is starting to guide this in a rational way (Khatri et al. 2014).

Rational approaches to designing compounds acting at the orthosteric binding site of glutamate receptors have also been employed (Demmer et al. 2015; Venskutonytė et al. 2014; Assaf et al 2013; Juknaitė et al 2012; Bunch and Krogsgaard-Larsen 2009; Sivaprakasam et al. 2009). An important goal at this point is to develop sub-type specific compounds, not necessarily as drugs, but as chemical probes. One of the main issues with targeting glutamate receptors is targeting the correct receptor subunit in right part of the brain, but at the present time this is hindered by a lack sub-type specific chemical probes.

In the next sections we briefly highlight some of the principles employed before going on to describe some case studies where these approaches have been used in on going drug-design projects.

12.1.1 In-sillico SBDD

Once one has the structure of a particular target, it can be used as the basis for SBDD and/or virtual screening (Forli and Olson 2015). This is often considered the most promising approach for the discovery of new ligands (Lounnas et al. 2013).

The presence of structural information in a drug-discovery program can lead to a step change in progress; Structure activity relationship (SAR) results can be rationalised, new avenues can be rationally explored and improved hypotheses generated.

In the absence of a structure, drug design typically precedes using ligand-based techniques, but these have well known limitations. Aside from the obvious limitation of the lack of information on how the ligand interacts with its target, it is often the case that any newly discovered molecules will have a similar chemotype to the parent molecule(s) and that can lead to intellectual property (IP) protection problems which render the process non-viable. Additional problems are that activity cliffs cannot be easily rationalized and it is not always possible to isolate effects of chemical modifications. If structural information is available with an important drug or lead compound, then new interactions between ligands and the target can be devised on simple chemical intuition and molecular mechanics principles. It is also easier to design molecules that are completely novel from an IP sense.

Structure-based virtual screening (SBVS) (Lionta et al. 2014) is a useful tool in terms of providing initial filtering of huge chemical libraries and to provide plausible suggestions (Cerqueira et al. 2015) that can be taken forward in a drug-discovery program (Shoichet 2004). However, the detailed prediction of precise modes of action and prediction of affinity, or even more difficult, efficacy (see below in Sect. 12.3.2), is still incredibly challenging and indeed over the years this aspect was arguably overhyped (Seddon et al. 2012). Despite these limitations, the low cost of virtual screening means that it is something employed in most drug design strategies to provide some initial assessment of potential binding modes.

SBVS has been around for a number of years and its development and use have been reviewed substantially (Lounnas et al. 2013; Lionta et al. 2014; Cheng et al. 2012). SBVS is based on high throughput docking where one has the structure of a target (protein usually but not necessarily) and a large (thousands to millions) compounds library. Compounds from this library are then docked to this target structure and ranked according to some kind of scoring function. SBVS relies on these functions, but whilst the ability of these methods to predict accurate poses (typically defined as within 2 Å RMSD of the crystallographic model) is quite reasonable, expert knowledge is often required to obtain that level of performance (Cross et al. 2009). Care must also be taken to ensure that due consideration is given to factors like explicit solvent and the flexibility of the binding pocket (Elokely and Doerksen 2013).

12.1.2 Homology Modelling of Related Targets

Although there has been a steady increase in the number of membrane protein structures being solved, there are still a large number of important targets for which no high resolution structure is available and thus if one wants to pursue any kind of structure-based approach, homology modelling is the only viable solution (Schmidt et al. 2014). Furthermore, even if structural information is available for one complex, if the timescale of obtaining that data is too slow, then homology modelling is often used instead. It is generally considered that in order for homology models to be useful for screening, there needs to be a high quality template with high sequence similarity, although it has been argued that for some approaches, such as off-target prediction, low-to moderate resolution structures can still be of value (Skolnick et al. 2013). Indeed it has also been argued for GPCRs that theoretical models can perform equally as well for some aspects of the drug discovery process (Tang et al. 2012). Similarly, a retrospective comparison for the Dopamine D3 receptor also concluded that well-built homology models can perform as well as crystal structures in terms of *in silico* docking (Levoin et al. 2011). Even in cases where the sequence identity to the target is very low, careful model building in conjunction with site-directed mutagenesis and binding assays can be very useful in aiding the future direction of a

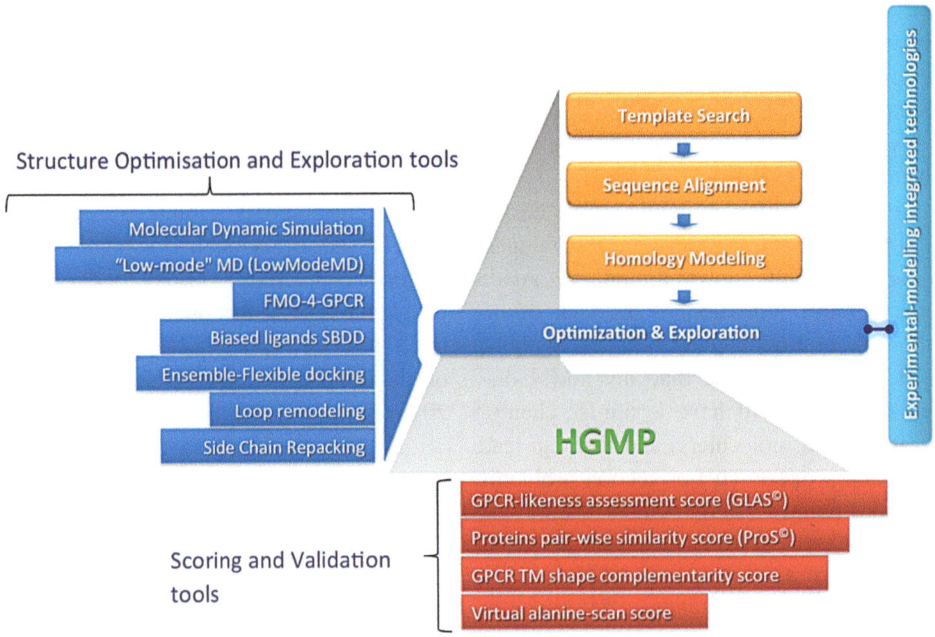

Fig. 12.1 A summary schematic of the Hierarchical GPCR Modelling Protocol (HGMP) developed at Evotec in collaboration with the University of Oxford, illustrating the various components that contribute to the modelling procedure

drug discovery program or indeed rationalizing existing SAR data (Lee et al. 2015).

These approaches can be incorporated into workflows. An example such workflow is the hierarchical GPCR modelling protocol (HGMP), which was developed to support SBDD programs (see Fig. 12.1). The HGMP generates a GPCR model and its potential complexes with small molecules by applying a series of computational methods in a workflow. The protocol makes use of homology modelling following by Molecular dynamics (MD) simulations and docking to predict binding poses and functions of ligands bound to GPCRs. The HGMP is equipped with GPCR-specific "plugins" including for example the GPCR-likeness assessment score (GLAS) to evaluate model quality (Heifetz et al. 2013a). The HGMP also includes a pairwise protein comparison method (ProS) used to cluster structural data and can distinguish between different activation sub-states (Heifetz et al. 2013a). Recently, the capabilities of HGMP have been extended by the addition of GPCR biased ligand tools. The HGMP has been used within real drug-discovery projects at the biotech company Evotec, some of which we briefly outline as case studies below.

12.1.3 SBDD Case Studies

12.1.3.1 Discovery of Selective 5-HT$_{2C}$ Agonists for the Treatment of Metabolic Disorders

In this project, which was performed prior to the crystal structures of 5-HT$_{2B}$ and 5-HT$_{1B}$ being solved, the challenge was to find novel 5-HT$_{2c}$ agonists that were selective in that they did not activate 5-HT$_{2A}$ and 5-HT$_{2B}$ receptors. The HGMP was applied to model both active and inactive receptor conformations, referred to as 5-HT$_{2C}^{active}$ and 5-HT$_{2C}^{inactive}$ respectively. Models were also built of the off targets, 5-HT$_{2A}$ and 5-HT$_{2B}$. Flexible docking was then applied to predict the binding modes of compounds with 5-HT$_{2A}$, 5-HT$_{2B}$ and 5-HT$_{2C}$. The binding site of 5-HT$_{2C}^{inactive}$ was proposed to be shallower compared to the binding site of 5-HT$_{2C}^{active}$ due to residues from TM3 and TM6 forming stabilizing inter-helical interactions in the $5\text{-}HT_{2C}^{inactive}$ binding site. It was proposed that these inter-helical interactions are broken in the active conformation of the $5\text{-}HT_{2C}$ receptor, which is stabilized by agonist molecules entering deeper into the binding site and compensating *via* interactions with various other residues, including $W324^{6.48}$, a key residue previously identified as a "transmission switch" residue (Han et al. 1997; Holst et al. 2010; Schwartz et al. 2006) and which may form part of a larger "hydrophobic hindering mechanism" recently suggested (Tehan et al. 2014). Agonists were suggested to interact with $W324^{6.48}$, thereby pushing the intracellular half of TM6 in to the active conformation. Agonists were proposed to interact simultaneously with both TM3 and TM6 in $5\text{-}HT_{2C}^{active}$ thus increasing the overall stability of $5\text{-}HT_{2C}^{active}$ and promoting activation. Furthermore, these modeling observations (which are directly supported by the published SDM data) were incorporated into the design of novel $5\text{-}HT_{2C}$ agonists (Tye et al. 2011). Hits were also assessed for hERG liability via docking to a model of the hERG channel. The result was the discovery of a novel compound (EC_{50} = 8.4, 762, 73 nM for $5\text{-}HT_{2C,\,2A,\,2B}$ and hERG inhibition of 11 % @ 10 CM). The whole design cycle for this project can be summarized in Fig. 12.2.

12.1.3.2 Fighting Obesity with a Sugar-Based Library

Obesity is an increasingly common disease. Although antagonism of the melanin-concentrating hormone-1 receptor (MCH-1R) has been widely reported as a promising therapeutic avenue for obesity treatment, no MCH-1R antagonists have reached the market. Discovery and optimization of new compounds targeting MCH-1R was hindered by low high throughput screening (HTS) success rates and a lack of structural information about the MCH-1R binding site. In this project, a conceptually pioneering approach that integrated GPCR modelling with design, synthesis and screening of a diverse library of

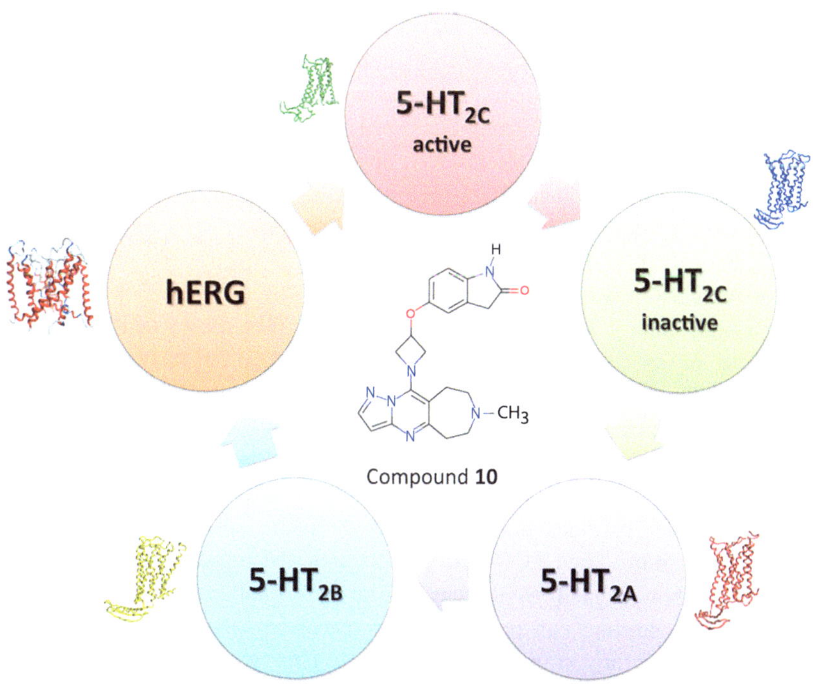

Fig. 12.2 Design cycle for potent, selective and hERG 'clean' $5HT_{2C}$ antagonists

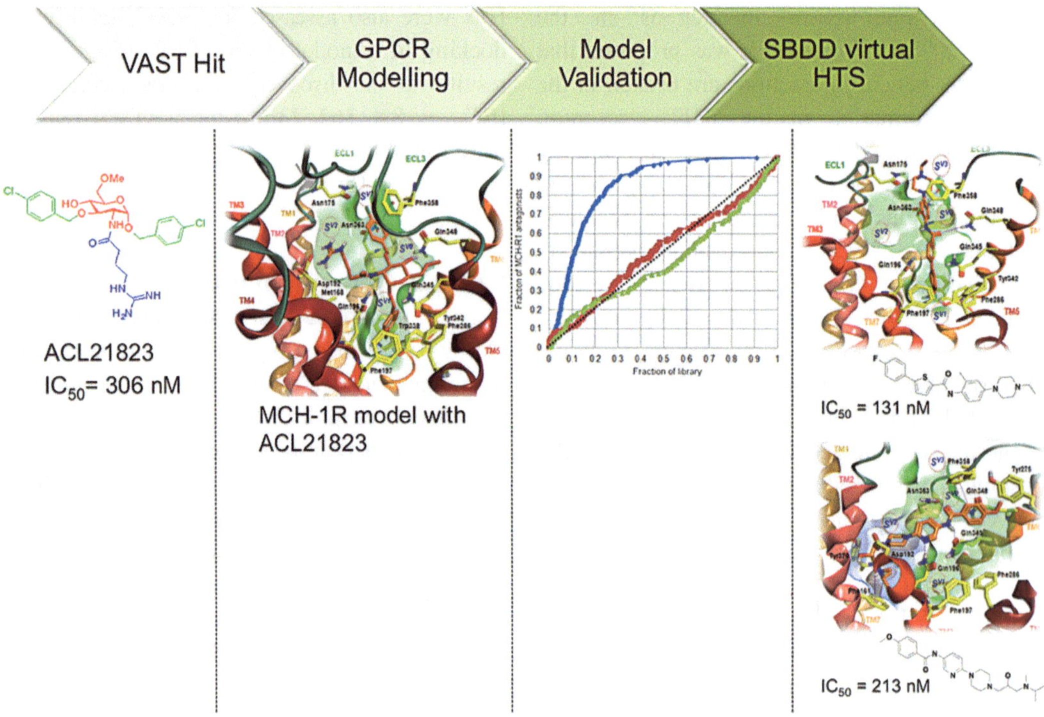

Fig. 12.3 Summary schematic of the VAST-GPCR modelling workflow which led to the discovery of new MCH-1R antagonists

sugar-based compounds from the VAST technology (Versatile Assembly on Stable Templates) was used, to provide structural insights on the MCH-1R binding site (Heifetz et al. 2013a, b, c). The 490 VAST compounds obtained from this library design were screened against MCH-1R, resulting in the discovery of a potent MCH-1R antagonist, ACL21823 (radioligand binding to MCH-1R gave an $IC_{50} = 306$ nM, see Fig. 12.3). The discovery of ACL21823 was utilized in the construction of a high quality MCH-1R model and the refinement of its antagonist binding site. The quality of the MCH-1R model was demonstrated by a virtual enrichment experiment and the model-driven structure-based expansion of ACL21823, which allowed the generation of a list of key MCH-1R residues potentially involved in antagonist binding. The GPCR-VAST method demonstrates how ligand SAR data, when combined with modelling, can provide a useful source of structural information on GPCR binding sites (Heifetz et al. 2013a, b, c). The usefulness of the GPCR-VAST method to drug discovery was demonstrated by a structure-based virtual screen, which achieved a hit rate of 14 % and yielded 10 new chemotypes of MCH-1R antagonists including EOAI3367472 ($IC_{50} = 131$ nM) and EOAI3367474 ($IC_{50} = 213$ nM). This approach creates a cost-efficient new avenue for structure-based drug discovery (SBDD) against GPCR targets.

12.1.3.3 Discovery of Potent and Selective OX_2 Receptor Antagonists

The orexin receptors (OX_1 and OX_2) are linked to a range of different physiological functions including the control of feeding, energy metabolism, modulation of neuro-endocrine function and regulation of the sleep-wake cycle. The key challenges of this project were to increase the OX_2 activity and selectivity over OX_1. This was particularly difficult as OX_1 and OX_2 receptors have over 80 % sequence similarity. This project was completed before the

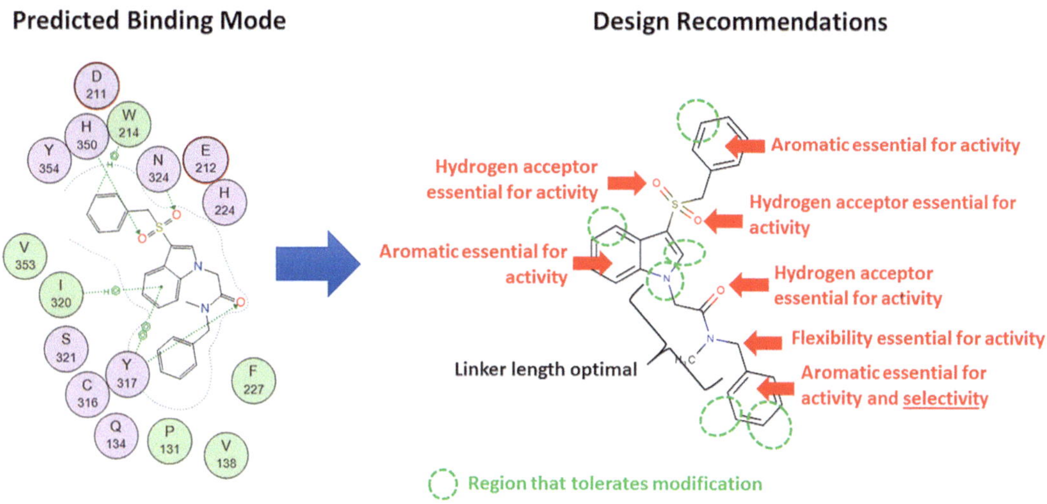

Fig. 12.4 Schematic summarizing how interaction maps derived from structural models and MD data can be used to provide synthesis recommendations

crystal structure of OX_2 bound to Suvorexant was solved. We used Molecular Dynamics (MD) simulations to study the OX_1 vs OX_2 selectivity (Heifetz et al. 2012). The MD process allowed refinement of the models that was not possible with static crystal structures or homology models alone. This study suggested that differences in intra-helical interactions resulted in differences in conformations of transmembranes (TMs) and differences in topology of the binding pocket. These are small differences but sufficient enough to design molecules with OX_2 selectivity. This rational design significantly decreased the amount of synthesis by focusing the effort to the relevant portion of the ligand structure as demonstrated in Fig. 12.4. The final compound, EP-009-0513, had inhibition constants, K_i, of 4363 and 5.7 nM for OX_1 and OX_2 respectively.

12.1.3.4 Discovery of Selective Dual Antagonists of H_3 and H_4 Receptors

An approach that integrates the HGMP with fragment based drug discovery (FBDD) had been applied for the discovery of selective and dual H_3 and H_4 histamine receptor antagonists (see Fig. 12.5). FBDD has emerged as a new tool for drug discovery in recent years and is typically aimed at a target for which a crystal structure can be determined in order to rationally guide fragment-hit expansion. While the majority of historical fragment screens have been focused towards biochemical targets, only a few examples have been published in which this method has been used to identify ligands for GPCRs. This is due to the current infeasibility of regularly crystallizing the GPCR-fragment complexes that are essential for further fragment expansion.

The HGMP was used to generate initial models of these receptors. Primary fragment screens yielded 44 H_3 selective, 21 H_4 selective and 20 dual selective fragment hits. These fragments were used to construct new high-quality H_3 and H_4 models followed by binding site exploration and a structure based virtual screen. Overall, 172 compounds were purchased for testing based on these virtual screening results. Of the 74 compounds predicted to have dual activity, 33 had activity against one or other of the two receptors (44 %), of which 17 had activity against both. Of the 19 compounds predicted to be H_3 selective, 13 were active against H_3 (68 %) and 10 of these also had selectivity over H_4. Of the 79 compounds predicted to be H_4 selective, 36 were active against H_4 (45 %) and 2 of these also had selectivity over H_3. This approach highlights a cost-efficient avenue for structure-based drug

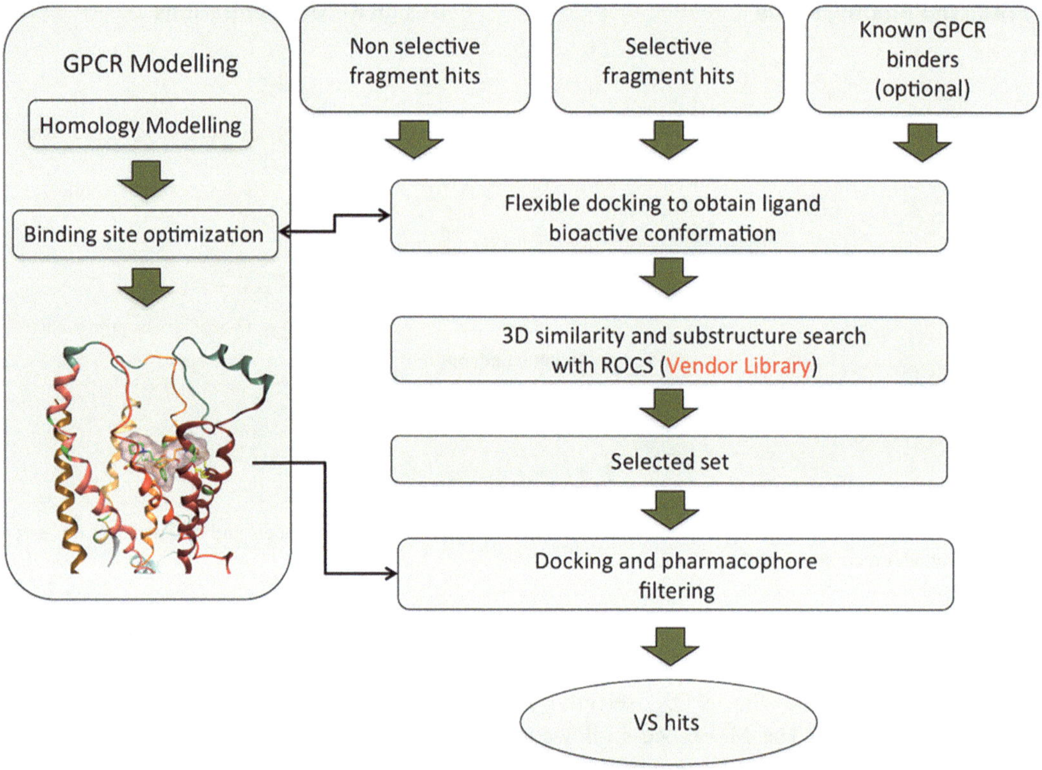

Fig. 12.5 Schematic of the HGMP-FBDD workflow to perform virtual screening

design (SBDD) against GPCR targets (Heifetz et al. 2013b).

12.2 Exploration of the Dynamic Nature of Membrane Proteins

Membrane proteins, by functional necessity, are often very dynamic entities. Indeed, this is one of the reasons why they have been recalcitrant to crystallization efforts. Ion channels have gating machinery for example, whilst GPCRs need to enable a wide variety of signalling pathways through changes in their response to different ligands. Crystal structures and modelling can provide a step-change in our understanding of membrane proteins and in the processes outlined above with SBDD. However, the structure is also the start point for one to be able to examine these dynamic responses and consequently address more complex questions about receptor flexibility and function. Molecular dynamics simulations can be used in a variety of ways including refinement of the structure in a more realistic membrane environment, the analysis of solvent and the discovery of so-called "cryptic sites" which may offer alternative pathways for therapy. We outline some of these aspects below.

12.2.1 Exploration of Membrane Protein Dynamics

Crystal structures are often obtained under conditions that are somewhat artificial and thus, care must be taken to interpret the data in a meaningful way (Wlodawer et al. 2008, 2013). MD simulations can provide refinement of homology models (as alluded to above) and can give information on lipid and solvent interactions for example. The dynamic and heterogeneous nature

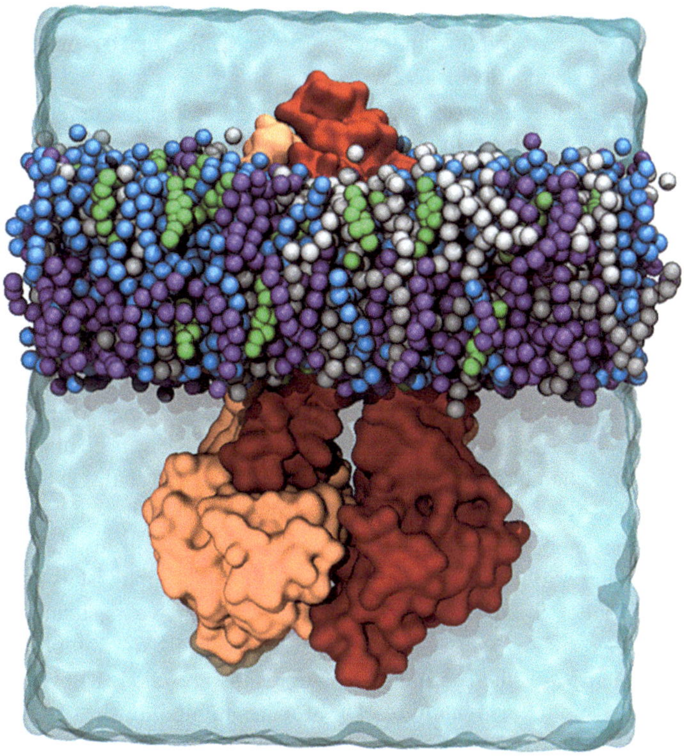

Fig. 12.6 Example simulation box. In case the protein is a homology model of human P-glycoprotein based on the refined mouse structure 4M1M (transmembrane helices 1–6 and 7–12 are *red* and *orange* respectively) embedded within a bilayer system comprised of 1-palmitoyl-2-oleoyl-sn-glycero-3-phosphocholine (POPC; *blue*), 1-palmitoyl-2-oleoyl-sn-glycero-3-phosphoethanolamine (POPE; *purple*) 1-palmitoyl-2-oleoyl-sn-glycero-3-phosphoserine (POPS; *dark grey*), sphingomeylin (*light grey*) and cholesterol (*green*). The lipids in front of the protein have been removed for clarity. It is usual for researchers to sodium and chloride ions in the solvent to a concentration that is representative of in vivo or in vitro studies (150 mM NaCl for example) (Figure courtesy of Laura Domicevica)

of lipid bilayers can make the analysis of interactions with membrane proteins difficult. In a typical approach the coordinates are extracted form the PDB and inserted into a lipid bilayer (see Fig. 12.6). Many methods have been published over the years to achieve this and the reader is referred to a specific review on this aspect (Biggin and Bond 2015).

Simulations can provide a unique and molecular view of the interaction of lipids with membrane proteins. Due to its abundance in mammalian membranes, cholesterol has been investigated in great depth (Grouleff et al. 2015) particularly for GPCRs (Sengupta and Chattopadhyay 2015), but as parameters for other lipids become available we can expect analysis of other important lipids and the interactions of more complex membrane systems (Goose and Sansom 2013) including negatively charged ones (Kalli et al. 2013). There is increasing evidence that the energetics of protein-lipid interactions can directly impact the functional properties of the protein (Mondal et al. 2014a, b). Simulations can also provide atomic level information on systems where it would be at best, very challenging if not impossible, to obtain information experimentally. For example, the response of the voltage sensor of potassium channels to a transmembrane voltage can be studied with MD simulations (Jensen et al. 2012) providing unique insight into these functionally important features at an atomistic level of detail.

MD simulations also allow one to explore the possibility of allosteric and cryptic binding pockets (Frembgen-Kesner and Elcock 2006). The latter are not exposed to bulk solvent all of the time and so may be hidden in certain crystallographic structures. MD allows these sites to manifest themselves (Lukman et al. 2014) and so permit docking and similar protocols to be followed in the usual manner. Simulations are also essential for understanding the mechanisms of allosteric modulation. This aspect of GPCRs has been investigated in terms of searching for "hidden" allosteric sites, which may be potential binding pockets (Miao et al. 2014; Ivetac and McCammon 2010). In ion channels, the effect of anesthetic on members of the cys-loop receptor has been postulated to be an allosteric effect, acting at the transmembrane region of the receptor (Murail et al. 2012; Murail et al. 2011 and Salari et al. 2014).

At a more fundamental level, there are questions of trying to understand the dynamics of protein targets as fully as possible (Micheletti 2013) with a view to comparing not just the structural similarity but also the dynamic similarity (Münz et al. 2010, 2012). The power of MD simulations to improve our understanding of function has meant growing interest in its potential in drug discovery (Durrant and McCammon 2011). It can be used in conjunction with many other tools including virtual screening to improve the prospects of finding a new compound (Nichols et al. 2011). The use of MD trajectories to generate an ensemble of possible receptor conformations is best highlighted by the so-called "relaxed complex scheme" (Amaro et al. 2008; Lin et al. 2002), where there has been some success in for soluble targets like HIV integrase (Schames et al. 2004) and the tumor suppressor, p53 (Wassman et al. 2013). The increasing viability of MD simulations with an ever-growing appreciation of the role of protein flexibility and solvent has meant that simulations are starting to attract the attention of the industrial community (Moroni et al. 2015).

12.2.2 The Role of Water Molecules in Receptor-Ligand Binding

Only high-resolution crystal structures will give any reliable indication as to the presence of waters molecules, yet it is known that water-mediated interactions between ligands and protein targets are extremely common (Lu et al. 2007). As it has been shown that the displacement of ordered water molecules can directly affect a ligand's binding affinity (Clarke et al. 2001; Lam et al. 1994), this has been the focus of many drug design strategies with the aim of designing compounds that can displace these waters (Lam et al. 1994; Chen et al. 1998; Wissner et al. 2000).

As one often does not have the presence of water molecules in key positions confirmed by experiment, the first task is the prediction of these sites. In recent years, many methods have been developed to tackle this problem and in general the results are good. Knowledge of the presence of water in binding sites can be useful in its own right and indeed several reports over the years have demonstrated how this is useful for membrane proteins including ligand-gated receptors (Sahai and Biggin 2011; Vijayan et al. 2010; Yu et al. 2014) and GPCRs (Mason et al. 2012).

The issue then is to compute whether it is likely that the water will be displaceable and indeed whether that displacement will give a favourable contribution to the overall free energy of binding. This latter aspect has proven surprisingly difficult to achieve reliable results for, though there have been some reported success mainly for soluble proteins (Mondal et al. 2014a, b; Pearlstein et al. 2013). The prediction of water molecule networks and their perturbation has also been examined in terms of the relationship to kinetics (and residence time – see Sect. 11.3.3 for a series of adenosine A2A receptor antagonists (Bortolato et al. 2013).

Running MD or Monte Carlo (MC) simulations and observing the peaks in water density (Henchman and McCammon 2002; Alvarez-Garcia and Barril 2014) can provide the location of water binding sites. However, these can be time-consuming to run, especially with buried cavities due to the long time it takes for water to permeate within the protein. Grand Canonical Monte Carlo methods can significantly reduce the length of the simulation, though even that can be quite demanding on resource. Thus, there have been several attempts to develop fast methods. JAWS for example is a grid-based MC method that estimates the free energy of displacing a water molecule into bulk (Michel and Essex 2010; Michel et al. 2009). An integral theory approach (3D-RISM) has also reported success in predicting solvation structure within ligand-binding sites (Imai et al. 2009) and protein cavities (Imai et al. 2007). Short molecular simulations can be used as the data for inhomogeneous fluid solvation theory (IFST), as reported by Lazaridis (1998a, b). This method has the distinct advantage that the free energy can be broken down into enthalpic and entropic components (Li and Lazaridis 2003, 2005a, b). IFST also forms the framework for WaterMap and has been used in number of cases (Abel et al. 2008; Robinson et al. 2010; Young et al. 2006) including glutamate receptors (Frydenvang et al. 2010), the ompC channel (Tran et al. 2013) and GPCRs (Higgs et al. 2010; Newman et al. 2012)

An even faster method, that exploits the docking program AutoDock Vina (Trott and Olson 2010) was found to reproduce water positions to a high degree of accuracy and could also predict whether a water molecule was displaced or conserved to an accuracy of 75 % (Ross et al. 2012). Figure 12.7 shows an example prediction for AMPA bound to ligand-binding domain of the GluA2 ionotropic glutamate receptor. This compromise between speed and accuracy may be desirable at the high-throughput stage of virtual screening.

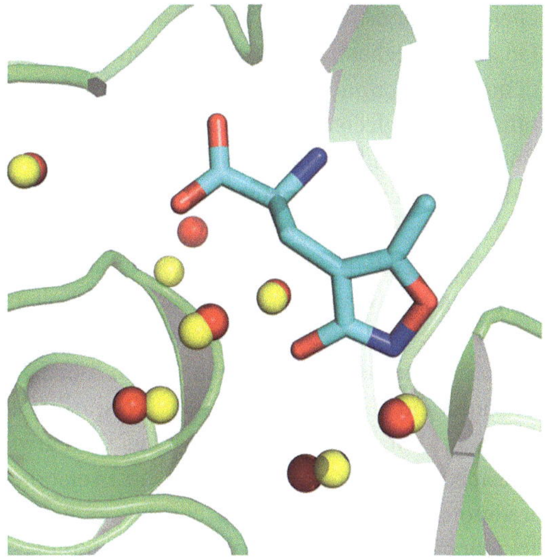

Fig. 12.7 An example of water position prediction from the WaterDock program performed on the ligand-binding domain of the GluA2 ionotropic glutamate receptor in complex with AMPA (shown in liquorice sticks). *Red spheres*: water molecules observed in at least two crystal structures. *Yellow spheres*: predicated water sites. WaterDock is able to predict all of the crystallographically observed water molecules (Figure taken from Ross et al. (2012))

12.3 Challenges for the Future

Experience has shown that the deepest insight can only be achieved when there is good interdisciplinary collaboration between experimental and theoretical groups. Challenges that will require this approach include understanding the conformational changes that are ligand dependent in GPCRs and how those are conveyed to the intracellular signalling cascades (see Bermudez and Wolber 2015 for a recent review). Properties such as accurate prediction of affinity, kinetics, the ever-increasing size and amount of data and the integration of structure into higher-order models are also areas of increasing interest. In this final section, we outline some of these challenges.

12.3.1 Deeper Understanding of Receptor-Ligand Interactions

As we have discussed, structural information of the target plays a central role in the rationalization, efficiency and cost-effectiveness of the drug design process. However, even with the crystal structure in hand, the simple molecule mechanics based approaches to rationalising affinity cannot always explain the full complexity of the chemical interactions between ligand and target. Quantum mechanics (QM) can provide the most complete description of the interactions including otherwise neglected components such as charge fluctuations and dynamic polarization that can make significant contributions to affinity. However, traditional QM methods are simply not feasible for large biological systems because of their huge computational cost. In recent years though, new QM based methods, such as the fragment molecular orbital (FMO) method developed by Fedorov and Kitaura offers a way forward (Fedorov and Kitaura 2007). The FMO method gives considerable computational speed up over other traditional QM methods and can be applied to membrane proteins and their ligand-complexes. Furthermore, the FMO method has the potential to contribute to the refinement process in terms of X-ray crystallographic data with drug complexes. This better understanding of the enthalpic contributions can help chemists in an intuitive way. However, the omission of entropic effects must be kept in mind and the prediction of the overall free energy of binding, ΔG, is still a major challenge as we discuss in the next section.

12.3.2 Affinity and Efficacy

Although docking programs generally do a reasonable job of pose-prediction, the correct prediction of binding affinity or even predicting the order of binding for a series of compounds, is much more error prone. The development of a generic scoring function that can successfully rank ligands across diverse targets is unlikely to be forthcoming in the near future and indeed it has been mathematically proven that specialized functions will always out-perform any generic scoring function (Ross et al. 2013). At the molecular level, a drug must associate with the receptor in order to cause a response, and the strength of such association is described by its affinity. The availability of structural data was thought to directly provide the information needed to interpret ligand-protein interactions and estimate the affinity of small molecules for any binding pocket (Beddell et al. 1976; Cohen 1977). It was soon realised however that while structural data is necessary, it is not sufficient on its own to describe drug-receptor association as it is in fact a complex process, with significant entropic and solvent effects in most instances that can hardly be explained by structure observation alone (Marshall 2012; Mobley and Dill 2009; Gilson and Zhou 2007). For these reasons, despite decades of efforts in computational studies on the effects of ligand binding to a receptor, the ability to predict affinity is still challenging. Nonetheless, in the last decade there have been continuous improvements in theory and computation that are improving binding affinity prediction methodologies (Chodera et al. 2011; Chipot 2014).

Currently, the most rigorous statistical mechanics approaches to estimate affinities rely upon Molecular Dynamics or Monte Carlo simulations for the sampling of the receptor, ligand and solvent conformations and their associated energies (Chipot 2014; Michel et al. 2010; Gohlke and Klebe 2002). The so called "alchemical" methods are based on a non-physical thermodynamic cycle, where binding free energy is computed as the sum of multiple steps during which the ligand is removed from the solution and inserted into the binding pocket. Steered or pulling methods follow instead a physical pathway, by applying a force that pulls the ligand away from the protein and calculating the work involved in this process (Chipot 2014). The advantage over scoring functions or implicit solvent approaches is that the full flexibility of protein and ligand is taken into account, as well as the discrete nature of the solvent.

While the prediction of absolute binding affinities still faces many challenges, the estimation of relative binding free energies, i.e. the difference in binding affinities between two ligands, appears to be more mature and ready to be applied to a wide range of biological targets (Shirts et al. 2007; Mobley and Klimovich 2012). Recent studies have shown how the prediction of relative affinities can guide medicinal chemistry efforts in lead optimisation (Jorgensen 2009). For instance, the Jorgensen lab has combined computational and medicinal chemistry in the development of potent HIV reverse transcriptase inhibitors (Bollini et al. 2011; Jorgensen et al. 2011). In one study, Bollini et al. used relative affinity calculations to identify the most promising modifications for an initial 5 μM affinity, which was later turned into a sub-nanomolar ligand. The authors demonstrated how the use of computational methods was pivotal for the identification of optimal substitution patters (Bollini et al. 2011). Similarly, Jorgensen et al. reported the evolution of three low affinity hits into potent inhibitor (<10 nM) of both the wild type and Y181C variant of HIV-1 reverse transcriptase. The use of free energy calculations allowed the identification of these potent and dual-target inhibitors while synthesizing only about 30 compounds (Jorgensen et al. 2011).

Wang et al. (2015) reported instead a large retrospective study with over 200 ligands and 10 protein targets, and using an improved force field. The authors showed how the predictions correlated extremely well (weighted average R-value of 0.75) and were mostly within 1.0 kcal/mol of the experimental values. Additionally, it was shown how the use of the calculations on two prospective studies involving IRAK4 and TYK2 allowed deprioritizing a large number of compounds and enriching the synthesis of tight-binding ligands.

Predicting absolute free energies prospectively, on the other hand, is still a challenge as shown by the latest SAMPL exercise, where participants were asked to predict affinities of HIV integrase inhibitors in the catalytic core domain (Mobley 2014). Whilst in relative affinity calculations sampling issues might have minor effects (Mobley and Klimovich 2012), these are more likely to have large repercussions when calculating absolute binding free energies. Mobley et al. (2007a) showed how under-sampling even a valine side-chain reorientation upon binding can lead to an error of several kcal/mol in the binding affinity prediction. For these reasons, prospective applications of absolute binding free energy are still rare and most efforts have focused on retrospective validation of the methodology. One of the most studied macromolecular systems has been the binding pocket of engineered T4 lysozymes. The binding of small fragments to such system has been calculated by Mobley et al. and Boyce et al., achieving root mean square errors just below 2 kcal/mol (Boyce et al. 2009; Mobley et al. 2007b). Another popular test system has been the FK506-binding protein (FKBP12). For this system the binding of drug-like molecules was tested computationally and RMS errors were still around 2 kcal/mol (Wang et al. 2006; Fujitani et al. 2005). More recently, Aldeghi et al. (2016) also achieved good success with absolute free energy calculations performed on a diverse set of drugs against a bromodomain. The RMS errors reported were of the order of 1 kca/mol.

While such level of accuracy would in fact be useful in a drug design context when screening compound libraries (Mobley and Klimovich 2012; Shirts et al. 2010), the performance of the approach still needs to be validated against more complex targets and small molecules. In fact, application to more complex proteins such as ion channels and GPCR where conformational changes play an important role in the function of the receptor (Jensen et al. 2012; Dror et al. 2015; Burg et al. 2015) amplify the challenges that need to be overcome in order to obtain reliable predictions. Similarly to the sampling issue of a side-chain rotation, slow degrees of freedoms like conformational changes upon binding cause significant sampling issues for the timescales currently accessible computationally.

Lin et al. (Lin and Roux 2013; Lin et al. 2013) have however, provided an elegant study that took a large conformational change into account when calculating binding affinity by dividing the calculation in two steps, showing how such calculations are still feasible given sufficient knowledge of the system at hand. Investigating the molecular reasons of Gleevec selectivity, the free energy change from DFG-in to –out conformations of Abl and c-Src kinases was first calculated, followed by the affinity calculation of Gleevec for both DFG-out conformations. The selectivity of Gleevec for Abl over c-Src was found to be a combination of conformational selection, due to the larger work needed to move the loop in c-Src, and differences in binding affinities to the DFG-out conformation (Lin and Roux 2013; Lin et al. 2013).

Considering the challenges that binding affinity predictions face, it is clear that predicting efficacy will be even more challenging at least in terms of separating the effects out from affinity. Additionally, efficacy is likely not to depend only on the thermodynamic quantity of affinity, but also the kinetics of the binding event, with residence times in the binding site playing a role in the biological response of the receptor to the bound drug. Experimentally, dissecting these various contributions out is also extremely difficult. However, for some ligand-gated ion channels such as the glycine receptor (Yu et al. 2014; Lape et al. 2008), where single channel behavior can be observed, there is the genuine prospect that progress can be made in this area, although there is no doubt that it will be extremely challenging.

Overall, predicting absolute ligand binding affinities is still a challenge for most systems and in particular large, complex membrane proteins. However, it is possible to foresee a point in the near future where binding affinity predictions will be routine and part of the valuable set of computational tools available to accelerate drug discovery. On the other hand, while prediction of efficacy is the natural next step, it will need a deeper mechanistic understanding of membrane protein systems before a solid foundation for such calculations can be laid.

12.3.3 Kinetics and Its Relationship to Structure

In recent years there has been growing interest in trying to relate structure to kinetics (Swinney 2009). The kinetics of drug-binding (Keserü and Swinney 2015) are increasingly being recognized as being important for the clinical effectiveness of drugs (Cusack et al. 2015). Indeed it has been shown experimentally that there is a positive correlation between functional efficacy and its so-called "residence time" at the receptor (Guo et al. 2012, 2014). A long residence time is thought to be important in many cases as it can extend the duration of pharmacodynamic activity during the body. It can thus not only increase the *in-vivo* efficacy but also reduce the potential of off-target effects (Cusack et al. 2015). For a recent review of how residence time has been considered in the development of some compounds the reader is referred to (Hoffmann et al. 2015).

Can we extract relationships between structures (of ligands and/or proteins) and their kinetics (so called Structure-kinetics-relationships, SKRs)? This area of research is still young, but there is tremendous interest in it, because the human body is anything but in equilibrium, and thus kinetics is presumably very important.

As with initial use of affinity, there was some naivety concerning assumptions (for example, that similar ligands would have similar k_{on} rates, etc). The situation has been shown, at least for soluble proteins (like kinases), to be more complex (Schneider et al. 2013) than first hoped. It is also hindered by the fact that there is not only a multitude of ways of performing the experiments but also the manner in which they are reported (Klebe 2015). Nevertheless, efforts have been made to systematically bring together observations across discrete families (Miller et al. 2012) and to analyse properties most likely to influence dissociation rates, with molecular weight apparently contributing the most (Miller et al. 2012). There have also been studies that have successfully developed new compounds against K_V 11.1 potassium channel targets with different off-rates (Yu et al. 2015). Such a consideration may be critical in the consideration of cardiotoxicity.

How do structural features of proteins dictate kinetics? To address that is going to be extremely challenging indeed and certainly much more difficult than the relationship to affinity. Nevertheless, some early attempts to examine this have been impressive in their ambition and the potential insight they can reveal; see (Cavalli et al. 2015). One of the biggest hurdles is simply ensuring that there are enough observations of both "on" and "off" rates and this ultimately comes down to a sampling problem. Various lines of approach have been devised ranging from brute force (i.e. simulate for a very long time using specialized hardware and simply count the on/off rates and residence times) as for example reported for ligand-GPCR binding.

A study of benzamine-trypsin (Buch et al. 2011) showed that 495 × 100 ns simulations could produce 187 binding events. By reconstructing the binding pathway, they were able to show that two intermediate binding states could be found between the solvent and the final bound state. Another study examined the binding of ligands to GPCRs and highlighted the specific role that dehydration can play in the ligand-binding process (Dror et al. 2011). The use of a large number of simulations for example via "the cloud" (Harvey and De Fabritiis 2015), together with Markov Modelling (Pande et al. 2010) and sophisticated sampling techniques (Doerr and De Fabritiis 2014) is a maturing field that is starting to show exciting results for ligand-binding events and indeed how they can be applied in terms of protein modulation (Shukla et al. 2015).

In addition to the approaches highlighted above, the use of metadynamics has also proven useful in the analysis of ligand-binding events both in term of binding (Limongelli et al. 2013) and unbinding (Tiwary et al. 2015). Metadynamics employs a sampling (in a non-systematic way) along a set of collective variables. A bias is added in a history-dependent fashion that adds Gaussian contributions to the potential energy surface to prevent the system visiting regions of the conformational space that have already been visited. The free energy surface can then be reconstructed as a function of the collective variables. The technique has the advantage that it can provide these insights on fewer computational resources than the Markovian reconstruction approach outlined above. However, the results tend to be sensitive to the choice of collective variables (Barducci et al. 2011).

12.3.4 Data, Pipelining and Unified Models

The previous section alluded to the fact that kinetics can be assessed via the use of a large number of trajectories. In fact this poses part of a more generic problem – the sheer volume of data and how to deal it. This is not a problem specific to simulation or computational methods per se, but is particularly acute for this field. Furthermore, the problem is more than just a data storage issue, managing data across several different compute systems, and in the cloud, requires some considerable strategic thought (Pronk et al. 2011).

However, once a strategy has been devised for simulation and data management more advanced protocols can be developed that can bring added-value to existing data. One such example is MemProtMD (Stansfeld et al. 2015), which is a simulation pipeline for predicting the location of membrane proteins within a lipid bilayer (Fig. 12.8). It will be interesting in the future to

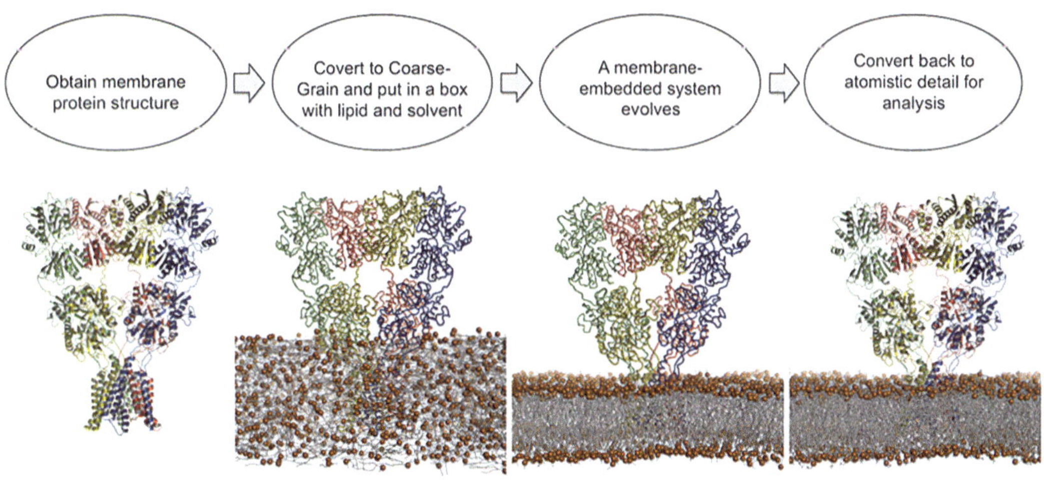

Fig. 12.8 The MemProtMD pipeline for inserting membrane proteins into bilayers. The first step is to detect the protein structures from the PDB, here shown for the GluA2 ionotropic glutamate receptor (PDB: 3KG2). The second is to set up a lipid, water, and protein simulation system. Coarse-grained (CG) simulations are then run (1 μs duration) to assemble and equilibrate a bilayer around each membrane protein structure. The CG simulation system is then converted to atomic resolution, allowing detailed analysis of lipid bilayer/protein interactions (Figure courtesy of Dr Phillip Stansfeld)

see how well computational pipelines can be integrated with high-throughput structural biology pipelines. Another challenge for the future will be how to stitch together the different levels of treatment both in terms of physical models (e.g. coarse-graining (Ayton et al. 2010)) and higher level models that are often called systems biology (Pei et al. 2014).

There are many complex challenges to be solved and even small steps of progress can give important insight for drug design. The changing landscape of the pharmaceutical sector also means that collaborations between academia and industry are likely to play central roles in moving the field forward in the coming decades (Heifetz et al. 2015).

Acknowledgments MA is supported by the EPSRC and Evotec via the Systems Approaches to Biomedical Sciences Doctoral Training Centre. Philip C. Biggin acknowledges support from the BBSRC and MRC.

References

Abel R, Young T, Farid R, Berne BJ, Friesner RA (2008) Role of the active-site solvent in the thermodynamics of factor Xa ligand binding. J Am Chem Soc 130:2817–2831

Aldeghi M, Heifetz A, Bodkin MJ, Knapp S, Biggin PC (2016) Accurate calculation of the absolute free energy of binding for drug molecules. Chem Sci 7:207–218

Alvarez-Garcia D, Barril X (2014) Molecular simulations with solvent competition quantify water displaceability and provide accurate interaction maps of protein binding sites. J Med Chem 57:8530–8539

Amaro R, Baron R, McCammon JA (2008) An improved relaxed complex scheme for receptor flexibility in computer-aided drug design. J Comput-Aided Mol Des 22:693–705

Armstrong N, Gouaux E (2000) Mechanisms for activation and antagonism of an AMPA-sensitive glutamate receptor: Crystal structures of the GluR2 ligand binding core. Neuron 28:165–181

Assaf Z, Larsen AP, Venskutonytė R, Han L, Abrahamsen B et al (2013) Chemoenzymatic synthesis of new 2,4-syn-functionalized (S)-glutamate analogues and structure–activity relationship studies at ionotropic glutamate receptors and excitatory amino acid transporters. J Med Chem 56:1614–1628

Ayton GS, Lyman E, Voth GA (2010) Hierarchical coarse-graining strategy for protein-membrane systems to access mesoscopic scales. Faraday Discuss 144:347–481

Barducci A, Bonomi M, Parrinello M (2011) Metadynamics. Wiley Interdiscip Rev Comp Mol Sci 1:826–843

Beddell CR, Goodford PJ, Norrington FE, Wilkinson S, Wootton R (1976) Compounds designed to fit a site of known structure in human haemoglobin. Br J Pharmacol 57:201–209

Bermudez M, Wolber G (2015) Structure versus function—the impact of computational methods on the discovery of specific GPCR–ligands. Bioorg Med Chem 23:3907–3912

Biggin PC, Bond PJ (2015) Molecular dynamics simulations of membrane proteins. Methods Mol Biol 1215:91–108

Bollini M, Domaoal RA, Thakur VV, Gallardo-Macias R, Spasov KA et al (2011) Computationally-guided optimization of a docking hit to yield catechol diethers as potent anti-HIV agents. J Med Chem 54:8582–8591

Bortolato A, Tehan BG, Bodnarchuk MS, Essex JW, Mason JS (2013) Water network perturbation in ligand binding: Adenosine A2A antagonists as a case study. J Chem Inf Model 53:1700–1713

Boyce SE, Mobley DL, Rocklin GJ, Graves AP, Dill KA et al (2009) Predicting ligand binding affinity with alchemical free energy methods in a polar model binding site. J Mol Biol 394:747–763

Buch I, Giorgino T, De Fabritiis G (2011) Complete reconstruction of an enzyme-inhibitor binding process by molecular dynamics simulations. Proc Natl Acad Sci U S A 108:10184–10189

Bunch L, Krogsgaard-Larsen P (2009) Subtype selective kainic acid receptor agonists: discovery and approaches to rational design. Med Res Rev 29:3–28

Burg JS, Ingram JR, Venkatakrishnan AJ, Jude KM, Dukkipati A et al (2015) Structural basis for chemokine recognition and activation of a viral G protein–coupled receptor. Science 347:1113–1117

Caldwell N, Harms JE, Partin KM, Jamieson C (2015) Rational design of a novel AMPA receptor modulator through a hybridization approach. ACS Med Chem Lett 6:392–396

Cavalli A, Spitaleri A, Saladino G, Gervasio FL (2015) Investigating drug–target association and dissociation mechanisms using metadynamics-based algorithms. Acc Chem Res 48:277–285

Cerqueira NM, Gesto D, Oliveira EF, Santos-Martins D, Brás NF et al (2015) Receptor-based virtual screening protocol for drug discovery. Arch Biochem Biophys, epub online

Chen JM, Xu SL, Wawrzak Z, Basarab GS, Jordan DB (1998) Structure-based design of potent inhibitors of scytalone dehydratase: displacement of a water molecule from the active site. Biochemistry 37:17735–17744

Chen H, Wang CZ, Ding C, Wild C, Copits B et al (2013) A combined bioinformatics and chemoinformatics approach for developing asymmetric bivalent ampa receptor positive allosteric modulators as neuroprotective agents. ChemMedChem 8: 226–230

Cheng T, Li Q, Zhou Z, Wang Y, Bryant SH (2012) Structure-based virtual screening for drug discovery: a problem-centric review. AAPS J 14:133–141

Chipot C (2014) Frontiers in free-energy calculations of biological systems. Wiley Interdiscip Rev Comput Mol Sci 4:71–89
Chodera JD, Mobley DL, Shirts MR, Dixon RW, Branson K et al (2011) Alchemical free energy methods for drug discovery: progress and challenges. Curr Opin Struct Biol 21:150–160
Clarke C, Woods RJ, Gluska J, Cooper A, Nutley MA et al (2001) Involvement of water in carbohydrate-protein binding. J Am Chem Soc 123:12238–12247
Cohen SS (1977) A strategy for the chemotherapy of infectious disease. Science 197:431–432
Cross JB, Thompson DC, Rai BK, Baber JC, Fan KY et al (2009) Comparison of several molecular docking programs: pose prediction and virtual screening accuracy. J Chem Inf Model 49:1455–1474
Cusack KP, Wang Y, Hoemann MZ, Marjanovic J, Heym RG et al (2015) Design strategies to address kinetics of drug binding and residence time. Bioorg Med Chem Lett 25:2019–2027
Demmer CS, Møller C, Brown PMGE, Han L, Pickering DS et al (2015) Binding mode of an α-amino acid-linked quinoxaline-2,3-dione analogue at glutamate receptor subtype GluK1. ACS Chem Neurosci 6:845–854
Doerr S, De Fabritiis G (2014) On-the-fly learning and sampling of ligand binding by high-throughput molecular simulations. J Chem Theor Comput 10:2064–2069
Dror RO, Pan AC, Arlow DH, Borhani DW, Maragakis P et al (2011) Pathway and mechanism of drug binding to G-protein-coupled receptors. Proc Natl Acad Sci U S A 108:13118–13123
Dror RO, Mildorf TJ, Hilger D, Manglik A, Borhani DW et al (2015) Structural basis for nucleotide exchange in heterotrimeric G proteins. Science 348:1361–1365
Durrant JD, McCammon JA (2011) Molecular dynamics simulations and drug discovery. BMC Biol 9:71–71
Elokely KM, Doerksen RJ (2013) Docking challenge: protein sampling and molecular docking performance. J Chem Inf Model 53:1934–1945
Fedorov DG, Kitaura K (2007) Extending the power of quantum chemistry to large systems with the fragment molecular orbital method. J Phys Chem A 111:6904–6914
Forli S, Olson A (2015) Computational challenges of structure-based approaches applied to HIV. In: Torbett BE, Goodsell DS, Richman DD (eds) The future of HIV-1 therapeutics.: Springer International Publishing, pp 31–51
Frembgen-Kesner T, Elcock AH (2006) Computational sampling of a cryptic drug binding site in a protein receptor: explicit solvent molecular dynamics and inhibitor docking to p38 MAP kinase. J Mol Biol 359:202–214
Frydenvang K, Pickering DS, Greenwood JR, Krogsgaard-Larsen N, Brehm L et al (2010) Biostructural and pharmacological studies of bicyclic analogues of the 3-isoxazolol glutamate receptor agonist ibotenic acid. J Med Chem 53:8354–8361
Fujitani H, Tanida Y, Ito M, Jayachandran G, Snow CD et al (2005) Direct calculation of the binding free energies of FKBP ligands. J Chem Phys 123(8):084108
Gilson MK, Zhou H-X (2007) Calculation of protein-ligand binding affinities. Annu Rev Biophys Biomol Struct 36:21–42
Gohlke H, Klebe G (2002) Approaches to the description and prediction of the binding affinity of small-molecule ligands to macromolecular receptors. Angew Chem Int Ed 41:2644–2676
Goose JE, Sansom MSP (2013) Reduced lateral mobility of lipids and proteins in crowded membranes. PLoS Comp Biol 9, e1003033
Grouleff J, Irudayam SJ, Skeby KK, Schiøtt B (2015) The influence of cholesterol on membrane protein structure, function, and dynamics studied by molecular dynamics simulations. Biochim Biophys Acta 1848:1783–1795
Guo D, Mulder-Krieger T, Ijzerman AP, Heitman LH (2012) Functional efficacy of adenosine A2A receptor agonists is positively correlated to their receptor residence time. Br J Pharm 166:1846–1859
Guo D, Hillger JM, Ijzerman AP, Heitman LH (2014) Drug-target residence time—a case for G protein-coupled receptors. Med Res Rev 34:856–892
Han M, Lou J, Nakanishi K, Sakmar TP, Smith SO (1997) Partial agonist activity of 11-cis-retinal in rhodopsin mutants. J Biol Chem 272:23081–23085
Harms JE, Benveniste M, MacLean JKF, Partin KM, Jamieson C (2013) Functional analysis of a novel positive allosteric modulator of AMPA receptors derived from a structure-based drug design strategy. Neuropharmacology 64:45–52
Harvey MJ, De Fabritiis G (2015) AceCloud: molecular dynamics simulations in the cloud. J Chem Inf Model 55:909–914
Heifetz A, Morris GB, Biggin PC, Barker O, Fryatt T et al (2012) Study of human Orexin-1 and −2 -G-protein coupled receptors with novel and published antagonists by modeling, molecular dynamics simulations, and site-directed mutagenesis. Biochemistry 51:3178–3197
Heifetz A, Barker O, Morris GB, Law RJ, Slack M et al (2013a) Toward an understanding of agonist binding to human Orexin-1 and Orexin-2 receptors with G-protein-coupled receptor modeling and site directed mutagenesis. Biochemistry 52:8246–8260
Heifetz A, Barker O, Verquin G, Wimmer N, Meutermans W, Pal S, Law RJ, Whittaker M (2013b) Fighting obesity with a sugar-based library: discovery of novel MCH-1R antagonists by a new computational-VAST approach for exploration of GPCR binding sites. J Chem Inf Model 53:1084–1099
Heifetz A, Mazanetz M, James T, Pal S, Law RJ et al (2013c) From receptors to ligands: fragment-assisted drug design for GPCRs applied to the discovery of H3 and H4 receptor antagonists. Med Chem 4:313–321
Heifetz A, Schertler GX, Seifert R, Tate C, Sexton P et al (2015) GPCR structure, function, drug discovery and crystallography: report from Academia-

Industry International Conference (UK Royal Society) Chicheley Hall, 1–2 September 2014. Naunyn-Schmiedeberg's Arch Pharmacol 388:883–903

Henchman RH, McCammon JA (2002) Extracting hydration sites around proteins from explicit water simulations. J Comput Chem 23:861–869

Higgs C, Beuming T, Sherman W (2010) Hydration site thermodynamics explain SARs for Triazolylpurines analogues binding to the A2A receptor. ACS Med Chem Lett 1:160–164

Hoffmann C, Castro M, Rinken A, Leurs R, Hill SJ et al (2015) Ligand residence time at GPCRs – why we should take our time to study it. Mol Pharm 88:552–560

Holst B, Nygaard R, Valentin-Hansen L, Bach A, Engelstoft MS et al (2010) A conserved aromatic lock for the tryptophan rotameric switch in TM-VI of seven-transmembrane receptors. J Biol Chem 285:3973–3985

Imai T, Hiraoka R, Kovalenko A, Hirata F (2007) Locating missing water molecules in protein cavities by the three-dimensional reference interaction site model theory of molecular solvation. Proteins Struct Func Genet 66:804–813

Imai T, Oda K, Kovalenko A, Hirata F, Kidera A (2009) Ligand mapping on protein surfaces by the 3D-RISM theory: toward computational fragment-based drug design. J Am Chem Soc 131:12430–12440

Isberg V, Vroling B, van der Kant R, Li K, Vriend G et al (2014) GPCRDB: an information system for G protein-coupled receptors. Nucleic Acids Res 42:D422–D425

Ivetac A, McCammon JA (2010) Mapping the druggable allosteric space of g-protein coupled receptors: a fragment-based molecular dynamics approach. Chem Biol Drug Des 76:201–217

Jamieson C, Maclean JKF, Brown CI, Campbell RA, Gillen KJ et al (2011) Structure based evolution of a novel series of positive modulators of the AMPA receptor. Bioorg Med Chem Lett 21:805–811

Jazayeri A, Dias JM, Marshall FH (2015) From G protein-coupled receptor structure resolution to rational drug design. J Biol Chem 290:19489–19495

Jensen M, Jogini V, Borhani DW, Leffler AE, Dror RO et al (2012) Mechanism of voltage gating in potassium channels. Science 336:229–233

Jorgensen WL (2009) Efficient drug lead discovery and optimization. Acc Chem Res 42:724–733

Jorgensen WL, Bollini M, Thakur VV, Domaoal RA, Spasov KA et al (2011) Efficient discovery of potent anti-HIV agents targeting the Tyr181Cys variant of HIV reverse transcriptase. J Am Chem Soc 133:15686–15696

Juknaitė L, Venskutonytė R, Assaf Z, Faure S, Gefflaut T et al (2012) Pharmacological and structural characterization of conformationally restricted (S)-glutamate analogues at ionotropic glutamate receptors. J Struct Biol 180:39–46

Kalli AC, Campbell ID, Sansom MSP (2013) Conformational changes in talin on binding to anionic phospholipid membranes facilitate signaling by integrin transmembrane helices. PLoS Comp Biol 9, e1003316

Keserü GM, Swinney DC (eds) (2015) Thermodynamics and kinetics of drug binding. Wiley-VCH, Verlag GmbH & Co. KGaA, Heidelberg

Khatri A, Burger PB, Swanger SA, Hansen KB, Zimmerman S et al (2014) Structural determinants and mechanism of action of a GluN2C-selective NMDA receptor positive allosteric modulator. Mol Pharm 86: 548–560

Klebe G (2015) The use of thermodynamic and kinetic data in drug discovery: decisive insight or increasing the puzzlement? ChemMedChem 10:229–231

Lam PY, Jadhav PK, Eyermann CJ, Hodge CN, Ru Y et al (1994) Rational design of potent, bioavailable, nonpeptide cyclic ureas as HIV protease inhibitors. Science 263:380–384

Lape R, Colquhoun D, Sivilotti LG (2008) On the nature of partial agonism in the nicotinic receptor superfamily. Nature 454:722–727

Lazaridis T (1998a) Inhomogeneous fluid approach to solvation thermodynamics. 1. Theory. J Phys Chem B 102:3531–3541

Lazaridis T (1998b) Inhomogeneous fluid approach to solvation thermodynamics. 2. Applications to simple fluids. J Phys Chem B 102:3542–3550

Lee J, Daniels V, Sands ZA, Lebon F, Shi J et al (2015) Exploring the interaction of SV2A with racetams using homology modelling, molecular dynamics and site-directed mutagenesis. PLoS ONE 10, e0116589

Levoin N, Calmels T, Krief S, Danvy D, Berrebi-Bertrand I et al (2011) Homology model versus x-ray structure in receptor-based drug design: a retrospective analysis with the dopamine D3 receptor. ACS Med Chem Lett 2:293–297

Li Z, Lazaridis T (2003) Thermodynamic contributions of the ordered water molecule in HIV-1 protease. J Am Chem Soc 125:6636–6637

Li Z, Lazaridis T (2005a) The effect of water displacement on binding thermodynamics: concanavalin A. J Phys Chem B 109:662–670

Li Z, Lazaridis T (2005b) Thermodynamics of buried water clusters at a protein at ligand binding interface. J Phys Chem B 110:1464–1475

Limongelli V, Bonomi M, Parrinello M (2013) Funnel metadynamics as accurate binding free-energy method. Proc Natl Acad Sci U S A 110:6358–6363

Lin Y-L, Roux B (2013) Computational analysis of the binding specificity of Gleevec to Abl, c-Kit, Lck, and c-Src tyrosine kinases. J Am Chem Soc 135:14741–14753

Lin JH, Perryman AL, Schames JR, McCammon JA (2002) Computational drug-design accomodating receptor flexibility: the relaxed complex scheme. J Am Chem Soc 124:5632–5633

Lin Y-L, Meng Y, Jiang W, Roux B (2013) Explaining why Gleevec is a specific and potent inhibitor of Abl kinase. Proc Natl Acad Sci U S A 110:1664–1669

Lionta E, Spyrou G, Vassilatis DK, Cournia Z (2014) Structure-based virtual screening for drug discovery:

principles, applications and recent advances. Curr Top Med Chem 14:1923–1938

Lounnas V, Ritschel T, Kelder J, McGuire R, Bywater RP et al (2013) Current progress in structure-based rational drug design marks a new mindset in drug discovery. Comput Struct Biotechnol J 5, e201302011

Lu Y, Wang R, Yang C-Y, Wang S (2007) Analysis of ligand-bound water molecules in high resolution crystal structures of protein-ligand complexes. J Chem Inf Model 47:668–675

Lukman S, Verma C, Fuentes G (2014) Exploiting protein intrinsic flexibility in drug design. In: Han K-l, Zhang X, Yang M-J (eds) Protein conformational dynamics. Springer International Publishing, pp 245–269

Marshall G (2012) Limiting assumptions in structure-based design: binding entropy. J Comput-Aided Mol Des 26:3–8

Mason JS, Bortolato A, Congreve M, Marshall FH (2012) New insights from structural biology into the druggability of G protein-coupled receptors. Trends Pharm Sci 33:249–260

Miao Y, Nichols SE, McCammon JA (2014) Mapping of allosteric druggable sites in activation-associated conformers of the M2 muscarinic receptor. Chem Biol Drug Des 83:237–246

Michel J, Essex JW (2010) Prediction of protein-ligand binding affinity by free energy simulations: assumptions, pitfalls and expectations. J Comput Aided Mol Des 24:639–658

Michel J, Tirado-Rives J, Jorgensen WL (2009) Energetics of displacing water molecules from protein binding sites: consequences for ligand optimization. J Am Chem Soc 131:15403–15411

Michel J, Foloppe N, Essex JW (2010) Rigorous free energy calculations in structure-based drug design. Mol Inf 29:570 578

Micheletti C (2013) Comparing proteins by their internal dynamics: exploring structure-function relationships beyond static structural alignments. Phys Life Rev 10:1–26

Miller DC, Lunn G, Jones P, Sabnis Y, Davies NL et al (2012) Investigation of the effect of molecular properties on the binding kinetics of a ligand to its biological target. Med Chem Commun 3:449–452

Mobley DL (2014) Blind prediction of HIV integrase binding from the SAMPL4 challenge. J Comput-Aided Mol Des 28:1–19

Mobley DL, Dill KA (2009) Binding of small-molecule ligands to proteins: "What you see" is not always "What you get". Structure 17:489–498

Mobley DL, Klimovich PV (2012) Perspective: alchemical free energy calculations for drug discovery. J Chem Phys 137:230901

Mobley DL, Chodera JD, Dill KA (2007a) Confine-and-release method: obtaining correct binding free energies in the presence of protein conformational change. J Chem Theory Comput 3:1231–1235

Mobley DL, Graves AP, Chodera JD, McReynolds AC, Shoichet BK et al (2007b) Predicting absolute ligand binding free energies to a simple model site. J Mol Biol 371:1118–1134

Mondal J, Friesner RA, Berne BJ (2014a) Role of desolvation in thermodynamics and kinetics of ligand binding to a kinase. J Chem Theory Comput 10: 5696–5705

Mondal S, Khelashvili G, Weinstein H (2014b) Not just an oil slick: how the energetics of protein-membrane interactions impacts the function and organization of transmembrane proteins. Biophys J 106: 2305– 2316

Moroni E, Paladino A, Colombo G (2015) The dynamics of drug discovery. Curr Top Med Chem 15:2043–2055

Münz M, Lyngsø R, Hein J, Biggin PC (2010) Dynamics-based alignment of proteins: An alternative approach to quantify dynamic similarity. BMC Bioinf 11:118

Münz M, Hein J, Biggin PC (2012) The role of flexibility and conformational selection in the binding promiscuity of PDZ domains. PLoS Comput Biol 8, e1002749

Murail S, Wallner B, Trudell James R, Bertaccini E, Lindahl E (2011) Microsecond simulations indicate that ethanol binds between subunits and could stabilize an open-state model of a glycine receptor. Biophys J 100:1642–1650

Murail S, Howard RJ, Broemstrup T, Bertaccini EJ, Harris RA et al (2012) Molecular mechanism for the dual alcohol modulation of cys-loop receptors. PLoS Comput Biol 8, e1002710

Newman AH, Beuming T, Banala AK, Donthamsetti P, Pongetti K et al (2012) Molecular determinants of selectivity and efficacy at the Dopamine D3 receptor. J Med Chem 55:6689–6699

Nichols SE, Baron R, Ivetac A, McCammon JA (2011) Predictive power of molecular dynamics receptor structures in virtual screening. J Chem Inf Model 51:1439–1446

Pande VS, Beauchamp K, Bowman GR (2010) Everything you wanted to know about Markov State Models but were afraid to ask. Methods 52:99–105

Partin KM (2015) AMPA receptor potentiators: from drug design to cognitive enhancement. Curr Opin Pharmacol 20:46–53

Pearlstein RA, Sherman W, Abel R (2013) Contributions of water transfer energy to protein-ligand association and dissociation barriers: Watermap analysis of a series of p38α MAP kinase inhibitors. Proteins Struct Func Bioinf 81:1509–1526

Pei J, Yin N, Ma X, Lai L (2014) Systems biology brings new dimensions for structure-based drug design. J Am Chem Soc 136:11556–11565

Pirotte B, Francotte P, Goffin E, de Tullio P (2013) AMPA receptor positive allosteric modulators: a patent review. Expert Opin Ther Pat 23:615–628

Pronk S, Larsson P, Pouya I, Bowman GR, Haque IS, et al (2011) Copernicus: a new paradigm for parallel adaptive molecular dynamics. In: Proceedings of 2011 international conference for high performance computing, networking, storage and analysis. ACM, Seattle, pp 1–10

Robinson DD, Sherman W, Farid R (2010) Understanding kinase selectivity through energetic analysis of binding site waters. ChemMedChem 5:618–627
Ross GA, Morris GM, Biggin PC (2012) Rapid and accurate prediction and scoring of water molecules in protein binding sites. PLoS ONE 7, e32036
Ross GA, Morris GM, Biggin PC (2013) One size does not fit all: the limits of structure-based models in drug discovery. J Chem Theory Comput 9:4266–4274
Sahai MA, Biggin PC (2011) Quantifying water-mediated protein-ligand interactions in a glutamate receptor; A DFT study. J Phys Chem 115:7085–7096
Salari R, Murlidaran S, Brannigan G (2014) Pentameric ligand-gated ion channels: insights from computation. Mol Sim 40:821–829
Schames JR, Henchman RH, Siegel JS, Sotriffer CA, Ni H et al (2004) Discovery of a novel binding trench in HIV integrase. J Med Chem 47:1879–1881
Schmidt T, Bergner A, Schwede T (2014) Modelling three-dimensional protein structures for applications in drug design. Drug Discov Today 19:890–897
Schneider EV, Böttcher J, Huber R, Maskos K, Neumann L (2013) Structure–kinetic relationship study of CDK8/CycC specific compounds. Proc Natl Acad Sci U S A 110:8081–8086
Schwartz TW, Frimurer TM, Holst B, Rosenkilde MM, Elling CE (2006) Molecular mechanism of 7tm receptor activation—a global toggle switch model. Ann Rev Pharmacol Toxicol 46:481–519
Seddon G, Lounnas V, McGuire R, van den Bergh T, Bywater RP et al (2012) Drug design for ever, from hype to hope. J Comp Aided Mol Des 26:137–150
Sengupta D, Chattopadhyay A (2015) Molecular dynamics simulations of GPCR–cholesterol interaction: an emerging paradigm. Biochim Biophys Acta 1848:1775–1782
Shirts MR, Mobley DL, Chodera JD (2007) Free energy calculations: ready for prime time? In: D. S, R. W (eds) Annu Rep Comput Chem, Elsevier, pp 41–59
Shirts MR, Mobley DL, Brown SP (2010) Free-energy calculations in structure-based drug design. In: Merz KM, Ringe D, Reynolds CH (eds) Drug design. Cambridge University Press, Cambridge, pp 61–86
Shoichet BK (2004) Virtual screening of chemical libraries. Nature 432:862–865
Shukla D, Hernández CX, Weber JK, Pande VS (2015) Markov state models provide insights into dynamic modulation of protein function. Acc Chem Res 48:414–422
Sivaprakasam M, Hansen KB, David O, Nielsen B, Traynelis SF et al (2009) Stereocontrolled synthesis and pharmacological evaluation of Azetidine-2,3-Dicarboxylic acids at NMDA receptors. ChemMedChem 4:110–117
Skolnick J, Zhou H, Gao M (2013) Are predicted protein structures of any value for binding site prediction and virtual ligand screening? Curr Opin Struct Biol 23:191–197
Spurny R, Debaveye S, Farinha A, Veys K, Vos AM et al (2015) Molecular blueprint of allosteric binding sites in a homologue of the agonist-binding domain of the α7 nicotinic acetylcholine receptor. Proc Natl Acad Sci U S A 112:E2543–E2552
Stansfeld Phillip J, Goose Joseph E, Caffrey M, Carpenter Elisabeth P, Parker Joanne L et al (2015) MemProtMD: automated insertion of membrane protein structures into explicit lipid membranes. Structure 23:1350–1361
Strong KL, Jing Y, Prosser AR, Traynelis SF, Liotta DC (2014) NMDA receptor modulators: an updated patent review (2013–2014). Exp Opin Ther Pat 24:1349–1366
Swinney DC (2009) The role of binding kinetics in therapeutically useful drug action. Curr Opin Drug Discov Dev 12:31–39
Tang H, Wang XS, Hsieh J-H, Tropsha A (2012) Do crystal structures obviate the need for theoretical models of GPCRs for structure based virtual screening. Proteins 80:1503–1521
Tate CG (2012) A crystal clear solution for determining G-protein-coupled receptor structures. Trends Biochem Sci 37:343–352
Tehan BG, Bortolato A, Blaney FE, Weir MP, Mason JS (2014) Unifying family A GPCR theories of activation. Pharm Res 143:51–60
Timm DE, Benveniste M, Weeks AM, Nisenbaum ES, Partin KM (2011) Structural and functional analysis of two new positive allosteric modulators of GluA2 desensitization and deactivation. Mol Pharm 80:267–280
Tiwary P, Limongelli V, Salvalaglio M, Parrinello M (2015) Kinetics of protein–ligand unbinding: predicting pathways, rates, and rate-limiting steps. Proc Natl Acad Sci U S A 112:E386–E391
Tran Q-T, Williams S, Farid R, Erdemli G, Pearlstein R (2013) The translocation kinetics of antibiotics through porin OmpC: insights from structure-based solvation mapping using WaterMap. Proteins Struct Func Bioinf 81:291–299
Trott O, Olson AJ (2010) AutoDock Vina: improving the speed and accuracy of docking with a new scoring function, efficient optimization, and multithreading. J Comput Chem 31:455–461
Tye H, Mueller SG, Prestle J, Scheuerer S, Schindler M et al (2011) Novel 6,7,8,9-tetrahydro-5H-1,4,7,10a-tetraaza-cyclohepta[f]indene analogues as potent and selective 5-HT2C agonists for the treatment of metabolic disorders. Bioorg Med Chem Lett 21: 34–37
Uteshev VV (2014) The therapeutic promise of positive allosteric modulation of nicotinic receptors. Eur J Pharm 727:181–185
Venkatakrishnan AJ, Deupi X, Lebon G, Tate CG, Schertler GF et al (2013) Molecular signatures of G-protein-coupled receptors. Nature 494:185–194
Venskutonytė R, Larsen AP, Frydenvang K, Gajhede M, Sagot E et al (2014) Molecular recognition of two 2,4-syn-functionalized (S)-glutamate analogues by the kainate receptor GluK3 ligand binding domain. ChemMedChem 9:2254–2259

Vijayan R, Sahai MA, Czajkowski T, Biggin PC (2010) A comparative analysis of the role of water in the binding pockets of ionotropic glutamate receptors. Phys Chem Chem Phys 12:14057–14066
Wang J, Deng Y, Roux B (2006) Absolute binding free energy calculations using molecular dynamics simulations with restraining potentials. Biophys J 91:2798–2814
Wang L, Wu Y, Deng Y, Kim B, Pierce L et al (2015) Accurate and reliable prediction of relative ligand binding potency in prospective drug discovery by way of a modern free-energy calculation protocol and force field. J Am Chem Soc 137:2695–2703
Ward SE, Harries M, Aldegheri L, Austin NE, Ballantine S et al (2010) Integration of lead optimization with crystallography for a membrane-bound ion channel target: discovery of a new class of AMPA receptor positive allosteric modulators. J Med Chem 54:78–94
Ward SE, Pennicott LE, Beswick P (2015) AMPA receptor-positive allosteric modulators for the treatment of schizophrenia: an overview of recent patent applications. Future Med Chem 7:473–491
Wassman CD, Baronio R, Demir Ö, Wallentine BD, Chen C-K et al (2013) Computational identification of a transiently open L1/S3 pocket for reactivation of mutant p53. Nat Commun 4:1407
Weeks AM, Harms JE, Partin KM, Benveniste M (2014) Functional insight into development of positive allosteric modulators of AMPA receptors. Neuropharm 85:57–66
Wissner A, Berger DM, Boschelli DH, Floyd MB Jr, Greenberger LM et al (2000) 4-Anilino-6,7-dialkoxyquinoline-3-carbonitrile inhibitors of epidermal growth factor receptor kinase and their bioisosteric relationship to the 4-anilino-6,7-dialkoxyquinazoline inhibitors. J Med Chem 43:3244–3256
Wlodawer A, Minor W, Dauter Z, Jaskolski M (2008) Protein crystallography for non-crystallographers, or how to get the best (but not more) from published macromolecular structures. FEBS 275:1–21
Wlodawer A, Minor W, Dauter Z, Jaskolski M (2013) Protein crystallography for aspiring crystallographers or how to avoid pitfalls and traps in macromolecular structure determination. FEBS 280:5705–5736
Young JW, Geyer MA (2013) Evaluating the role of the alpha-7 nicotinic acetylcholine receptor in the pathophysiology and treatment of schizophrenia. Biochem Pharm 86:1122–1132
Young RJ, Campbell M, Borthwick AD, Brown D, Burns-Kurtis CL et al (2006) Structure- and property-based design of factor Xa inhibitors: pyrrolidin-2-ones with acyclic alanyl amides as P4 motifs. Bioorg Med Chem Lett 16:5953–5957
Yu R, Hurdiss E, Greiner T, Lape R, Sivilotti L et al (2014) Agonist and antagonist binding in human glycine receptors. Biochemistry 53:6041–6051
Yu Z, van Veldhoven JPD, Louvel J, 't Hart IME, Rook MB et al (2015) Structure–affinity relationships (SARs) and structure–kinetics relationships (SKRs) of Kv11.1 blockers. J Med Chem 58:5916–5929

ERRATUM TO

The Next Generation in Membrane Protein Structure Determination

Isabel Moraes

I. Moraes (ed.), *The Next Generation in Membrane Protein Structure Determination*,
Advances in Experimental Medicine and Biology 922, DOI 10.1007/978-3-319-35072-1

DOI 10.1007/978-3-319-35072-1_13

Chapter 4 in: I. Moraes (ed.), The Next Generation in Membrane Protein Structure Determination, DOI 10.1007/978-3-319-35072-1_4

Chapter 5 in: I. Moraes (ed.), The Next Generation in Membrane Protein Structure Determination, DOI 10.1007/978-3-319-35072-1_5

Chapter 6 in: I. Moraes (ed.), The Next Generation in Membrane Protein Structure Determination, DOI 10.1007/978-3-319-35072-1_6

Chapter 8 in: I. Moraes (ed.), The Next Generation in Membrane Protein Structure Determination, DOI 10.1007/978-3-319-35072-1_8

Chapters 4, 5, 6 and 8 were originally published © Springer International Publishing Switzerland, but have now been made available © The Author(s) under a CC BY 4.0 license.

Updated versions of the original chapters can be found at

http://dx.doi.org/10.1007/978-3-319-35072-1_4

http://dx.doi.org/10.1007/978-3-319-35072-1_5

http://dx.doi.org/10.1007/978-3-319-35072-1_6

http://dx.doi.org/10.1007/978-3-319-35072-1_8

The updated original online version for this book can be found at
DOI 10.1007/978-3-319-35072-1

I. Moraes (ed.), *The Next Generation in Membrane Protein Structure Determination*,
Advances in Experimental Medicine and Biology 922, DOI 10.1007/978-3-319-35072-1_13

Index

I. Moraes (ed.), *The Next Generation in Membrane Protein Structure Determination*, Advances in Experimental Medicine and Biology 922, DOI 10.1007/978-3-319-35072-1

MIX
Papier aus verantwortungsvollen Quellen
Paper from responsible sources
FSC® C105338

If you have any concerns about our products,
you can contact us on
ProductSafety@springernature.com

In case Publisher is established outside the EU,
the EU authorized representative is:
Springer Nature Customer Service Center GmbH
Europaplatz 3, 69115 Heidelberg, Germany

Printed by Libri Plureos GmbH
in Hamburg, Germany